BIOLOGICAL MONITORING OF MARINE POLLUTANTS

ACADEMIC PRESS RAPID MANUSCRIPT REPRODUCTION

Proceedings of a Symposium on Pollution and Physiology of Marine Organism, held in Milford, Connecticut, November 7–9, 1980. Sponsored by National Oceanic and Atmospheric Administration.

BIOLOGICAL MONITORING OF MARINE POLLUTANTS

Edited by

F. JOHN VERNBERG

Belle W. Baruch Coastal Research Institute
University of South Carolina
Columbia, South Carolina

ANTHONY CALABRESE

FREDERICK P. THURBERG

National Marine Fisheries Service
Middle Atlantic Coastal Fisheries Center
Milford Laboratory
Milford, Connecticut

WINONA B. VERNBERG

College of Health
University of South Carolina
Columbia, South Carolina

ACADEMIC PRESS 1981
A Subsidiary of Harcourt Brace Jovanovich, Publishers

New York   Toronto   London   Sydney   San Francisco

ACADEMIC PRESS, INC.
111 Fifth Avenue, New York, New York 10003

United Kingdom Edition published by
ACADEMIC PRESS, INC. (LONDON) LTD.
24/28 Oval Road, London NW1 7DX

Library of Congress Cataloging in Publication Data
Main entry under title:

Biological monitoring of marine pollutants.

    "Papers ... presented at a symposium entitled
'Pollution and Physiology of Marine Organisms' held at
the National Marine Fisheries Service, Milford Laboratory,
Milford, Connecticut, November 7-9, 1979 ... sponsored
jointly by the School of Public Health and the Belle W.
Baruch Institute for Marine Biology and Coastal Research
both at the University of South Carolina, and the Northeast
Fisheries Center, National Marine Fisheries Service" --Pref.
    Includes index.
    1. Aquatic organisms--Effect of water pollution on--
Congresses. 2. Marine pollution--Environmental aspects--
Congresses. I. Vernberg, F. John, Date. II. University of
South Carolina. School of Public Health. III. Belle W. Baruch
Institute for Marine Biology and Coastal Research. IV. United
States. Northeast Fisheries Center.
QH545.W3B57   628.1'686162   81-3583
ISBN 0-12-718450-3   AACR2

CONTENTS

v

Section III   PETROLEUM HYDROCARBONS

Section IV   PHYSIOLOGICAL MONITORING

CONTRIBUTORS

Numbers in parentheses indicate the pages on which authors' contributions begin.

James E. Armstrong (449), Department of Environmental Regulation, Tallahassee, Florida 32301

Pete Benville, Jr. (483), National Marine Fisheries Service, Southwest Fisheries Center, Tiburon Laboratory, Tiburon, California 94920

J. R. Bridge (357), Battelle Pacific Northwest Laboratories, Marine Research Laboratory, Sequim, Washington 98382

John A. Calder (449), NOAA, Bering Sea-Gulf of Alaska Project, Juneau, Alaska 99801

Angela C. Cantelmo (37), Ramapo College of New Jersey, 505 Ramapo Valley Road, Mahwah, New Jersey 07430

Judith M. Capuzzo (405), Woods Hole Oceanographic Institution, Woods Hole, Massachusetts 02543

Neil G. Carmichael (145), National Institute of Environmental Health Sciences, Laboratory of Organ Function and Toxicology, Research Triangle Park, North Carolina 27709

R. Scott Carr (73), Department of Biology, Texas A&M University, College Station, Texas 77843

Philip J. Conklin (37), Department of Biology, The University of West Florida, Pensacola, Florida 32504

John D. Costlow, Jr. (241), Duke University Marine Laboratory, Beaufort, North Carolina 28516

Geraldine M. Cripe (21), U.S. Environmental Protection Agency, Environmental Research Laboratory, Gulf Breeze, Florida 32561

Margaret A. Dawson (335), National Marine Fisheries Service, Milford Laboratory, Milford, Connecticut 06460

A. S. Drum (357), Battelle Pacific Northwest Laboratories, Marine Research Laboratory, Sequim, Washington 98382

Maxwell B. Eldridge (483), National Marine Fisheries Service, Southwest Fisheries Center, Tiburon Laboratory, 3150 Paradise Drive, Tiburon, California 94920

David W. Engel (127, 145), National Marine Fisheries Service, Southeast Fisheries Center, Beaufort Laboratory, Beaufort, North Carolina 28516

Steven W. G. Fehler (449), National Oceanic and Atmospheric Administration Environment Assessment Division, Washington, D.C. 20233

Bruce A. Fowler (127, 145), Laboratory of Environmental Toxicology, National Institute of Environmental Health Science, Research Triangle Park, North Carolina 27709

Ferris R. Fox (37), Department of Biology, The University of West Florida, Pensacola, Florida 32504

Walter Galloway (335), U.S. Environmental Protection Agency, South Ferry Road, Narragansett, Rhode Island 02882

Teresa K. Goochey (389), Southern California Coastal Water Research Project, Long Beach, California 90806

Edith Gould (335, 377), National Marine Fisheries Service, Milford Laboratory, Milford, Connecticut 06460

Timothy L. Hamaker (3, 21), U.S. Environmental Protection Agency, Environmental Research Laboratory, Gulf Breeze, Florida 32561

K. Kanungo (107), Biology Department, Western Connecticut State College, Danbury, Connecticut 06810

Bruce A. Lancaster (405), Woods Hole Oceanographic Institution, Woods Hole, Massachusetts 02543

William H. Lang (165), U.S. Environmental Protection Agency, South Ferry Road, Narragansett, Rhode Island 02882

Richard F. Lee (323), Skidaway Institute of Oceanography, Savannah, Georgia 31406

Wen Yuh Lee (467), The University of Texas, Marine Science Institute, Port Aransas, Texas 78373

M. Marcy (165), Environmental Protection Agency, Environmental Research Laboratory, South Ferry Road, Narragansett, Rhode Island 02882

E. Matthews (3), ERL, Environmental Protection Agency, Sabine Island, Gulf Breeze, Florida 32561

Charles L. McKenney, Jr. (205, 241), Biology Department, Texas A&M University, College Station, Texas 77843

Alan J. Mearns (389), Southern California Coastal Water Research Project, Long Beach, California 90806

D. C. Miller (165, 265), U.S. Environmental Protection Agency, South Ferry Road, Narragansett, Rhode Island 02882

J. C. Moore (3), ERL, Environmental Protection Agency, Sabine Island, Gulf Breeze, Florida 32561

Jerry M. Neff (73, 205), Department of Biology, Texas A&M University, College Station, Texas 77843

J. A. C. Nicol (467), The University of Texas Marine Science Institute, Port Aransas Marine Laboratory, Port Aransas, Texas 78373

DelWayne R. Nimmo (3, 21), ERT, P.O. Box 2105, 1716 Heath Parkway, Fort Collins, Colorado 80522

Philip S. Oshida (389), Southern California Coastal Water Research Project, Long Beach, California 90806

Jose J. Pereira (107), Department of Biology, University of Bridgeport, Bridgeport, Connecticut 06810

Donald K. Phelps (335), U.S. Environmental Protection Agency, Environmental Research Laboratory, South Ferry Road, Narragansett, Rhode Island 02882

K. Ranga Rao (37), University of West Florida, Pensacola, Florida 32504

Stanley D. Rice (425), National Marine Fisheries Service, Northwest and Alaska Fisheries Center, P.O. Box 155, Auke Bay, Alaska 99821

P. J. Ritacco (165), Environmental Protection Agency, Environmental Research Laboratory, South Ferry Road, Narragansett, Rhode Island 02882

G. Roesijadi (357), Battelle Pacific Northwest Division, Marine Research Laboratory, Sequim, Washington 98382

Akella N. Sastry (265), University of Rhode Island, Graduate School of Oceanography, Narragansett, Rhode Island 02882

Sara Singer (323), Skidaway Institute of Oceanography, Savannah, Georgia 31406

Katherine S. Squibb (145), National Institute of Environmental Health Sciences, Laboratory of Organ Function and Toxicology, Research Triangle Park, North Carolina 27709

James Stolzenbach (323), Skidaway Institute of Oceanography, Savannah, Georgia 31406

William G. Sunda (127), National Marine Fisheries Service, NOAA, Southeast Fisheries Center, Beaufort Laboratory, Beaufort, North Carolina 28516

Kenneth R. Tenore (323), Skidaway Institute of Oceanography, Savannah, Georgia 31406

Peter Thomas (73), Department of Biology, Texas A&M University, College Station, Texas 77843

Robert E. Thomas (425), Chico State University, Department of Biological Sciences, Chico, California 95926

Frederick P. Thurberg (335), National Marine Fisheries Service, Milford Laboratory, Milford, Connecticut 06460

Sandra L. Vargo (295), University of Rhode Island, Graduate School of Oceanography, Narragansett, Rhode Island 02882

Jeannette A. Whipple (483), National Marine Fisheries Service, Tiburon Laboratory, Southeast Fisheries Center, 3150 Paradise Drive, Tiburon, California 94920

PREFACE

The papers in this volume were presented at a symposium entitled Pollution and Physiology of Marine Organisms held at the National Marine Fisheries Service, Milford Laboratory, Milford, Connecticut, November 7–9, 1979. This symposium, which was sponsored jointly by the School of Public Health and the Belle W. Baruch Institute for Marine Biology and Coastal Research, both at the University of South Carolina, and the Northeast Fisheries Center, National Marine Fisheries Service, is the fourth one organized around the theme of physiological effects of pollutants on marine organisms.

There has been considerable laboratory research performed on the physiological effects of pollutants on marine organisms, with a more recent trend toward applying laboratory techniques to field studies. The objective of this volume is to show how these techniques may be used in field-oriented biological monitoring programs, a small number of which are now underway in various parts of the world. The combination of laboratory studies with field studies should give us a better understanding of the effects of pollutants on marine life.

We are indebted to the authors for their cooperation, to the reviewers of all manuscripts for sharing their views with us, and to the staff of the University of South Carolina for their assistance in many ways.

F. John Vernberg

Anthony Calabrese

Frederick P. Thurberg

Winona B. Vernberg

Section I

SYNTHETIC ORGANICS

AN OVERVIEW OF THE ACUTE AND CHRONIC EFFECTS
OF FIRST AND SECOND GENERATION PESTICIDES
ON AN ESTUARINE MYSID

D. R. Nimmo

Environmental Research and Technology, Inc.
Post Office Box 2105
Fort Collins, Colorado

T. L. Hamaker
E. Matthews
J. C. Moore

ERL, Environmental Protection Agency
Sabine Island
Gulf Breeze, Florida

INTRODUCTION

The mysid shrimp, *Mysidopsis bahia,* was introduced as a
laboratory organism for life-cycle toxicity testing in 1977
(Nimmo *et al*. 1977). This species has since been used
experimentally to determine (1) the toxicity of dredge
materials (Shuba *et al*. 1978), (2) the toxicities of several
priority pollutants (U.S. EPA 1978), (3) the acute and
chronic effects of a metal (Nimmo *et al*. 1978a) and (4) pesticides (Nimmo *et al*. 1977; Nimmo *et al*. 1979; Schimmel *et al*.
1979; and Nimmo *et al*. 1980). In addition, *M. bahia* has been
used successfully to determine possible long-term effects due
to dredging in estuaries (Nimmo *et al*. in press).

Because some of the research completed with pesticides has
not been published elsewhere, this paper will briefly
summarize data from eleven studies in which pesticides were
used as test toxicants. Additional purposes of this paper are
to summarize criteria for determining chronic effects,
where possible, to compare the sensitivity of mysids with
other species exposed to the same chemicals and, finally, to

discuss acute versus chronic test results with regard to
estimated "safe" concentrations of a chemical throughout an
entire life cycle.

MATERIALS AND METHODS

Detailed procedures for culturing the mysids and testing
protocols have been published previously (Nimmo *et al.* 1978b
and 1978c). Mysids, obtained from Santa Rosa Sound near
Pensacola, Florida, were cultured by two methods: (1)
flowing sea water supplied to 38-liter glass aquaria or a
150-liter circular fiberglass tank, and (2) a static recirculating method in which a biological filter was established in the substratum of a 38-liter aquarium. To provide circulated water, an air-lift tube set in its filter
base, was attached to a small glass chamber (vented to the
atmosphere) and mounted directly above the aquarium. Currents
created by the water flow appeared to aid adult mysids in
obtaining food. In all culture devices, larval brine shrimp,
Artemia salina, were fed to the mysids *ad libitum.*
The mysids were exposed to the eleven pesticides in
flowing saltwater; the carrier solvent was triethylene glycol.
All tests were conducted in systems delivering intermittent
flows from a diluter (Mount and Brungs 1967) or continuous
flows with each toxicant added by an infusion pump (Bahner
et al. 1975). Previous studies have shown these methods to be
comparable because approximately equal volumes of water flow
through the test aquaria and the desired pesticide concentrations are easily maintained with either system.
The mysids were observed through a life cycle and, in most
cases, two life cycles for a total of 28 days. In a temperature range of 22 to 25°C, female mysids usually release
multiple broods with the number of juveniles per brood
varying from about 3 to 12 per female. Thus, the number of
young per female (or reproductive success) is an easily
observed effect. Other effects, such as delay in release of
young, growth, abnormal behavior, effects on feeding activity
and death, were also noted.
The pesticides used in this study are listed in Table 1.
DEF® and trifluralin and herbicides, the remaining nine are
insecticides. Diazionon®, EPN, leptophos, Methyl parathion®
and phorate and organophosphates; Kepone® and Toxaphene®
are chlorinated hydrocarbons; Sevin, a carbamate; and
Dimilin®, a benzoylphenyl urea.

TABLE 1. *Pesticides used in this study and their chemical names.*

COMMON NAME OR REGISTERED TRADEMARK	CHEMICAL NAME
DEF	S, S, S-Tributyl phosphorotrithioate
Diazinon	O, O-Diethyl O-(2-isopropyl-6-methyl-4-pyrimidinyl) phosphorothioate
Dimilin	N-[[4-Chlorophenyl) amino]carbonyl] 2, 6-difluorobenzamide
EPN	O-Ethyl O-p-nitrophenyl phenylphos-phonothioate
Kepone	Decachloro octahydro-1, 3, 4-metheno-2H-cyclobuta [cd] pentalene-2-one
Leptophos (Phosvel)	O-(4-Bromo-2, 5-dichlorophenyl)O-methylphenylphosphonothioate
Methyl Parathion	O, O-Dimethyl O-p-nitrophenyl phos-phorothioate
Phorate (Thimet)	O, O-Diethyl S-(ethylthio)methyl phosphorodithioate
Carbaryl (Sevin)	N-Naphthyl-N-methylcarbamate
Toxaphene (Camphechlor)	Technical chlorinated camphene
Trifluralin (Treflan)	α, α, α-Trifluoro-2, 6-dinitro-N, N-dipropyl-p-toluidine

Test aquaria were sampled to determine concentrations of pesticide. Samples from each test concentration were analyzed at least twice during a 96-hour acute test and at least four times during a life-cycle test of 28 days duration. Results showed a variance of less than 20% for each concentration. One liter samples were extracted with two 100-ml portions of an organic solvent. Petroleum ether was used to extract most of the compounds; methylene chloride was used in Kepone and Sevin analyses; and 1:1 diethyl ether-petroleum ether (V/V) in Dimilin analyses. The extracts were combined and concentrated on a steam table prior to analysis.

Two chromatographic methods were used to separate and quantify the pesticides: high-pressure liquid chromatography (HPLC) with UV-absorption detection and gas-liquid chromatography (GLC) with electron capture and nitrogen phosphorus flame ionization detectors (NPFID). Sevin and Dimilin concentrations were determined by the HPLC method. Phorate concentrations were determined by the GLC method with a NPFID and the others were determined by the GLC method, using electron caputre detectors (Table 2).

STATISTICAL ANALYSES

The LC50s and 95% confidence limits were calculated by linear regression analysis after probit transformation. Abbot's correction factor for adjusting control mortality was used in the statistical program.

The maximum acceptable toxicant concentration (MATC) was calculated as the geometric mean between the lowest concentration that had a significant effect and the highest concentration that had no significant effect in the life-cycle test. The most sensitive criterion for effects in the life-cycle test was used to calculate the MATC; i.e., survival, reproduction, or growth. Selection of the most sensitive criterion was based on the results of the statistical analysis of each set of data. The two sample "t" test was used to compare controls and treatment with respect to growth and survival. Analysis of Variance (ANOVA) was used to determine significant differences among the six treatments (five test concentrations and control) with respect to the number of young per female. If the results of the ANOVA showed significant difference among those treatments, multiple comparisons (*post hoc*) tests were made to distinguish between "no effect" and "effect" groups of concentrations by determining nonsignificant group. Student-Newman-Keuls (SNK), Duncan's, Dunnett's, and Bonferrone's tests were used to

TABLE 2.   Detection limits and recoveries of pesticides from water.

Compound	Method of Detection	Column and Temperature	Detection Limit (μg/1)	Fortified Recovery (%)	Reference
DEF	GLC/EC[1]	2% SP2100 210°C	0.010	>85	Unpublished
Diazinon	GLC/EC	2% SP2100 175°C	0.020	>90	Goodman 1979
Dimilin	UV-ABSOR[2]	μ NH$_2$[3]	0.40	>85	Nimmo et al. 1980
EPN	GLC/EC	2% SP2100 210°C	0.20	85	Schimmel et al. 1979
Kepone	GLC/EC	2% SP2100 200°C	0.20	>85	Schimmel and Wilson 1977
Leptophos	GLC/EC	2% SP2100 210°C	0.025	>85	Schimmel et al. 1979
Methyl Parathion	GLC/EC	2% SP2100 210°C	0.10	>85	Unpublished
Phorate	GLC/NPFID	2% SP2100 200°C	0.020	>85	Unpublished
Sevin	UV-ABSOR	μ NH$_2$	3.0	>95	Unpublished
Toxaphene	GLC/EC	2% SP2100 200°C	0.020	>85	Schimmel et al. 1977
Trifluralin	GLC/EC	5% QF-1 175°C	0.020	>85	Unpublished

[1]GLC/EC, Gas- Liquid Chromatography/Electron-Capture Detector

[2]UV-ABSOR, UV Absorbance Detection

[3]μ NH$_2$, Micro NH$_2$ Column

compare treatments and controls with respect to total brood.
These statistical tests are similar and have been modified for
use in computer programs. Since these studies spanned a
period of four years, differing tests were employed as
computational equipment and programs became available. Significant differences ($\alpha=0.05$) were attributed to the effects
of the respective pesticide being tested.

RESULTS AND DISCUSSION

Acute Toxicities

     All pesticides were acutely toxic to *Mysidopsis bahia,*
except trifluralin, which had a 96-hour LC50 greater than
136.0 µg/l (Table 3). Greater test concentrations were not
possible due to low solubility in water. Phorate was the
most acutely toxic chemical, 96-hour LC50=0.33 µg/l. Therefore, the range of LC50s for these pesticides was three
orders of magnitude.
     Although the criterion of effect in acute tests was death,
observations of the animals exposed to some chemicals indicated stress before death. In tests with Phorate and
Methyl parathion, loss of equilibrium, reduced feeding
activity and reduced growth (development) occurred. Reduced
growth of the juveniles was particularly conspicuous in the
higher concentrations of Kepone and DEF. An effect observed
only in the Dimilin tests was failure of the mysids to produce
a chitinous exoskeleton during the molt.
     In acute tests, differences in mortality were noted
between juveniles and adults and, in general, juveniles showed
the greater sensitivity. Exceptions were that Diazinon and
EPN were more acutely toxic to adults than to juveniles;
however, these differences were less than a factor of 1/2 or
1/3. The reason for using juveniles instead of adult animals
for acute toxicity tests was the minimal effort to stand-
ardize the age and sizes of the animals prior to beginning
the test. Juveniles 48 hours old or younger, released from
adult females, were used to begin each acute and chronic
test (Nimmo *et al.,* 1978c).
     Two comparisons can be made between the sensitivites of
mysids and other species tested with the same pesticide, the
first being a comparison between mysids and decapods tested
with the pesticides used in this study (Table 4). The 48- or
96-hour LC50s of penaeid shrimp were available with one
exception: the stone crab, *Menippe mercenaria,* was exposed
to Dimilin from the hatching of the larvae to the first crab

TABLE 3. Summary of toxicity tests using Mysidopsis bahia and eleven pesticides.

Compound	96-Hour LC50 (μg/l) 95% Confidence Limits (juveniles)	MATC Limits (μg/l) (juvenile through adult)	Application Factor[1]
DEF	4.55 (4.21 - 4.92)	<0.34[2]	<0.08
Diazinon	4.82 (4.11 - 5.87)	>1.15 < 3.27	0.54
Dimilin	1.97 (1.73 - 2.24)	<<0.4[2]	<<0.2
EPN	3.01 (2.46 - 3.75)	>0.44 < 4.13	0.45
Kepone	10.29 (8.07 -12.57)	>0.026< 0.34	0.01
Leptophos	3.31 (2.77 - 4.04)	>0.64 < 1.77	0.32
Methyl Parathion	0.77 (0.64 - 0.98)	>0.11 < 0.16	0.17
Phorate	0.33 (0.27 - 0.43)	>0.09 < 0.14	0.33
Sevin	>7.7 --	--	--
Toxaphene	2.67 (1.94 - 4.05)	>0.07 < 0.14	0.04
Trifluralin	>136.4 --	--	--

[1]Derived by dividing the geometric mean of the maximum acceptable toxicant limits by the 96-hour LC50.

[2]A no-effect concentration has not been determined.

TABLE 4. Comparative toxicities of pesticides used in this study to juvenile mysids and some representative marine decapods.

Pesticide	Mysid 96-Hr.LC50 (μg/l)	Species	48-Hr.LC50 (μg/l)	Reference
Phorate	0.33	Penaeus aztecus	0.46	Butler 1964
Methyl Parathion	0.77	Penaeus aztecus	3.2	Butler 1964
Dimilin	1.97	Menippe mercenaria	0.5[2]	Costlow 1979
Toxaphene	2.67	Penaeus aztecus	2.7	Lowe, personal communication[1]
EPN	3.01	Penaeus duorarum	0.29[3]	Schimmel et al. 1979
Leptophos	3.31	Penaeus duorarum	1.88[3]	Schimmel et al. 1979
DEF	4.55	Penaeus aztecus	28.0	Butler 1965
Diazinon	4.82	Penaeus aztecus	28.0	Lowe, personal communication[1]
Sevin	>7.70	Penaeus aztecus	1.5	Butler 1963
Kepone	10.29	Penaeus aztecus	68.0	Butler 1963
Trifluralin	>136.0	Penaeus duorarum	>210.0[3]	Schimmel, personal communication[1]

[1]Personal communication from J. I. Lowe and S. G. Schimmel. Information available from ERL, EPA, Sabine Island, Gulf Breeze, Florida 32561.

[2]Decreased survivorship of larvae.

[3]96-Hr. LC50.

stage (Costlow 1979). The LC50s for the decapods, most of
them tested as sub-adults, were within an order of magnitude
of those for juvenile mysids. Although these data are not
conclusive because of differing procedures, exposure times,
and small sample size, they indicate fair agreement in
results. One obvious advantage of using mysids rather
than juveniles or sub-adults of decapod crustaceans is the
mysid's availability throughout the year.

Other investigations showed the mysid's sensitivity
compared to other species commonly tested against a variety
of chemicals. These studies (U.S. EPA 1978) were conducted
to fill informational gaps in water quality criteria documents
for both fresh-water and marine species. Results showed that
M. bahia was equal to, or was the most sensitive species, 52%
of the time based on 29 priority pollutants for which comparable data existed. The fresh water species in these tests
were the alga, *Selenastrum capricornutum*; a daphnid, *Daphnia
magna*; and the bluegill sunfish, *Lepomis macrochirus*. The
marine species were: an alga, *Skeletonema costatum*; the
mysid, *Mysidopsis bahia;* and the sheepshead minnow, *Cyprinodon
variegatus*. The order of decreasing species sensitivity
(expressed as LC50s for the crustaceans and fishes and EC50s
for the algae) was: *M. bahia* (the most sensitive), followed
by *S. costatum, L. macrochirus, D. pulex, S. capricornutum*
and *C. variegatus* (the least sensitive).

LIFE CYCLE TESTS

The data listed in Table 3 illustrate the differences in
sensitivities of this mysid to chemicals tested for 96 hours
compared to tests through an entire life cycle. Few estuarine
animals have, historically, been available to biologists for
life-cycle testing and the primary purpose of these studies
was to generate data for comparisons of acute versus chronic
tests with chemicals, and ultimately to predict long-term
effects from short-term tests. The Maximum Acceptable
Toxicant Concentration (MATC) (Table 3) was calculated for
each chemical to provide an estimated concentration considered
to be "safe" or incapable of providing an observable or
measurable effect throughout an animal's life cycle (Eaton
1973). The Application Factor (AF) (Table 3) is the ratio of
the MATC divided by the LC50 and provides a number by which
estimates of chronic toxicity can be made for chemicals for
which only acute toxicity data exists. It should be pointed
out that initially the application factor concept was used

primarily for fresh water fishes because adequate data bases
existed for these species.

In our studies with mysids, the end point of effect in
the acute tests was death but the criteria to determine the
MATC depended upon such responses as reduced growth, decreased
number of young per female, or chronic toxicity (Table 5).

In tests using measured concentrations of pesticide,
Kepone was the most toxic chemical throughout an entire life
cycle and the MATC was based on a measurement of growth
(Table 3). Female mysids exposed to 0.072 µg/l Kepone grew
about 6% less than control mysids, whereas growth of females
exposed to 0.026 µg/l was not significantly different from
controls. Also, the average number of young per female was
8.9 in the 0.39 µg/l, compared to 15.3 in the controls.
Therefore, the effects of Kepone in reducing both growth
and reproduction in our tests were apparently consistent with
the findings of Jensen (1958) whose studies with 13 species
of malacostracans, including five species of mysids, established that the number of eggs produced per female was a
direct function of body length.

The most toxic pesticide, based on nominal concentrations,
in a life cycle was Dimilin. This test is distinguished from
the test with Kepone because the effects of Dimilin on mysids
were observed in concentrations below the limit of detection
of our analytical capability. Only 13.5 young per female were
produced in an estimated concentration of 0.075 µg/l, whereas
21.4 and 20.9 young per female were produced in controls
(Table 6). With Dimilin we observed a significant decrease in
reproduction with increasing concentrations of the chemical.
Survival was greater than 62% in all concentrations, but the
number of juveniles per female was significantly less. Therefore, we have not been able to arrive at a "no effect"
concentration of Dimilin.

EPN and Methyl Parathion produced significant effects on
both survival and reproduction in the same concentration of
pesticide. For example, with continuous exposure to EPN at a
concentration of 4.13 µg/l (Table 7), survival was only 32.5%,
a significant reduction compared to the 89.7 and greater in
all other test concentrations and controls. Also, the number
of juveniles per female was only 3.5 compared to the 14.6 and
greater in all other tests. Thus, the total numbers of mysids
in the test concentration of 4.13 µg/l EPN were reduced by
two factors acting in concert: (1) greater mortality of
parental stock and (2) decreased production of juveniles by
surviving females.

TABLE 5. Criteria used to establish MATC[1] limits for nine pesticides.

TABLE 5. Criteria used to establish MATC[1] limits for nine pesticides.

Compound	Criteria[2]
Kepone	Growth (Development)
DEF	# Young/Female
Dimilin	# Young/Female
Toxaphene	# Young/Female
Leptophos	# Young/Female
Diazinon	Chronic Toxicity
Phorate	Chronic Toxicity
Methyl Parathion	# Young/Female and Chronic Toxicity
EPN	# Young/Female and Chronic Toxicity

[1]MATC, Maximum Acceptable Toxicant Concentration.

[2]Growth, length measurement of mysids from tip of carapace to end of uropod; number young/female, number of young per producing female; chronic toxicity, mortality of parental stock throughout a life cycle.

TABLE 6. Effect of Dimilin® on reproductive success
and survival of *Mysidopsis bahia* at 22–26°C and 22–28%
salinity.

Nominal Test Concentrations (μg/l)	Percentage Survival Of Adults (Day 28)	Juveniles Per Female (Day 28)
Seawater Control	97.5	21.4
TEG[1] Control	94.5	20.9
0.075	75.5	13.5^2
0.25	81.5	10.2^2
0.5	71.0	7.2^2
0.75	62.0	2.4^2

[1]Triethylene Glycol Control.

[2]Significantly different from controls at α = 0.05 (Dunnett's Test).

TABLE 7. Survival and reproductive success of *Mysidopsis bahia* chronically exposed to EPN in flowing seawater for 28 days at 23 to 24° and 22 to 26% salinity.

Measured Test Concentrations (μg/1)	Percentage Survival Of Adults (Day 28)	Juveniles Per Female (Day 28)
Seawater	89.7	16.1
Triethylene Glycol Control	97.4	14.6
0.057	100.0	18.5
0.076	92.5	14.8
0.44	92.5	15.8
4.13	32.5[1]	3.5[2]

[1]Significantly different from controls (Student's t) α = 0.05.

[2]Significantly different from controls at α = 0.05 (Dunnett's test).

Application Factors (the ratio of MATC divided by the LC50) ranged from 0.54 for Diazinon to 0.01 for Kepone, indicating that for some of the chemicals such as Diazinon, EPN, leptophos and phorate, a 96-hour acute test is close approximation of the MATC.   Interestingly, all but Kepone above are organophosphates with "quick kill" properties.   However, making estimates or predictions about the long-term effects from 96-hour LC50 values for some of the other pesticides (DEF, Kepone and Toxaphene), may cause the observer to underestimate the long-term chronic effects on mysids. Research with mysids to date suggests that caution should be taken when estimating application factors without data from laboratory studies.   Perhaps with more laboratory experiments, using a variety of chemicals to arrive at variance estimates, a single generic application factor can be computed.

SUMMARY AND CONCLUSIONS

Results of laboratory experiments conducted at the Environmental Research Laboratory for several years indicate that the mysid shrimp, *Mysidopsis bahia,* can be used as a practical test species for life-cycle studies to determine the sublethal effects of toxicants.   Ratios of the acute 96-hour lethal concentrations to the maximum acceptable toxicant concentration (MATC) ranged from 0.01 to 0.54 indicating that, in some instances, acute tests would be predictive of chronic effects.   Criteria used to establish MATC limits were effects on growth, reduced reproduction, chronic toxicity and, with two chemicals, reproduction and chronic toxicity, concurrently.   Though the data base is incomplete, the acute tests with *M. bahia* and eleven pesticides show sensitivities within a factor of 10 to some decapod crustaceans--most of them being penaeid shrimp. Acute studies conducted elsewhere show *M. bahia* to be as sensitive to, or in many instances, more sensitive than five fresh water and marine species commonly used in toxicity testing.

LITERATURE CITED

Bahner, L. H., C. D. Craft, and D. R. Nimmo. 1975. A salt
    water flow-through bioassay method with controlled temper-
    ature and salinity. Progressive Fish-Culturist 37: 126-
    129.

Butler, P. A. 1963. Commercial Fishery Investigations. U.S.
    Fish and Wildlife Service Circular No. 167: 18.

Butler, P. A. 1964. Commercial Fishery Investigations. U.S.
    Fish and Wildlife Service Circular No. 199: 24.

Butler, P. A. 1965. Commercial Fishery Investigations. U.S.
    Fish and Wildlife Service Circular No. 226: 70.

Costlow, J. C., Jr. 1979. Effect of Dimilin on development
    of larvae of the stone crab *Menippe mercenaria,* and blue
    crab *Callinectes sapidus.* In: W. B. Vernberg, F. P.
    Thurberg, A. Calabrese, and F. J. Vernberg (eds.) Marine
    Pollution: Functional Responses, Academic Press, Inc.,
    New York.

Eaton, J. G. 1973. Recent developments in the use of lab-
    oratory bioassays to determine "safe" levels of toxicants
    for fish. G. E. Glass (ed.). Ann Arbor Science Pub-
    lishers, Inc., Ann Arbor, Michigan. Pages 107-115.

Goodman, L. 1979. Chronic toxicity to and brain acetyl-
    cholinesterase inhibition in the sheepshead minnow,
    Cyprinodon variegatus. Trans. Am. Fish. Soc. 108: 479-
    488.

Jensen, J. P. 1958, The relation between body size and
    number of eggs in marine malacostrakes. Meddelelser Fra
    Danmarks Fiskeri-og Harundensogelser 2 (19): 1-25.

Mount, D. I. and W. Brungs. 1967. A simplified dosing
    apparatus for fish toxicology studies. Water Research
    2: 21-29.

Nimmo, D. R., L. H. Bahner, R. A. Rigby, J. M. Sheppard and
    A. J. Wilson, Jr. 1977. *Mysidopsis bahia:* an estuarine
    species suitable for life-cycle toxicity tests to de-
    termine the effects of a pollutant. In: F. L. Mayer and
    J. L. Hamelink (eds) Aquatic Toxicology and Hazard
    Evaluation, ASTM STP 634. Pages 109-116.

Nimmo, D. R., R. A. Rigby, L. H. Bahner and J. M. Sheppard.
    1978a. The acute and chronic effects of cadmium on the
    estuarine mysid, *Mysidopsis bahia*. Bull. Environ.
    Contam. Toxicol. 12: 80-85.

Nimmo, D. R., T. L. Hamaker, and C. A. Sommers. 1978b.
    Culturing the mysid (*Mysidopsis bahia*) in flowing seawater
    or a static system. In: Bioassay Procedures for the
    Ocean Dispodal Permit Program. ERL USEPA, Gulf Breeze,
    Florida 32561, EPA-600/9-78-010: 59-60.

Nimmo, D. R., T. L. Hamaker, and C. A. Sommers. 1978c.
    Entire life cycle toxicity test using mysids (*Mysidopsis
    bahia*) in flowing water. In: Bioassay Procedures for the
    Ocean Disposal Permit Program. ERL USEPA, Gulf Breeze,
    Florida 32561, EPA-600/9-78-010: 64-68.

Nimmo, D. R., T. L. Hamaker, J. C. Moore, and C. A. Sommers.
    1979. Effect of diflubenzuron on an estuarine crustacean.
    Bull. Environ. Contamin. Toxicol. 22: 767-770.

Nimmo, D. R., T. L. Hamaker, J. C. Moore and R. A. Wood.
    1980. Acute and chronic effects of Dimilin on survival
    and reproduction of *Mysidopsis bahia*. Aquatic Toxicology,
    ASTM STP, J. G. Eaton, P. R. Parrish, and A. C. Hendricks,
    Eds., Am. Soc. Testing and Materials. Pages 366-376.

Nimmo, D. R., T. L. Hamaker, E. Matthews, and W. T. Young.
    In press. The long-term effects of suspended sediment on
    survival and reproduction of the mysid shrimp, *Mysidopsis
    bahia,* in the laboratory. In: Ecological Stress and the
    New York Bight: Science and Management. NOAA, MESA
    Publications.

Schimmel, S. C., T. L. Hamaker, and J. Forester. 1979.
    Toxicity and bioconcentration of EPN and Leptophos to
    selected estuarine animals. Contrib. Mar. Sci. 22: 193-
    203.

Schimmel, S. C., J. M. Patrick, and J. Forester. 1977. Up-
    take and toxicity of toxaphene in several estuarine
    organisms. Arch. Environ. Contam. Toxicol. 5: 353-367.

Schimmel, S. C., and A. J. Wilson, Jr. 1977. Acute toxicity
    of Kepone to four estuarine animals. Chesapeake Science
    18 (2): 224-227.

Shuba, P. J., H. E. Tatem, and J. H. Carroll. 1978. Bio-
    logical assessment methods to predict the impact of open
    water disposal of dredged material. Tech. Report D-78-
    50. Environmental Laboratory, U. S. Army Engineer Water-
    ways Exp. Sta., P. O. Box 631, Vicksburg, Miss. 39180.
    86 pages.

U.S.E.P.A. 1978. In-depth studies on health and environ-
    mental impacts of selected water pollutants. Contract
    No. 68-01-4646.

EFFECTS OF TWO ORGANOPHOSPHATE PESTICIDES ON SWIMMING STAMINA OF THE MYSID *MYSIDOPSIS BAHIA*

Geraldine M. Cripe
DelWayne R. Nimmo [1]
Timothy L. Hamaker

Gulf Breeze Environmental Research Laboratory
EPA, Gulf Breeze, Florida

INTRODUCTION

Swimming responses of fishes have been used to document effects of altered environmental factors. Laboratory-measured swimming performance is influenced by reduced dissolved oxygen concentrations (Katz *et al.*, 1959; Dahlberg *et al.*, 1968), and environmental temperature changes (Brett, 1964; Otto and Rice, 1974; Rulifson, 1977). Swimming speed measurements are used to compare wild and domesticated brook trout (Vincent, 1960), and naturally and artificially propagated fry of sockeye salmon (Bams, 1967), as well as to observe effects of bleached kraft pulpmill effluent (Howard, 1975), hydrogen sulfide (Oseid and Smith, 1972), and fenitrothion (Peterson, 1974) on fish populations.

Most studies of crustacean activity in flowing water have concentrated on determining the environmental cues used for directional movement. Using a portunid crab, *Macropipus holsatus*, Venema and Creutzberg (1973) found an inverse relationship between swimming activity and salinity in laboratory tests. Hughes (1969, 1972) investigated the response of postlarval and juvenile *Penaeus duorarum* to salinity as a possible tidal transport mechanism. Changes in phototaxis and geotaxis

[1] *present address: P. O. Box 2105, 1716 Heath Parkway, Ft. Collins, Colorado*

of some larval crustaceans have been related to salinity fluctuations (Latz and Forward, 1977). Allen (1978) found that
when *Neomysis americana* were placed in a transparent circular
container within a rotating white cylinder, the mysids maintained their position with respect to the vertical lines on
the cylinder wall.

Mysidopsis bahia exhibit positive rheotaxis in both field
and laboratory (personal observation) when attempting to maintain position in a current with respect to their environment.
This mysid, therefore, is suitable for experiments designed to
test swimming performance.

Mysids are an important factor in the estuarine community.
They constitute 52% of the stomach contents of the bay
anchovy, *Anchoa mitchilli* (Darnell, 1958). Copepods and
mysids form 96% of the stomach contents of the juvenile leath-
erjacket, *Oligoplites saurus*, and are the principal zooplankton in the diets of 15 or 20 species of juvenile fish
studied by Carr and Adams (1973).

Mysidopsis bahia are easily cultured in the laboratory
(Nimmo *et al.*, 1977). About 14 days after release from the
brood pouch, the young (approximately 1 cm in length) become
sexually mature. Within the next 14 days, the new adults can
produce at least two broods of young (Nimmo *et al.*, 1980).

M. bahia are used routinely to determine acute (96-hr
LC50, or Median Lethal Concentration (Sprague, 1969) and
chronic life-cycle effects (i.e., reproduction rates, growth,
and survival) of pollutants (Nimmo *et al.*, 1980). A 96-hr
LC50 concentration of a particular toxicant does not necessarily indicate the concentration at which long-term effects
become apparent, as do life-cycle tests with *M. bahia* that
require 28 days. Thus, useful data could be obtained in a
much shorter time if a 96-hr test is used to correlate a
behavioral response (e.g., swimming ability) with specific
concentrations of a toxicant that cause adverse effects on
the life cycle.

Organophosphate insecticides, in tests with aquatic
organisms, have reduced fish brain acetylcholinesterase
activity (Coppage, 1972; Coppage and Matthews, 1974), prevented reproduction, and also decrease survival of fishes
(Carlson, 1971; Hansen and Parrish, 1977) and mysids (Nimmo *et
al.*, 1977; Nimmo *et al.*, 1978).

The purpose of our study was to test the ability of *M.
bahia* to maintain its position in a current and to determine
the feasibility of a behavioral bioassay based on short-term
toxicant exposure and rheotactic response in a stamina
tunnel. This was accomplished by (1) observing the maximum
sustained speed (MSS) of various age groups of mysids
throughout a life cycle, (2) comparing the performance of

laboratory-reared *M. bahia* at specific ages with young of feral *M. bahia* and (3) comparing the MSS of laboratory-reared control *M. bahia* with that of mysids that had been exposed for 96 hr to different concentrations of the organophosphate insecticides, methyl parathion[2] or phorate[2].

MATERIALS AND METHODS

Source and Culture of Test Mysids

Laboratory-reared *Mysidopsis bahia* were obtained from a 3-yr-old continuous culture maintained in filtered flowing seawater at the Gulf Breeze Laboratory, near Pensacola, Florida. Feral *M. bahia* were collected at the Laboratory from shallow man-made ponds constantly supplied with unfiltered seawater. Animals cultured for testing were held in a 76-liter aquarium with filtered recirculating water of $21\pm1°C$, 20 °/oo, and constant light. They were fed 48-hr-old *Artemia salina* larvae daily.

Gravid females were maintained in an aerated 4-liter glass beaker with a nylon screen collar (Nitex[3], mesh number 210). A recirculating aquarium filter maintained a continuous water flow in the beaker. Within 24 hr of release from the brood pouch, juvenile mysids were collected in groups of 45 to 60 and placed in petri dish culture containers described by Nimmo et al. (1977). When a group of mysids attained the desired age of 2, 4, 7, 12, 14, 17 or 22 days, 35 to 45 animals from that age group were divided randomly into three samples for testing in the tunnel. Stamina measurements from each age group were pooled for analysis. Laboratory-reared (hereafter called "lab stock") and feral mysids ("feral stock") were obtained and confined separately in this manner. Of the feral stock, only first generation offspring raised in the laboratory were used in this study. Stamina measurements were taken to determine differences between sexes in a group of 17-day-old mysids.

[2]*Methyl parathion, (O,O-Dimethyl-O-p-nitrophenyl-phosphorothioate); phorate, (O,O-Dimethyl-O-[4-(methylthio)-n-tolyl] phosphorothioate).*

[3]*Registered trademark, Tobler, Ernst and Trabor, Inc., Murray St., New York, NY. Mention of tradename does not constitute endorsement by the U.S. Environmental Protection Agency or the University of West Florida.*

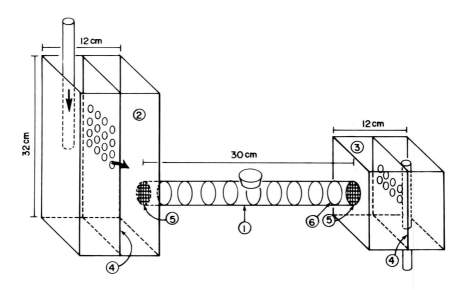

FIGURE 1. Stamina tunnel used to measure the ability of M. bahia to maintain its position in a current. Arrows indicate water flow. 1, experimental chamber; 2, incurrent headbox; 3, excurrent headbox; 4, baffle; 5, retaining screens; 6, benchmark.

Stamina Measurement Apparatus

The stamina tunnel (Figure 1) was modified from Thomas *et al.* (1964). Water flowed through siphons from an overhead storage tank to an incurrent chamber. The incurrent and ex-current chambers were attached to opposite ends of the 5-cm (i.d.) stamina tunnel that was 30-cm long and constructed from 3-mm thick, clear plexiglass. Test animals were intro-duced through a 2.5-cm-diameter hole in the top center of the tunnel. The hole was closed by a rubber stopper trimmed flush with the inside surface. A plexiglass baffle with 4-mm-diameter holes was placed in each chamber to aid in obtaining rectilinear flow through the tunnel. Retaining screens of Nitex (mesh number 210) confined the animals to the test area of the tunnel, as well as provided microturbulence for laminar flow in the tunnel. Dye injection was used to demonstrate microturbulence. Water velocity was calculated by dividing the total volume of water flowing into the incurrent chamber by the cross-sectional area of the tunnel. Water exited through the standpipe into a holding tank, then was pumped to the overhead tank and recirculated. The water in the system was changed and the tunnel cleaned before each age group was

tested and before mysids exposed to either pesticide were
tested. The water was maintained at $24.5 \pm 0.5°C$ by heat-
exchange tubing from a MGW Lauda-Brinkman Model RC3 cooler
and heater during testing; salinity was the same in the tunnel
and culture.

Black lines (2-mm wide) that encircled the tunnel at 2.5-
cm intervals served as potential visual cues for the mysids
and reference points for the observer. The line immediately
adjacent to the excurrent chamber was chosen as the benchmark.
Only the speed of the animals located upstream from the benchmark at the beginning of the test was recorded. An opaque
curtain encircling the stamina apparatus isolated the mysids
from outside stimuli. A 5-cm observation hole was cut in the
curtain near the benchmark, and diffused light from a 100-watt
incandescent bulb provided illumination for the tests.

Test Procedures

Stamina tunnel. For each test, 10 to 15 mysids were
pipetted into the completely filled tunnel and left undisturbed in still water for 15 minutes; then a very slow flow
of water was begun to acclimate the animals. Step increases
in water velocity were made according to the regime in Table
1.

The number of animals upstream from the benchmark was recorded at the beginning and end of the periods of acclimation
and at each increment of velocity. The velocity of the water
and length of time each mysid swam before it was swept past
the benchmark was recorded and used to compute its maximum
sustained speed. The animals were then discarded.

Calculation of Maximum Sustained Speed. The calculation
of the maximum sustained speed (MSS) at which the mysids
could maintain their position against the current is similar
to that used by Brett (1964) for critical swimming speeds of
fish. MSS is calculated by:

$$MSS = HV + VI\ (\ TE/TT\ ),$$

where

MSS = maximum sustained speed in cm/sec,
HV = highest velocity in cm/sec that was maintained by
the mysid for the prescribed time period,
VI = velocity increment in cm/sec between HV and the
final velocity,
TE = time in sec endured at final velocity,
TT = prescribed time increment in sec.

TABLE 1. Water velocity (cm/sec) regime for testing the stamina of various ages of M. bahia.

	2-days-old		4-days-old		7-, 12-, 14-, 17-, and 22-days-old	
	Velocity Increment	Cumulative Velocity	Velocity Increment	Cumulative Velocity	Velocity Increment	Cumulative Velocity
Acclimation (15 min.)	0	0	0	0	0	0
Acclimation (15 min.)	3.0	3.0	1.8	1.8	3.0	3.0
Increment 1 (5 min.)	1.8	4.8	1.2	3.0	3.0	6.0
Increment 2 (5 min.)	1.8	6.6	3.0	6.0	6.2	12.2
Increment 3 (5 min.)	1.8	8.4	6.2	12.2	6.2	18.4
Increment 4 (5 min.)	3.0	11.4	6.2	18.4	3.9	22.3
Increment 5 (5 min.)	3.1	14.5	3.9	22.3	3.6	25.9
Increment 6 (5 min.)	3.1	17.6	-	-	-	-

Pesticide Exposure Tests. Laboratory-reared mysids for the pesticide exposure tests were obtained in a manner similar to that for testing age and sex differences; 40 to 45 mysids, less than 24-hr-old, were exposed for 96 hr to each treatment of a phorate[4] flow-through toxicity test. Stock solutions, prepared by dissolving the pesticide in a carrier, triethylene glycol (TEG), were metered into filtered seawater via a diluter (Mount and Brungs, 1967) to obtain the desired concentrations. Test mysids were exposed to seawater and TEG carrier control, as well as to 0.18, 0.78, and 0.045 µg phorate/ℓ (measured concentrations). These concentrations are below the 96-hr LC50 value of 0.33 µg phorate/ℓ (95% Confidence Limits: 0.27-0.43 µg/ℓ) reported by Nimmo *et al*. (this volume). During exposure, the animals were maintained at 25°C and 18 to 22 °/∘∘ salinity and in constant light from 40-watt fluorescent tubes. After exposure, mysids from each treatment were divided randomly into three groups per concentration and placed in beakers of pesticide-free seawater for one hour before stamina measurements were taken. The velocity regime for 4-day-old mysids (Table 1) was used to measure the MSS. Mysids from all treatments of each pesticide exposure were tested on the same day. The tunnel salinity was identical to that of the exposure water (21 °/∘∘).

Effects of methyl parathion[5] on stamina measurements were quantified by exposing and testing other groups of mysids in the same manner. The mysids were exposed for 96 hr to concentrations of 0.58, 0.31, and 0.10 µg methyl parathion/ℓ (measured concentrations), TEG carrier and seawater controls. These concentrations are below the 96-hr LC50 value of 0.77 µg methyl parathion/ℓ (95% Confidence Limits: 0.64-0.98 µg/ℓ) determined by Nimmo *et al*., (1978). Throughout this exposure, the animals were maintained at 23 to 24 °/∘∘ salinity and 24°C.

Experimental Design. A two-way factorial design was used to compare age-group performances of laboratory and feral stocks. The factors compared were source of animals (laboratory vs. feral), and age (2, 4, 7, 12, 14, 17, or 22 days). A general linear model for analysis of variance (Barr *et al*., 1976) and Duncan's multiple range test was used to determine statistical differences. Interactions between the two factors were analyzed by the Student-Newman-Keuls test. All statistical analyses were performed at $p < 0.05$.

[4] *Source, American Cyanamid, Princeton, NJ.*
[5] *Source, Chem Service, West Chester, PA.*

A one-way analysis of variance was used to compare males vs. females and pesticide-treated mysids to control animals. Duncan's multiple range tests were performed to determine which pesticide concentrations were statistically different.

RESULTS

Orientation in the Tunnel

When placed in the tunnel and left undisturbed, the mysids distributed themselves randomly on the tunnel surfaces, often with the body axis perpendicular to the long axis of the tube. During the acclimation to flowing water, the mysids moved throughout the tube by orienting to the current. Some test animals did not respond to the flowing water and remained downcurrent from the benchmark throughout the test. These were not included in the MSS calculations (see Table II). At various test velocities, the mysids moved to the surface of the tube where the velocity was lowered by the drag of the moving water: 2-day-old, 6.60 cm/sec; 4-day-old, 6.02 cm/sec; 7- through 22-day-old, 12.16 cm/sec. The mysids then oriented into the current with thoracic appendages extended and in contact with the tube, thus maintaining their position. Those swept downstream used quick swimming movements and abdominal flexions to regain their previous position. Eventually, the water velocity overcame all efforts, and each animal was swept past the benchmark.

Comparison of Cultured and Feral Young

Averages and standard deviations of the MSS measurements from laboratory-cultured and feral mysids are listed in Table II. No significant difference in MSS (analysis of variance, $p \leq 0.05$) was indicated when pooled measurements of laboratory stock were compared to pooled measurements of feral mysids. With two exceptions, MSS of the various laboratory and feral age groups were not significantly different from each other. MSS of 14-day-old laboratory-reared mysids differed significantly from those of 7- through 17-day-old feral mysids (Student-Newman-Keuls test, $p = 0.05$).

TABLE II. Comparison of the Maximum Sustained Speed (MSS) in cm/sec of laboratory-reared and feral stocks of Mysidopsis bahia as determined in a stamina tunnel. (Only those mysids that responded by remaining upstream of the benchmark were used for MSS calculations.)

| | | | | MSS | |
Stock	Age (Days)	Number Tested	Number Responding	Mean (cm/sec)	Standard Deviation
Laboratory	2	45	16	11.49	2.53
	4	45	21	11.42	4.97
	7	45	23	10.23	3.67
	12	45	24	7.76	3.81
	14	45	19	18.24	4.49
	17	45	21	12.19	4.29
	22	40	23	11.82	4.60
Feral	2	45	31	11.21	3.94
	4	45	23	10.70	4.20
	7	38	18	13.50	6.38
	12	35	15	12.80	5.23
	14	45	22	14.76	5.08
	17	35	17	12.90	5.12
	22	45	23	10.69	4.12

Differences Due to Sex

The swimming performances of three groups of five 17-day-old males and three groups of five 17-day-old females from laboratory stocks were not significantly different (Analysis of Variance, $p \geq 0.05$).

Differences Due to Pesticide Treatment

The MSS of animals exposed for 96 hr to phorate differed significantly (analysis of variance, $p \leq 0.001$) between treatments. Mean velocities sustained were inversely related to the concentration of phorate (Figure 2). The MSS of mysids from the seawater control, TEG control and 0.045 µg phorate/ℓ were not significantly different from each other, but the MSS of mysids exposed to 0.078 µg/ℓ differed from that of the seawater control and 0.045 µg phorate/ℓ (Student-Newman-Keuls test, $p = 0.05$). The MSS of mysids exposed to 0.18 µg

phorate/ℓ differed significantly from those of all other
treatments (Student-Newman-Keuls, p = 0.05). Therefore, the
threshold concentration for observing an effect of phorate on
the mysids' Maximum Sustained Speed lies between 0.078 and
0.18 µg/ℓ

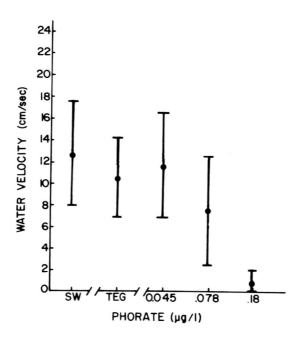

*FIGURE 2. Means and standard deviations of Maximum Sus-
tained Speed (MSS) of Mysidopsis bahia after 96 hr exposure to
phorate. SW, seawater control; TEG, triethylene glycol
control.*

Effects of treatment with different concentrations of
methyl parathion differed significantly (analysis of variance,
p ≤ 0.01). Mean velocities and standard deviations of the MSS
for mysids are shown in Figure 3. The effects of seawater
control, TEG control, 0.10 or 0.31 µg methyl parathion/ℓ, were
not significantly different (Duncan's multiple range test,
p ≥ 0.05), but in the highest concentration tested (0.58 µg/ℓ)

performance was significantly less than in all other
treatments. The threshold for the effects on the MSS of *M.
bahia* is between 0.31 and 0.58 µg methyl parathion/ℓ.

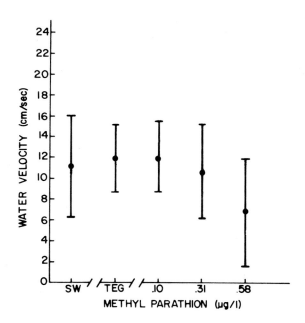

*FIGURE 3. Means and standard deviation of Maximum Sus-
tained Speeds (MSS) of* Mysidopsis bahia *after 96 hr exposure
to methyl paration. SW, seawater control; TEG, triethylene
glycol control.*

DISCUSSION

We cannot satisfactorily explain the low MSS of 12-day-
old mysids nor the high MSS of 14-day-old mysids from laboratory stocks. MSS measurements should have been comparable
to those of 12- and 14-day old feral mysids, because there was
no significant difference between the MSS of the two stocks of
mysids in other equivalent age groups. The laboratory culture
had been maintained for three years without an addition of
feral *Mysidopsis bahia*. This might have produced: (1) a

shift in the mean MSS values of laboratory mysids, or (2) an
increase in the standard deviation of the MSS for each age
group because of the survival of weaker organisms in the absence of predators. Nevertheless, means and standard deviations of MSS for most age groups of laboratory-reared and
feral mysids were similar.

The swimming ability exhibited by mysids may allow them
to remain in favorable environmental conditions (i.e., temperature, salinity, and light), catch suspended food particles, and avoid predators. For these reasons, the impact of
toxicants on the movement of mysids is a concern.

Effects of organophosphate pesticides on crustacean
behavior have been studied. Farr (1977) reported a decreased
degree of tail-snap escape response of grass shrimp, *Palaemonetes pugio*, exposed to methyl or ethyl parathion. He also
found that the toxicants impaired the grass shrimp's ability
to escape predation by gulf killifish, *Fundulus grandis*. Reduction in the MSS of mysids in the highest concentrations of
phorate and methyl parathion shows that these compounds inhibit a mysid's ability to maintain a stationary position in
flowing water. Although the physiological reason for reduction of mysid stamina by organophosphate insecticides is
not clear, the behavioral effect was quite evident. The MSS
measurements are used here as a behavioral indicator of the
organisms' physiological state.

The life-cycle bioassay is frequently used to measure
physiological stress in aquatic organisms. "Safe" toxicant
levels determined in such tests are often expressed as MATC
(Maximum Acceptable Toxicant Concentration) range (Mount and
Stephan, 1967; Eaton, 1970). This range consists of the
maximum concentration tested that produces no effect on mortality, reproduction, growth, or viability of eggs and larvae
and the minimum concentration at which an effect was observed.
The MATCs found by exposing mysids for 28 days to the organophosphates used in this study were 0.11 to 0.16 µg methyl
parathion/ℓ, and 0.09 to 0.14 µg phorate/ℓ (Nimmo, *et al.*,
this volume). These concentrations are very similar to the
MSS threshold of effect concentrations determined in the
stamina tunnel (0.31 to 0.58 µg methyl parathion/ℓ and 0.078
to 0.18 µg phorate/ℓ). However, the MSS threshold concentrations were determined by exposing mysids to the toxicant
for only 96 hr. Thus a MSS measurement after 96 hr exposure
may indicate concentrations which should be tested for chronic
effects.

SUMMARY

1. No significant difference in MSS (analysis of variance, p $\geq$ 0.05) was indicated when pooled measurements of
laboratory stock were compared to pooled measurements of
feral mysids.
2. The MSS measurements of laboratory-reared mysids were
recorded after a 96-hr exposure to the organophosphate insecticides, methyl parathion or phorate. The results indicated a
threshold of effect concentration for reduction of swimming
stamina between 0.31 and 0.56 µg methyl parathion/ℓ, and
between 0.078 and 0.18 µg phorate/ℓ.
3. The effect and no-effect concentrations found from the
MSS measurements after 96 hr exposure were very similar to
those obtained in a 28-day life-cycle test that exposed mysids
to the same compounds (Nimmo, *et al*., this volume). This
suggests a potential for use of the MSS measurement as a
rapid-screening biossay.

ACKNOWLEDGMENTS

The authors thank S. Foss for his assistance with the
illustrations. Gulf Breeze Contribution Number 387.

LITERATURE CITED

Allen, D. M. 1978. Population dynamics, spatial and temporal
    distributions of mysid crustaceans in a temperate marsh
    estuary. Ph.D. Thesis. Lehigh Univ. 157 pp.

Bainbridge, R. 1960. Speed and stamina in three fish. J.
    Exp. Biol. 37: 129-153.

Bams, R. A. 1967. Differences in performance of naturally
    and artificially propagated sockeye salmon migrant fry,
    as measured with swimming and predation tests. J. Fish.
    Res. Bd. Canada. 24: 1117-1153.

Barr, A. J., Goodnight, J. P. Sall, and J. T. Helwig. *A
    User's Guide to SAS 76*. 494 pp. SAS Institute, Raleigh,
    N.C. 1976.

Brett, J. R. 1964. The respiratory metabolism and swimming performance of young sockeye salmon. J. Fish. Res. Bd. Canada. 21: 1184-1226.

Carr, W. E. S., and C. A. Adams. 1973. Food habits of juvenile marine fish occupying seagrass beds in the estuarine zone near Crystal River, Florida. Trans. Amer. Fish. Soc. 102: 511-540.

Carlson, A. R. 1971. Effects of long-term exposure to carbaryl (Sevin) on survival, growth, and reproduction of the fathead minnow (*Pimephales promelas*). J. Fish. Res. Bd. Canada. 29: 583-587.

Coppage, D. L. 1972. Organophosphate pesticides: Specific level of brain AChE inhibition related to death in sheepshead minnows. Trans. Amer. Fish. Soc. 101: 534-536.

Coppage, D. L., and E. Matthews. 1974. Short-term effects of organophosphate pesticides on cholinesterases of estuarine fishes and pink shrimp. Bull. Environ. Contam. Toxicol. 11: 438-488.

Dahlberg, M. L., D. L. Shumway, and P. Doudoroff. 1968. Influence of dissolved oxygen and carbon dioxide on swimming performance of largemouth bass and coho salmon. J. Fish. Res. Bd. Canada. 25: 49-70.

Darnell, R. M. 1958. Food habits of fishes and larger invertebrates of Lake Pontchartrain, Louisiana, an estuarine community. Pub. Inst. Mar. Sci., Univ. Texas. 5: 353-416.

Eaton, J. G. 1970. Chronic malathion toxicity to the bluegill (*Lepomis macrochirus* Rafinesque). Water Res. 4: 673-684.

Farr, J. A. 1977. Impairment of antipredator behavior in *Palaemonetes pugio* by exposure to sublethal doses of parathion. Trans. Amer. Fish Soc. 106: 287-290.

Hansen, D. J., and P. R. Parrish. 1977. Suitability of sheepshead minnows (*Cyprinodon variegatus*) for life-cycle toxicity tests. Aquatic Toxicology and Hazard Evaluation, ASTM STP 634, F. L. Mayer and J. I. Hamelick, (Eds,), Amer. Soc. Test . Mater. pp. 117-126.

Howard, T. E. 1975. Swimming performance of juvenile coho
    salmon (*Oncorhynchus kisutch*) exposed to bleached kraft
    pulpmill effluent. J. Fish. Res. Bd. Canada. 32: 789-793.

Hughes, D. A. 1969. Responses to salinity change as a tidal
    transport mechanism of pink shrimp, *Penaeus duorarum*.
    Biol. Bull. 136: 45-53.

Hughes, D. A. 1972. On the endogenous control of tide-assoc-
    iated displacement of pink shrimp, *Penaeus duorarum*
    Burkenroad. Biol. Bull. 142: 271-280.

Katz, M., A. Pritchard, and C. E. Warren. 1959. Ability of
    some salmonids and a centrarchid to swim in water of
    reduced oxygen content. Trans. Amer. Fish. Soc. 88: 88-
    95.

Latz, M. E., and R. B. Forward. 1977. The effect of salinity
    upon phototaxis and geotaxis in a larval crustacean. Biol.
    Bull. 153: 163-179.

Mount, D. I., and W. A. Brungs. 1967. A simplified dosing
    apparatus for fish toxicology studies. Water Res. 1: 21-
    29.

Mount, D. I., and C. E. Stephan. 1967. A method for estab-
    lishing acceptable toxicant limits for fish--malathion and
    the butoxyethanol ester of 2,4-D. Trans. Amer. Fish. Soc.
    96(2): 185-193.

Nimmo, D. R., L. H. Bahner, R. A. Rigby, J. M. Sheppard, and
    A. J. Wilson, Jr. 1977. *Mysidopsis bahia*: an estuarine
    species suitable for life-cycle toxicity tests to
    determine the effects of a pollutant. Aquatic Toxicology
    and Hazard Evaluation, ASTM STP 634, F. L. Mayer and J. I.
    Hamelink, (Eds.), Amer. Soc. Test. Mater. pp. 109-116.

Nimmo, D. R., E. Matthews, and T. L. Hamaker. 1978. Phys-
    iology. *In:* Research Review 1977, B. P. Jackson, (Ed.),
    U.S. EPA, Environmental Research Laboratory, Gulf Breeze,
    FL., Ecol. Res. Ser. EPA-600/9-78-014, p. 7.

Nimmo, D. R., T. L. Hamaker, J. C. Moore, and R. A. Wood.
    1980. Acute and chronic effects of Dimilin on survival
    and reproduction of *Mysidopsis bahia*. Aquatic Toxicology.
    ASTM STP 707, J. G. Eaton, P. R. Parrish, and A. C.
    Hendricks, (Eds.), Amer. Soc. Test. Mater. pp. 366-376.

Nimmo, D. R., T. L. Hamaker, E. Matthews, and J. C. Moore. In press. Acute and chronic effects of eleven pesticides on the mysid shrimp *Mysiodopsis bahia*. Symposium on Pollution and Physiology of Marine Organisms, November 6-9, 1977, National Marine Fisheries Service, Milford, Conn.

Oseid, D., and L. L. Smith., Jr. 1972. Swimming endurance and resistance to copper and malathion of bulegills treated by long-term exposure to sublethal levels of hydrogen sulfide. Trans. Amer. Fish. Soc. 4: 620-625.

Otto, R. G., and J. O. Rice. 1974. Swimming speeds of yellow perch (*Perca flavescens*) following an abrupt change in environmental temperature. J. Fish. Res. Bd. Canada. 31: 1731-1734.

Petersen, R. H. 1974. Influence of fenitrothion on swimming velocities of brook trout (*Salvelinus fontinalis*). J. Fish. Res. Bd. Canada. 31: 1757-1762.

Rulifson, R. A. 1977. Temperature and water velocity effects on the swimming performances of young-of-the-year striped mullet (*Mugil cephalus*), spot (*Leiostomus xanthurus*), and pinfish (*Lagodon rhomboides*). J. Fish. Res. Bd. Canada. 34: 2316-2322.

Sprague, J. B. 1969. Measurement of pollutant toxicity to fish. I. Bioassay methods for acute toxicity. Water Res. 3: 793-821.

Thomas, A. E., R. E. Burrows, and H. H. Chenoweth. 1964. A device for stamina measurement of fingerling salmonids. U. S. Fish Wildl. Ser. Res. Rep. 67. 15 pp.

Venema, S. C., and F. Cruetzberg. 1973. Seasonal migration of the swimming crab *Macropipus* holsatus in an estuarine area controlled by tidal streams. Neth. J. Sea Res. 7: 103-111.

Vincent, R. E. 1960. Some influences of domestication upon three stocks of brook trout (*Salvelinus fontinalis* Mitchill). Trans. Amer. Fish. Soc. 8: 35-52.

COMPARATIVE TOXICOLOGY AND PHARMACOLOGY
OF CHLOROPHENOLS: STUDIES ON THE GRASS SHRIMP,
PALAEMONETES PUGIO

K. Ranga Rao
Ferris R. Fox
Philip J. Conklin
Angela C. Cantelmo[1]

Department of Biology
The University of West Florida
Pensacola, Florida

INTRODUCTION

Chlorophenols such as 2,4-dichlorophenol, 2,4,5-trichloro-
phenol, 2,4,6-trichlorophenol, 2,3,4,6-tetrachlorophenol, and
pentachlorophenol are broad spectrum biocides. Although
chlorophenols are used mainly for wood preservation and treatment, their bactericidal, fungicidal, algicidal, herbicidal,
insecticidal, molluscicidal, and disinfectant properties have
led to a widespread application of chlorophenol formulations.
Additionally, some of the chlorophenols are used in the manufacture of other biocides. For example, 2,4-dichlorophenol
is used in the manufacture of 2,4-D (2,4-dichlorophenoxyacetic
acid), while 2,4,5-trichlorophenol is used in the manufacture
of 2,4,5-T (2,4,5-trichlorophenoxyacetic acid), hexachloro-
phene, and trichlorophene (Doedens, 1964; Rao, 1978; Buikema
et al., 1979).
    Because of this widespread biocidal and industrial usage,
chlorophenols often find their way into the aquatic
environment. A recent review of phenolics in aquatic systems

[1]*present address: Department of Biology, Ramapo College
of New Jersey, Mahwah, New Jersey.*

(Buikema *et al.*, 1979) shows that chlorophenols are detectable in sewage effluents, industrial discharges, and in natural surface waters. Some of the chlorophenols present in drinking waters and wastewaters, however, are thought to be chlorination products of natural phenolics (Jolley *et al.*, 1975; Rockwell and Larsen, 1978). Chlorophenols formed from natural phenolics may include 4-chlorophenol, 2,4-dichlorophenol, 2,4,6-trichlorophenol, 3,5-dichloro-4-hydroxy benzoic acid, 5-chlorovanillic acid, and 4-chloro-2-methoxyphenol (Rockwell and Larsen, 1978).

Whatever the causes of chlorophenolic contamination of the aquatic ecosystems may be, the ubiquitous distribution of several different chlorophenols and their potential impact on aquatic organisms are of concern. Thus, many investigations have been undertaken to evaluate the toxicology and pharmacology of chlorophenols (for detailed bibliographies, see Bevenue and Beckman, 1967; Rao, 1978; Rao *et al.*, 1979; Buikema *et al.*, 1979). Most of these investigations have dealt with pentachlorophenol. With the exception of the studies on an unidentified fish (Ingolls *et al.*, 1966), and on goldfish (Kobayashi, 1979), *Daphnia magna* (Klopperman *et al.*, 1974; Bringmann and Kuhn, 1977), *Astacus fluviatilis* (Kaila and Saarikoski, 1977), *Crangon septemspinosa* and *Mya arenaria* (McLeese *et al.*, 1979), little has been done to evaluate the comparative toxicity of di-, tri-, and tetrachlorophenols to aquatic animals.

Therefore, we have conducted a series of experiments with the grass shrimp, *Palaemonetes pugio*, in order to assess the toxicity of several chlorophenols: 2,4-dichlorophenol, 2,4,5-trichlorophenol, 2,4,6-trichlorophenol, 2,3,4,5-tetrachlorophenol, 2,3,4,6-tetrachlorophenol, 2,3,5,6-tetrachlorophenol, and pentachlorophenol. These experiments were designed to determine whether the grass shrimp would exhibit a cyclic variation in susceptibility to various chlorophenols in relation to the molt cycle. Previous studies (Conklin and Rao, 1978) have shown that the grass shrimp is most susceptible to sodium pentachlorophenate at the time of molting. Because previous studies have shown that sodium pentachlorophenate inhibits regenerative limb growth in grass shrimp (Rao *et al.*, 1978), we conducted additional experiments to determine whether other chlorophenols would exert similar effects.

Since little is known of the fate of chlorophenols in crustaceans (See Bose and Fujiwara, 1978), we studied the uptake, tissue distribution, and depuration of ^{14}C-2,4,5-trichlorophenol and ^{14}C-pentachlorophenol in the grass shrimp.

Additionally, utilizing thin layer chromatography and auto-radiography, we attempted to identify the metabolites of ^{14}C-pentachlorophenol in the grass shrimp.

MATERIALS AND METHODS

Grass Shrimp

Adult grass shrimp, *Palaemonetes pugio*, were collected from grass beds in Santa Rosa Sound, Gulf Breeze, Florida. They were used in experiments within two weeks after collection. The average length of the shrimp used was 25 mm, with a range of 22 to 28 mm. Shrimp were maintained in sea-water (10 $^{\circ}/_{\circ\circ}$ salinity) at 20 ± 1° C under 12 hr light: 12 hr dark conditions. They were fed live brine shrimp nauplii on alternate days during the period of holding in laboratory stock aquaria, as well as during long-term (> 96 hr) experiments. They were not fed during the 96-hr acute toxicity tests or during the experiments with labelled chlorophenols.

Prior to usage in experiments, the grass shrimp were examined microscopically so as to determine their stage in the molt cycle. As described earlier (Conklin and Rao, 1978), epidermal retraction and progress in neosetogenesis in the uropods were the external criteria for recording the onset and progress of premolt (proecdysial) preparations. The shrimp in which the epidermis did not retract from the cuticle were considered to be in the intermolt stage (Stage C) of the molt cycle; typically, such shrimp required 9 to 14 days to complete premolt preparations and undergo molting. The shrimp which completed neosetogenesis were considered to be in late premolt stages (Stages D_2-D_4), and they usually molted within the next two or three days.

Acute Toxicity Tests

The chlorophenols used in these tests were obtained as follows. Analytical grade (99% pure) 2,4-dichlorophenol, 94% pure 2,4,5-trichlorophenol, 98% pure 2,4,6-trichlorophenol, 98% pure 2,3,4,5-tetrachlorophenol, 98% pure 2,3,5,6-tetra-chlorophenol, and 99% pure pentachlorophenol were obtained from Aldrich Chemical Company. Technical grade 2,3,4,6-tetra-chlorophenol was obtained from Fluka AG.

Each chlorophenol was initially dissolved in ethanol; then, suitable amounts of this stock solution were added to filtered (5 μm) seawater (10 $^{\circ}/_{\circ\circ}$ salinity) to obtain the desired test concentrations. The stock solutions as well as

the diluted solutions of the test compound were prepared
fresh, just prior to their use.

In separate experiments, each chlorophenol was tested
on intermolt (Stage C) shrimp and on late premolt (Stage
D_2-D_4) shrimp. In each experiment, a minimum of 20 shrimp
were exposed to each of the five test concentrations and the
two control media (seawater and ethanol-containing seawater).
To prevent cannibalism, the shrimp were maintained individually in glass jars; each jar contained 250 ml of the
appropriate medium. The experimental and control media were
renewed daily with freshly prepared solutions. The shrimp
were examined daily for evidence of molting and/or death.

The 96-hr median lethal concentrations (96-hr LC_{50} were
computed by probit analysis (Finney, 1971) using a statistical
computer program (Barr *et al.*, 1976)

Limb Regeneration

Experiments were conducted to determine the effects of
2,4,5-trichlorophenol, 2,4,6-trichlorophenol, 2,3,4,5-tetra-
chlorophenol, 2,3,4,6-tetrachlorophenol, and pentachlorophenol
on limb regeneration in the grass shrimp, *Palaemonetes pugio*.
Each chlorophenol was tested at three concentrations; the
highest concentration was equal to, or less than, the LC_{10}
value computed from acute toxicity tests with intermolt
(Stage C) grass shrimp. A minimum of 25 shrimp were exposed
to each of the test concentrations and to the control media.

The experimental methods were similar to those described
earlier by Rao *et al.* (1978). Intermolt (Stage C) grass
shrimp were employed in these experiments. The left fifth
pereiopod was removed from each shrimp with a fine-tipped
jeweler's forceps. The shrimp were placed in the experimental
or control media on the day of limb removal, and maintained in
these media until the shrimp molted. During this period, the
media were renewed daily. The shrimp were microscopically
examined on the 4th, 7th, 9th, and 11th day after limb removal
to assess limb bud growth and progress in premolt preparations. The regeneration index, R value, was calculated
according to the method of Bliss (1956):

$$R \text{ value} = \frac{\text{length of limb bud}}{\text{length of carapace}} \times 100$$

The R values were utilized for comparing the limb bud
growth in control and experimental shrimp. The data were
subjected to ANOVA treatment. The concentration of chlorophenol required for causing 50% inhibition of limb bud growth
(EC_{50}) was computed using probit analysis (Finney, 1951).

Accumulation and Distribution of Labelled
2,4,5-Trichlorophenol in Grass Shrimp Tissues

Uniformly labelled ^{14}C-2,4,5-trichlorophenol (specific activity: 12.42 mCi/mM), obtained from Pathfinder Laboratories Inc., St. Louis, was utilized in this study. In separate experiments, intermolt (Stage C) grass shrimp and newly molted (Stage A; within 5 minutes after molting) grass shrimp were exposed to filtered seawater containing 1.0 ppm 2,4,5-tri-chlorophenol (0.118 ppm labelled + 0.882 ppm unlabelled) for a period of one hour. After this exposure period, the shrimp were transferred to a seawater wash for two minutes, then blotted dry, and weighed individually. Some of the shrimp were processed as a whole; others were dissected, and each of the several tissues - abdomen, hepatopancreas, digestive tract, and cephalothorax - was weighed and processed separately. The shrimp (or their tissues) were placed in scintillation vials containing up to 1.0 ml of Protosol® tissue solubilizer (New England Nuclear), then macerated and digested overnight in a shaking water bath at 50°C. Following digestion, samples were neutralized with acetic acid; then, 15 ml of Aquasol® (New England Nuclear) was added to each sample. The radioactivity of these samples was determined by using a Beckman LS-133 liquid scintillation counter. Background and quench corrections were performed using control (unexposed) animals of a similar weight range as the experimental animals. Quench correction for counting efficiency was made by the internal standard method.

Additional experiments were conducted with intermolt (Stage C) grass shrimp to determine the influence of the duration of exposure to 2,4,5-trichlorophenol-containing media on the extent of accumulation of this chlorophenol in grass shrimp tissues. The shrimp were exposed to 1.0 ppm 2,4,5-trichlorophenol (0.118 ppm labelled + 0.882 ppm unlabelled) for varying durations (15 min, 30 min, 1 hr, 3 hr, 6 hr, or 12 hr) and then processed for liquid scintillation counting as described above.

In the next experiment, grass shrimp were exposed to 1.0 ppm 2,4,5-trichlorophenol (0.118 ppm labelled + 0.882 ppm unlabelled) for 12 hours and then transferred to seawater. At the end of 1, 2, 3, 6, 12, 18 and 24 hour(s) after transfer to seawater, shrimp were removed in groups of 4 to 8, and processed individually for liquid scintillation counting. This experiment provided information on the ability of grass shrimp to eliminate the accumulated 2,4,5-trichlorophenol.

Accumulation and Distribution of Labelled
Pentachlorophenol in Grass Shrimp Tissues

Uniformly labelled ^{14}C-pentachlorophenol (specific
activity: 10.04 mCi/mM; 10.62 mCi/mM), obtained from Pathfinder Laboratories Inc., St. Louis, was utilized in these
experiments. The accumulation of pentachlorophenol in intermolt and newly molted grass shrimp, the influence of the
duration of exposure on the extent of pentachlorophenol
accumulation in intermolt grass shrimp, and the ability of
these shrimp to eliminate the accumulated pentachlorophenol
were examined following the procedures described above for
2,4,5-trichlorophenol. The major difference was that the
shrimp were exposed to 2.0 ppm pentachlorophenol (0.266 ppm
labelled + 1.734 ppm unlabelled). Additionally, some of the
experiments were conducted with berried females; these shrimp
and their eggs were processed separately.
        Details of the number of shrimp processed in each ex-
periment are given at appropriate places in the text figures.

Metabolism of [14]*C-pentachlorophenol in the*
Grass Shrimp

In this series of experiments, grass shrimp (10/group)
were exposed to 0.228 ppm ^{14}C-pentachlorophenol in seawater
(10°/oo salinity) for periods of 6, 24, or 72 hours. To avoid
photodecomposition of pentachlorophenol (see Wong and Crosby,
1978), these exposures were done in the dark. The processing
of shrimp tissues was done under dim light. Each shrimp was
dissected so as to separate the abdominal muscle, cephalothorax, digestive tract, and hepatopancreas. Samples of a
given tissue dissected from a group of shrimp were pooled and
processed as follows.
        Utilizing a ground glas homogenizer, the tissues were
homogenized in 0.1 N H_2SO_4; this was followed by extraction in
cyclohexane/isopropanol (85/15; v/v). The tissues were extracted three times using a ten to one volume to weight ratio.
The extracts were dried over calcium chloride and evaporated
to dryness. The dried samples were stored in the cold until
used.
        The experimental media were also extracted with cyclo-
hexane/isopropanol (3 volumes of this solvent: 1 volume of the
medium). The extract was dried over ammonium sulfate and
evaporated to dryness.
        The dried extracts prepared from shrimp tissues and the
medium were processed as follows. Each sample was dissolved

in ethanol and chromatographed on silicic acid thin layer plates (Gelman SA) using two solvent systems: benzene alone, or a mixture of hexane/acetone (80/20; v/v). The R_f values of the radioactive materials were determined by incubating the chromatographed plates with electron microscope film (Kodak 4489) for periods of two to six weeks. The film was then developed and laid over the chromatography plates to locate the spots. The R_f values of these materials were compared with those of known standards (obtained from Aldrich Chemical Co.) chromatographed on silicic acid plates containing a fluorescent indicator.

For quantitative studies of the radioactive metabolites of ^{14}C-pentachlorophenol, the chromatographic plates were cut into sections, 5 mm high and 10 mm wide. The sections were placed in scintillation vials, two drops of water and 15 ml Aquasol were added prior to liquid scintillation counting.

In order to detect the occurrence of pentachlorophenol conjugates, some samples were subjected to acid and enzymatic hydrolysis. Acid hydrolysis was done by treating an aliquot of the sample with 6N HCl. The hydrolysate was neutralized, and then extracted with cyclohexane/isopropanol. The extract was processed and chromatographed as described above.

For enzymatic hydrolysis, an aliquot of the sample was placed in 10 mM phosphate buffer (pH 7.0) containing β-glucoronidase (50 U/ml), and incubated overnight at 25°C in a sealed tube. The hydrolysate was then extracted with cyclohexane/isopropanol, and processed for chromatography.

In some cases, enzymatic hydrolysis was done directly on the chromatograph plate. The extract of the desired tissue was chromatographed in one dimension. Then the material at the origin which was suspected to be the conjugated form of pentachlorophenol, was treated overnight with β-glucuronidase. Then, the chromatography was allowed to proceed in the second dimension. As in the case of previous experiments, autoradiography and liquid scintillation were employed for identification and quantification of the metabolites.

RESULTS

Toxicity of Chlorophenols

It can be seen from Table 1 that, with the exception of 2,4-dichlorophenol, the chlorophenols were more toxic to the molting grass shrimp (late premolt individuals which molted during the test period) than to the intermolt shrimp (which did not molt during the test period). This difference

TABLE 1. Toxicity of Chlorophenols to the Grass Shrimp, *Palaemonetes pugio*, during 96-hour Tests (20$^+_-$ 1°C; 10‰ Salinity; 7.6-7.7 pH).

Chlorophenol	Molecular Weight[a]	Ionization Constant[a] pKa	96-hour LC$_{50}$ (95% fiducial limits) mg/l; ppm	
			Intermolt Shrimp[b]	Molting Shrimp[c]
2,4-Dichlorophenol	163.00	7.67	2.55 (2.28-2.86)	2.16 (1.49-2.73)
2,4,6-Trichlorophenol	197.45	7.43	3.95 (3.28-4.95)	1.21 (1.11-1.31)
2,4,5-Trichlorophenol	197.45	7.42	1.12 (0.92-1.43)	0.64 (0.36-0.80)
2,3,4,5-Tetrachlorophenol	231.89	7.04	0.86 (0.73-0.98)	0.37 (0.35-0.39)
2,3,4,6-Tetrachlorophenol	231.89	6.62	3.70 (2.98-5.25)	0.81 (0.64-0.89)
2,3,5,6-Tetrachlorophenol	231.89	5.48	4.10 (3.30-5.31)	1.17 (1.08-1.27)
Pentachlorophenol	266.34	5.07	2.50 (1.91-3.29)	0.44 (0.18-0.67)

[a] Data from Buikema *et al*. (1979).

[b] At the beginning of the test, the shrimp were in Stage C of the intermolt cycle; they did not molt during the 96-hour test period.

[c] At the beginning of the test, the shrimp were in late premolt stages (D_2-D_4); they molted during the 96-hour test period.

between toxicity to intermolt and molting shrimp (i.e., increased toxicity to molting shrimp) was most pronounced in the case of pentachlorophenol, which exhibited a 5.68-fold increase in toxicity to molting shrimp. For other chlorophenols, this increase in toxicity to molting shrimp was: 4.56X (2,3,4,6-tetrachlorophenol); 3,5X (2,3,5,6-tetrachlorophenol); 3.26X (2,4,6-tetrachlorophenol); 2.32X (2,3,4,5-tetrachlorophenol), and 1.64X (2,4,5-trichlorophenol). However, unlike these chlorophenols, 2,4-dichlorophenol was equally toxic to intermolt and molting grass shrimp; its LC_{50} for intermolt shrimp was not significantly different from that for molting shrimp.

In general, the toxicity of chlorophenols was not proportional to the number of chlorine residues in these compounds. For example, in tests with intermolt shrimp, 2,4,5-trichlorophenol and 2,3,4,5-tetrachlorophenol were more toxic than pentachlorophenol. In tests with molting shrimp, 2,4,5-trichlorophenol and 2,3,4,5-tetrachlorophenol were as toxic as pentachlorophenol.

The relative toxicities of the various chlorophenols did not appear to be related to the known ionization constants (pK_a values) of these compounds (see Table 1). For example, 2,4,5-trichlorophenol was more toxic than 2,4,6-trichlorophenol, althouth its pK_a was not significantly different from that of the latter trichlorophenol. Among the tetrachlorophenols, however, there seemed to be a correlation between the observed toxicity and the known pK_a values. Indeed, 2,3,4,5-tetrachlorophenol, which has a higher pK_a than 2,3,4,6-tetrachlorophenol and 2,3,5,6-tetrachlorophenol, was the most toxic tetrachlorophenol.

Effects of Chlorophenols on Limb Regeneration

The R-values calculated from the limb bud measurements made on the 9th day following limb removal (Fig. 1) illustrate the effects of chlorophenols on limb regeneration in the grass shrimp, *Palaemonetes pugio*. Neither 2,4,6-trichlorophenol nor 2,4,5-trichlorophenol caused dose-dependent inhibition of limb regeneration, although the latter compound caused significant ($p<0.05$) inhibition of limb regeneration at concentrations of 0.5 and 0.75 ppm. Unlike these trichlorophenols, the tetrachlorophenols (2,3,4,5-tetrachlorophenol and 2,3,4,6-tetrachlorophenol) and pentachlorophenol were able to inhibit limb regeneration in a dose-dependent manner. At the concentrations tested, these chlorophenols did not alter the duration of the intermolt and premolt stages of the intermolt cycle of the grass shrimp.

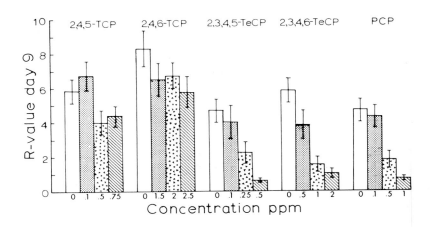

*FIGURE 1. Effects of trichlorophenols (TCP), tetra-
chlorophenols (TeCP), and Pentachlorophenol (PCP) on regeneration of the left fifth pereiopod in the grass shrimp,
Palaemonetes pugio. The R values (mean ± standard errors)
were derived from the measurements of limb bud size on the 9th
day after limb removal. The control shrimp (open bars) were
maintained in seawater; the experimental shrimp were exposed
to chlorophenol-containing seawater beginning the day of limb
removal until the completion of ecdysis. The shrimp were fed
during these experiments.*

The median effective concentrations - EC_{50} values along
with 95% fiducial limits - of chlorophenols causing inhibition
of limb regeneration were: 0.30 ppm (0.15-0.54 ppm) for
2,3,4,5-tetrachlorophenol; 0.78 ppm (0.63-0.94 ppm) for
2,3,4,6-tetrachlorophenol; 0.57 ppm (0.42-0.75 ppm) for
pentachlorophenol. The EC_{50} value for pentachlorophenol was
not significantly different from that reported previously
(Rao *et al.*, 1978) for sodium pentachlorophenate: 0.56 ppm
(0.45-0.71 ppm).
    It is pertinent to recall that, for each of these chloro-
phenols, the EC_{50} value for inhibition of limb regeneration
was computed from the results of tests with grass shrimp
which did not molt during the 9-day period following limb
removal. For 2,3,4,5-tetrachlorophenol, 2,3,4,6-tetrachloro-
phenol and pentachlorophenol, the EC_{50} values for inhibition
of limb regeneration were lower than the 96-hr LC50 values
derived from tests with intermolt (non-molting) grass shrimp.
However, these EC_{50} values for inhibition of limb regeneration

were well within the 95% fiducial limits of the 96-hr LC_{50} values derived from tests with molting grass shrimp (for comparison, see Table 1).

In other words, the developing limb bud tissues (which have little protective covering during the early phases of development) and the newly molted shrimp (which have a relatively thin cuticle, and highly permeable epithelia) were more sensitive to chlorophenols in the medium than were inter-molt shrimp (which have a relatively thicker cuticle, and less permeable epithelia). This led to the question of whether the observed differences in sensitivity to chloro-phenols were attributable to differences in the permeability of epithelia and to differences in the bioaccumulation of chlorophenols. In an effort to answer this question, we studied the bioaccumulation and distribution of two chloro-phenols (2,4,5-trichlorophenol and pentachlorophenol) in grass shrimp tissues.

Accumulation and Distribution of ^{14}C-2,4,5-Trichlorophenol in Grass Shrimp Tissues

It can be seen from Figure 2 that newly molted shrimp accumulated 2.5 times more 2,4,5-trichlorophenol than did the intermolt shrimp. This trend of increased trichlorophenol accumulation during the period immediately following molting was also noted among the various tissues examined. The most pronounced increase in trichlorophenol accumulation was noted in the digestive tract (includes stomach, midgut, and hind-gut).

2,4,5-Trichlorophenol bioconcentration (expressed as Bio-concentration Index-BI; amount in the tissue divided by amount in the medium) was higher in the digestive tract (BI: 154 and 516, for tissues from intermolt and newly molted shrimp, respectively) and hepatopancreas (BI: 104 and 265) than in cephalothorax (BI: 13 and 27) and the abdomen (BI: 6.5 and 12). For the whole shrimp, the BI values were 13 (for intermolt shrimp) and 32 (for newly molted shrimp). These BI values are for shrimp which have been exposed to 1.0 ppm 2,4,5-trichlorophenol for a period of one hour.

When the shrimp were exposed to 1.0 ppm 2,4,5-trichloro-phenol for longer periods, up to 12 hours, the magnitude of bioconcentration increased (see Figs. 3 and 4). For example, at the end of a 12-hr exposure, the bioconcentration indices for intermolt shrimp were: 32 (for whole shrimp), 396 (for the digestive tract), 276 (for the hepatopancreas), 44 (for the cephalothorax), and 19 (for the abdomen). When such shrimp were transferred to clean seawater, however, approxi-

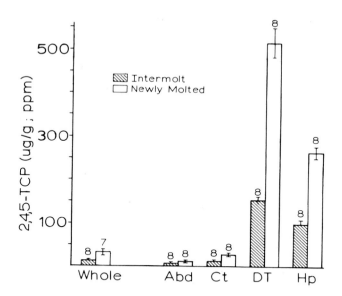

FIGURE 2. Accumulation of 2,4,5-trichlorophenol (TCP) in the whole body and in the abdomen (Abd), cephalothorax (ct), digestive tract (DT; includes the stomach, midgut, and hindgut), and hepatopancreas (Hp) of intermolt (Stage C) and newly molted (within 5 minutes after molting) grass shrimp, Palaemonetes pugio. The shrimp were exposed to 1.0 ppm 2,4,5-trichlorophenol (0.118 ppm ^{14}C-2,4,5-TCP plus 0.882 ppm unlabelled 2,4,5-TCP) for a period of one hour. The amounts of TCP accumulated in the whole body and in the various tissues were estimated on the basis of radioactivity measurements. Each bar represents the mean ± standard errors. The number of animals (or tissue samples) used is shown on top of each bar.

mately 87% of the accumulated trichlorophenol was lost from the shrimp within 12 hours. By the end of a 24-hour period of depuration, nearly 96% of the accumulated trichlorophenol was lost from the shrimp (Fig. 3). This trend of rapid depuration was also evident in the various tissues examined, although in some tissues (digestive tract and hepatopancreas) the trichlorophenol loss was somewhat less rapid than in the others (cephalothorax and abdomen; see Fig. 4).

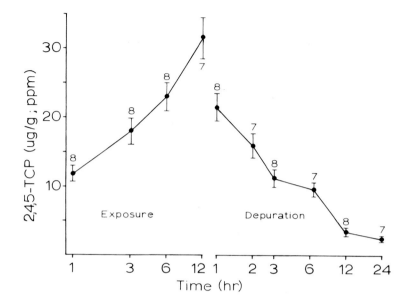

FIGURE 3. The levels of 2,4,5-trichlorophenol (TCP) in intermolt (Stage C) grass shrimp, Palaemonetes pugio, exposed to 1.0 ppm 2,4,5-TCP (0.118 ppm ^{14}C-2,4,5-TCP plus 0.882 ppm unlabelled 2,4,5-TCP) for a 12-hour period, and then trans-ferred to seawater for a 24-hour period of depuration. The levels of TCP in the grass shrimp were estimated on the basis of radioactivity measurements. Each data point shows the mean ± standard errors. The number of shrimp used is shown near each data point.

Accumulation and Distribution of
[14]C-Pentachlorophenol in Grass Shrimp Tissues

When exposed to a medium containing 2.0 ppm pentachloro-phenol for one hour, the amount of pentachlorophenol accumu-lated by newly molted shrimp was 20 times higher than that accumulated by intermolt shrimp (Fig. 5). This trend of increased pentachlorophenol accumulation during the period immediately following molting was also seen among the various tissues examined. It should be noted, however, that the magnitude of the relative increase in pentachlorophenol accumulation in newly molted shrimp was much greater than that

observed with 2,4,5-trichlorophenol accumulation (Compare
Figs. 2 and 5).

Because of this, the newly molted shrimp bioconcentrated
pentachlorophenol (BI: 150) to a greater extent than 2,4,5-
trichlorophenol (BI: 32), although the intermolt shrimp bioconcentrated 2,4,5-trichlorophenol (BI: 13) to a greater
extent than pentachlorophenol (BI: 7.5).

The magnitude of pentachlorophenol bioaccumulation in

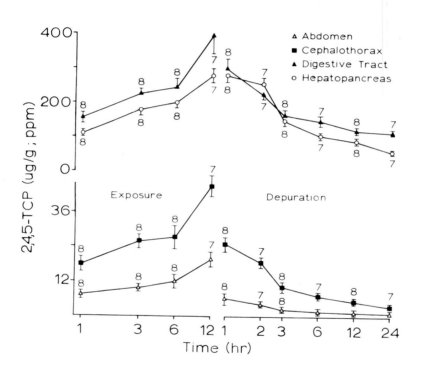

FIGURE 4. *The levels of 2,4,5-trichlorophenol (TCP) in
the tissues of intermolt (Stage C) grass shrimp, Palaemonetes
pugio, exposed to 1.0 ppm 2,4,5-TCP (0.118 ppm ^{14}C-TCP plus
0.882 ppm unlabelled 2,4,5-TCP) for a 12-hour period, and
then transferred to seawater for a 24-hour period of
depuration. The levels of TCP in the grass shrimp tissues
were estimated on the basis of radioactivity measurements.
Each data point shows the mean ± standard errors. The
number of tissue samples used is shown near each data point.*

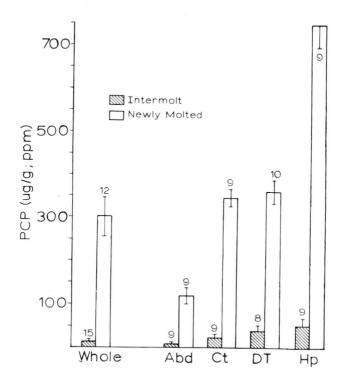

FIGURE 5. Accumulation of pentachlorophenol (PCP) in the whole body and in the abdomen (Abd), cephalothorax (Ct), digestive tract (DT), and hepatopancreas (Hp) of intermolt (Stage C) and newly molted (within 5 minutes after molting) grass shrimp, Palaemonetes pugio. The shrimp were exposed to 2.0 ppm PCP (0.226 ppm ^{14}C-PCP plus 1.734 ppm unlabelled PCP - for a period of one hour. The amounts of PCP accumulated in the whole body and in the various tissues were estimated on the basis of radioactivity measurements. Each bar represents the mean ± standard errors. The number of shrimp (or tissue samples) used is shown on top of each bar.

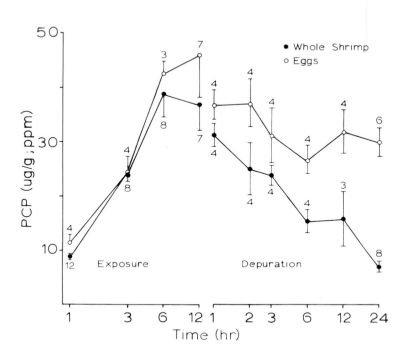

FIGURE 6. The levels of pentachlorophenol (PCP) in the body and in the eggs (attached to the pleopods) of berried grass shrimp, Palaemonetes pugio, exposed to 2.0 ppm penta-chlorophenol (0.266 ppm ^{14}C-PCP plus 1.734 ppm unlabelled PCP) for a 12-hour period, and then transferred to seawater for a 24-hour period of depuration. The levels of PCP in the whole body and in the eggs were estimated on the basis of radio-activity measurements. Each data point shows the mean ± standard errors. The number of berried shrimp used is shown near each data point.

intermolt grass shrimp (Fig. 6) depended on the duration of exposure to the chlorophenol-containing medium. During the first six-hour period of exposure, the pentachlorophenol accumulation in shrimp increased with the duration of exposure.

Experiments with berried grass shrimp showed that penta-chlorophenol accumulated in the eggs (attached to the pleo-pods) as well as in the body of the shrimp (Fig. 6). In fact, by the end of a 12-hour exposure to 2.0 ppm pentachlorophenol,

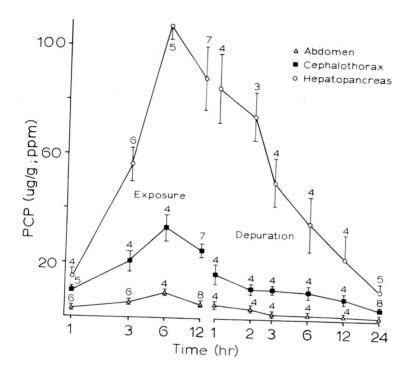

FIGURE 7. *The levels of pentachlorophenol (PCP) in the abdomen, cephalothorax, and hepatopancreas of intermolt (Stage C) grass shrimp, Palaemonetes pugio, exposed to 2.0 ppm pentachlorophenol (0.266 ppm ^{14}C-PCP plus 1.734 ppm unlabelled PCP) for a 12-hour period, and then transferred to seawater for a 24-hour period of depuration. The levels of PCP in the grass shrimp tissues were estimated on the basis of radioactivity measurements. Each data point shows the mean ± standard errors. The number of tissue samples used is shown near each data point.*

the bioconcentration of pentachlorophenol in the eggs (BI: 23) was somewhat higher than in the shrimp body (BI: 18.5). Furthermore, when such pentachlorophenol-exposed berried shrimp were transferred to clean seawater, the eggs lost only 20% of the accumulated pentachlorophenol by the end of a 24-hour period of depuration, although the body of the shrimp -including the hepatopancreas, abdomen, and cephalothorax- lost 80 to 88% of the accumulated pentachlorophenol (Fig. 7).

The rapid loss of pentachlorophenol from shrimp tissues
during the period of depuration in clean seawater was similar
to that observed with 2,4,5-trichlorophenol (Fig. 4).

Metabolism of ^{14}C-Pentachlorophenol
in the Grass Shrimp

The ^{14}C-pentachlorophenol employed in this study appeared
to be homogeneous in thin layer chromatography (Fig. 8). In
media without the grass shrimp, ^{14}C-pentachlorophenol was not
degraded during an exposure period of 6 to 72 hours in
darkness. But in media containing grass shrimp, radioactive
metabolites were present in the medium and in the shrimp
tissues. By means of thin layer chromatography, it was
possible to detect eight metabolites of ^{14}C-pentachlorophenol.
Two of these metabolites (one, tentatively identified as
2,4,6-trichlorophenol; and the other, designated as unknown
compound-III) were present in the shrimp tissues, but not in
the medium. The metabolite designated as unknown compound-II
was present in the medium, but not in the shrimp tissues. The
metabolites tentatively identified as pentachloranisole,
tetrachlorohydroquinone, 2,3,4,5-tetrachlorophenol, and
2,3,4,6-tetrachlorophenol, as well as the one designated as
unknown compound-I, were present in the medium and in some of
the shrimp tissues (Fig. 8).
Upon treatment with β-glucuronidase, the metabolite
designated as the unknown compound-I dissociated into three
components; one of these had the same R_f value as pentachlorophenol (Fig. 9). This suggests that some of the material
designated as unknown compound-I may be a glucuronide conjugate of pentachlorophenol.
The qualitative distribution of the various metabolites,
and their occurrence in shrimp tissues and in the medium in
relation to the duration of exposure to ^{14}C-pentachlorophenol
are summarized in Table 2. It can be readily seen that most
of the metabolites were detectable by the end of a six-hour
exposure period. Some of the metabolites in certain tissues
(e.g., 2,3,4,6-tetrachlorophenol in the hepatopancreas and
abdomen; pentachloroanisole in the hepatopancreas and cephalothorax), however, were detectable only at the end of 24 and/
or 72 hours of exposure to ^{14}C-pentachlorophenol.
The relative distribution of radioactivity among the
various metabolites in shrimp tissues and in the medium at
the end of the 72-hour exposure period is shown in Figure 10.
It can be seen that some of the radioactive material in the
tissues and in the medium was not extractable; the nature of
this material remains to be determined. In the abdomen and

TABLE 2. Radioactive Metabolites extracted from Grass Shrimp exposed for 6, 24, or 72 hours to seawater containing 0.228 ppm ^{14}C-Pentachlorophenol.

Metabolites	R_f[a]	Medium[b]			Hepatopancreas			Abdomen			Digestive Tract			Cephalothorax		
		6	24	72	6	24	72	6	24	72	6	24	72	6	24	72
Unknown-I	0.01	+	+	+[c]	+	+	+	+	+	+	+	+	+	+	+	+
Unknown-II	0.16	+	+	+	-	-	-	-	-	-	-	-	-	+	+	-
Tetrachlorohydro-quinone	0.24	+	+	+	+	+	+	-	-	-	+	+	+	-	-	-
Unknown-III	0.32	-	-	-	+	+	+	+	+	+	+	+	+	+	+	+
2,3,4,5-Tetrachloro-phenol	0.46	+	+	+	+	+	+	+	+	+	+	+	+	+	+	+
Pentachlorophenol	0.58	+	+	+	+	+	+	+	+	+	+	+	+	+	+	+
2,3,4,6-Tetrachloro-phenol	0.66	+	+	+	-	+	+	-	+	+	+	+	+	+	+	+
2,4,6-Trichlorophenol	0.77	-	-	-	+	+	+	+	+	+	+	+	+	+	+	+
Pentachloroanisole	0.97	+	+	+	-	-	+	-	-	-	+	+	+	-	+	+

[a] R_f values derived from thin layer chromatography on Gelman SA plates (polysilicic acid gel impregnated glass fiber sheets); solvent = 100% benzene.

[b] Hours of exposure.

[c] +, present; -, absent.

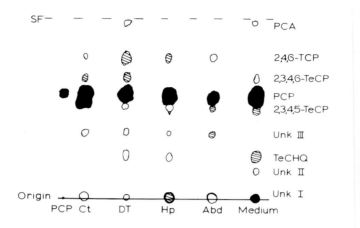

FIGURE 8. Diagrammatic representation of the separation
of radioactive metabolites of ^{14}C-pentachlorophenol (PCP) by
thin layer chromatography on Gelman SA plates, utilizing 100%
benzene as a solvent. The radioactive spots were localized
by autoradiography. Some of the metabolites were tentatively
identified by comparing their R_f values with those of
reference compounds. The samples chromatographed were: ^{14}C-
PCP; the extracts of cephalothorax (Ct), digestive tract (DT),
hepatopancreas (Hp), and abdomen (Abd) of the grass shrimp
exposed for six hours to 0.228 ppm ^{14}C-PCP; and an extract of
the exposure medium. The tentatively identified metabolites
were: pentachloroanisole (PAC), 2,4,6-trichlorophenol (TCP),
2,3,4,6-tetrachlorophenol (TeCP), 2,3,4,5-tetrachlorophenol,
and tetrachlorohydroquinone (TeCHQ); other metabolites were
designated as unknown compounds *Unk I, II, and III).
SF: solvent front.

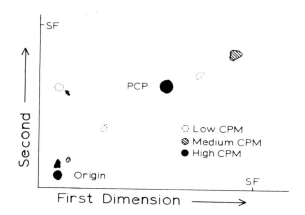

FIGURE 9. Two-dimensional thin layer chromatography of an extract of the hepatopancreas from grass shrimp exposed for six hours to 0.228 ppm ^{14}C-PCP. Chromatography in both dimensions was done utilizing 100% benzene as the solvent. Upon completion of chromatography in the first dimension, the material near the origin (Unknown I) was treated overnight with β-glucuronidase; then, the chromatography was allowed to proceed in the second dimension. It can be seen that, upon treatment with β-glucuronidase, the material near the origin dissociated into three components; one of these had the same R_f as PCP (short arrow). The radioactive spots were localized by autoradiography. The various spots were shaded differently to illustrate the low (100-300), medium (500-800) or high (> 2000) CPM as revealed by liquid scintillation counting. SF: solvent front.

in the cephalothoracic tissues (other than the digestive tract and hepatopancreas), 50 to 53% of the radioactivity was associated with pentachlorophenol. But in the case of hepato-pancreas, the digestive tract, and the medium, most of the radioactivity was associated with the metabolites of penta-chlorophenol and certain non-extractable material(s); only 18 to 28% of the radioactivity was attributable to pentachloro-phenol.

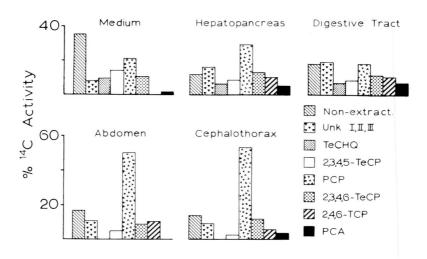

*FIGURE 10. The relative distribution of radioactivity
(^{14}C) among the various metabolites of ^{14}C-pentachlorophenol
detected in the exposure medium and in the various tissues of
the grass shrimp exposed to 0.228 ppm ^{14}C-pentachlorophenol
for a period of 72 hours. The extracts were subjected to
thin layer chromatography, and the zones containing the
appropriate metabolites were then processed for liquid
scintillation counting. The metabolites in the extracts were:
unknowns I, II, and III; tetrachlorohydroquinone (TeCHQ);
2,3,4,5-tetrachlorophenol (TeCP); 2,3,4,6-tetrachlorophenol;
2,4,6-trichlorophenol (2,4,6-TCP), and pentachloroanisole
(PCA). Certain amount of the radioactivity was associated
with material(s) which could not be extracted with cyclo-
hexane/isopropanol (85/15: v/v).*

DISCUSSION

Comparative Toxicity of Chlorophenols

     The toxicity of phenolic compounds varies with the type,
position, and number of substitutions on the parent molecule.
A survey of the available toxicity data shows that halogenated
phenols are more toxic than nitrophenols (see Buikema *et al.*,
1979). Among the halogenated phenols, iodophenols and bromo-
phenols appear to be more toxic than chlorophenols (Ingolls

et al., 1966). Nevertheless, chlorophenols are of particular concern because of (a) their widespread usage, (b) their ubiquitous occurrence in the environment (Buikema *et al.*, 1979), (c) their accumulation in human (Bevenue and Beckman, 1967; Daugherty, 1978; Kutz *et al.*, 1978) and animal tissues (Landner *et al.*, 1977; Kopperman *et al.*, 1976; Pierce and Victor, 1978), and (d) their impurities such as chlorinated phenoxyphenols, diphenyl ethers, dibenzodioxins, dibenzo-furans, and dihydroxybiphenyls (Nilsson and Renberg, 1974; Buser, 1975; Buser and Bosshart, 1976; Levin *et al.*, 1976; Nilsson *et al.*, 1978).

Studies on rats (Farquharson *et al.*, 1958; Ahlborg and Larsson, 1978), and fishes (Ingolls *et al.*, 1966; Kobayashi, 1979) showed that the toxicity of chlorophenols increases with an increase in chlorination. For example, the median lethal doses (LC_{50} values) for chlorophenols administered intraperitoneally to rats were: 276-372 mg/kg for four different isomers of trichlorophenols, 130 mg/kg for 2,3,4,6-tetrachlorophenol, and 56 mg/kg for pentachlorophenol (Farquharson *et al.*, 1958). Acute toxicity tests with an unidentified fish yielded the following LC_{50} values: 14-58 ppm (mg/l) for monochlorophenols, 5.2-13.7 ppm for 2,4-di-chlorophenol, and 3.2 ppm for 2,4,6-trichlorophenol (Ingolls *et al.*, 1966). Tests with goldfish (Kobayashi, 1979) yielded the following LC_{50} values: 60 ppm (phenol), 16 ppm (o-chlorophenol), 9 ppm (p-chlorophenol), 7.8 ppm (2,4-dichloro-phenol), 10 ppm (2,4,6-trichlorophenol), 1.7 ppm (2,4,5-tri-chlorophenol), 0.75 ppm (2,3,4,6-tetrachlorophenol), and 0.27 ppm (pentachlorophenol). The LC_{50}s of pentachlorophenol for many other fishes were reported to be in the range of 0.038 to 0.6 ppm (see review, Buikema *et al.*, 1979; Borthwick and Schimmel, 1978; Schimmel *et al.*, 1978). Somewhat higher LC_{50}s of pentachlorophenol (1.2 and 1.74 ppm) were reported for zebrafish and flagfish, respectively (Fogels and Sprague, 1977).

As in the case of fishes, pentachlorophenol seems to be more toxic (LC_{50}: <1.0 ppm; Weber, 1965) to *Daphnia magna* than are monochlorophenols (4.8-7.4 ppm; Kopperman *et al.*, 1974) and 2,4-dichlorophenol (2.6 ppm., Kopperman *et al.*, 1974; 11.0 ppm, Bringman and Kuhn, 1977). Tests on *Astacus fluviatilis* (Kaila and Saarikoski, 1977) revealed, however, that 2,3,6-trichlorophenol is more toxic (LC_{50} at pH 7.5: 19 ppm; at pH 6.5: 5.4 ppm) than pentachlorophenol (LC_{50} at pH 7.5: 53 ppm; at pH 6.5: 19 ppm). Similarly, our studies on intermolt grass shrimp showed that certain lower chlorinated phenols (i.e., 2,4,5-trichlorophenol and 2,3,4,5-tetrachlorophenol; LC_{50} values: 1.12 and 0.86 ppm, respectively) are more toxic than pentachlorophenol (LC_{50}:

2.5 ppm). Thus, factors other than the degree of chlorination
seem to influence the toxicity of chlorophenols to some
organisms.

In addition to the degree of chlorination, factors such as
the position(s) of chlorination, the pK_a (ionization constant)
of the compound (Blackman *et al.*, 1955; Crandall and Goodnight, 1959; Kaila and Saarikoski, 1977), and the octanol/
water partition coefficient of the compound (McLeese *et al.*,
1979) are thought to influence the toxicity of chlorophenols.
The relative influence of these factors may depend on the mode
of chlorophenol application, and on the species tested. For
example, in tests with *Astacus fluviatilis* (Kaila and Saarikoski, 1977), 2,3,6-trichlorophenol was more toxic than pentachlorophenol when these chlorophenols were added to the
medium; but pentachlorophenol was more toxic than 2,3,6-tri-
chlorophenol when they were injected into the crayfish. The
higher toxicity of 2,3,6-trichlorophenol in the aquatic
medium was thought to be attributable to the trichlorophenol's
ability to cross biological membranes more easily than can
pentachlorophenol. The pH of the test medium (7.5 or 6.5)
was closer to the pK_a of 2,3,6-trichlorophenol than to the
pK_a of pentachlorophenol; the former chlorophenol would be
less ionized under these conditions and, because of this, it
would be able to cross biological membranes more easily. This
phenomenon might, at least in part, account for the higher
toxicity of some of the lower chlorinated phenols to intermolt
grass shrimp (Table 1).

Furthermore, among the tetrachlorophenols tested, there
seemed to be a correlation between the pK_a of the compound and
the toxicity to grass shrimp. The order of their pK_a values
(from high to low) was the same as the order of their toxicity
(from high to low): 2,3,4,5-tetrachlorophenol > 2,3,4,6-
tetrachlorophenol >2,3,5,6-tetrachlorophenol. Interestingly,
these tetrachlorophenols exhibited a similar order of toxicity
when administered orally to gerbils (Ahlborg and Larson,
1978). But when administered orally or intraperitoneally to
rats, the order of toxicity was: 2,3,5,6-tetrachlorophenol >
2,3,4,6-tetrachlorophenol > 2,3,4,5-tetrachlorophenol (Ahlborg
and Larson, 1978). This points out the importance of considering species differences in sensitivity in evaluating the
toxicity of chlorophenols.

The transport of chlorophenols across biological membranes
may not always be related to the pK_a values. For example,
goldfish bioconcentrated 2,4,5-trichlorophenol to a greater
extent than 2,4,6-trichlorophenol (Kobayashi, 1979),
although the pK_a value of the former trichlorophenol is
essentially the same as that of the latter (Table 1). Because
of this difference in bioaccumulation, 2,4,5-trichlorophenol

is more toxic to goldfish than is 2,4,6-trichlorophenol (Kobayashi, 1979). Interestingly, our tests on grass shrimp have also shown that 2,4,5-trichlorophenol is more toxic than 2,4,6-trichlorophenol; whether this is due to differences between the bioaccumulation of these two trichlorophenols in grass shrimp remains to be investigated.

The overall sensitivity of the grass shrimp to various chlorophenols seems to be quite comparable to that exhibited by the goldfish. In tests with intermolt and molting grass shrimp, the LC_{50} values for the various chlorophenols ranged between 0.37 and 4.10 ppm (Table 1). For these same chlorophenols, the LC_{50} values for goldfish ranged between 0.27 and 10 ppm (Kobayashi, 1979).

Toxicity of Chlorophenols to Intermolt and Molting Grass Shrimp

Because of the periodic shedding of their exoskeleton, crustaceans exhibit cyclic changes in epithelial permeability. That such cyclic permeability changes might lead to cyclic variations in the susceptibility of crustaceans to aquatic pollutants has long been suspected (Anderson, 1948). Our previous studies with sodium pentachlorophenate (Conklin and Rao, 1978), and our current studies with trichlorophenols, tetrachlorophenols and pentachlorophenol show that these compounds are more toxic to molting grass shrimp than to non-molting, intermolt grass shrimp. Radiotracer studies with ^{14}C-2,4,5-trichlorophenol (Fig. 2) and ^{14}C-pentachlorophenol (Fig. 5) showed that newly molted shrimp accumulated these chlorophenols to a greater extent than did the intermolt shrimp. This increased bioaccumulation of chlorophenols during the period shortly after molting may account for the greater susceptibility of newly molted grass shrimp to chlorophenols in the medium.

Among the chlorophenols which exhibited an increase in toxicity to grass shrimp at the time of molting, pentachlorophenol showed the greatest increase (5.68-fold) in toxicity, while 2,4,5-trichlorophenol showed the least increase (1.64-fold). Interestingly, this difference can be correlated with the differences between the extents of increase in the bioaccumulation of these chlorophenols in newly molted shrimp. In comparison with intermolt shrimp, the newly molted shrimp accumulated pentachlorophenol to a greater extent than 2,4,5-trichlorophenol. This provides additional support to the view that an increased bioaccumulation of chlorophenols contributes to the greater toxicity of these compounds to molting shrimp.

Recent studies on the toxicity of chromium to *Daphnia
pulex* (Lee and Buikema, 1979), of copper and zinc to *Crangon
crangon* (Price and Uglow, 1979), and of chlorine to *Penaeus
kerathurus* (Saroglia and Scarano, 1979) have also revealed an
increased susceptibility of crustaceans to aquatic pollutants
at the time of, or shortly after, molting; but none of these
studies has examined pollutant bioaccumulation in relation to
the molt cycle.

*Effects of Chlorophenols on Limb
Regeneration in the Grass Shrimp*

Sodium pentachlorophenate inhibits the early phases of
limb regeneration (wound healing, mitosis, and early de-
differentiation) in the grass shrimp, *Palaemonetes pugio* (Rao
et al., 1978, 1979). The present study shows that 2,3,4,5-
tetrachlorophenol, 2,3,4,6-tetrachlorophenol, and pentachlorophenol inhibit limb regeneration in the grass shrimp in a
dose-dependent manner (Fig. 1). The inhibitory effects of
chlorophenols on limb regeneration do not seem to be related
to the acute toxicity of chlorophenols. For example, 2,4,5-
trichlorophenol failed to cause a dose-dependent inhibition
of limb regeneration, although it was one of the most toxic
chlorophenols tested.

Since the tetrachlorophenols and pentachlorophenol were
able to inhibit limb regeneration without causing alterations
in the duration of the intermolt cycle, it would appear
likely that they exert their effects directly on the target
tissue (developing limb bud) rather than acting indirectly
via the neuroendocrine system of the grass shrimp. Unlike
these chlorophenol effects on grass shrimp, aquatic pollutants
such as DDT (Weis and Mantel, 1976) and heavy metals (Weis,
1976, 1977; Weis and Weis, 1979) seem to influence regeneration as well as molting in fiddler crabs. The latter
pollutants seem to exert their effects not only on the
developing limb bud but also on the neuroendocrine system of
the fiddler crab.

*Accumulation and Depuration of
Chlorophenols*

Our studies show that grass shrimp bioconcentrate pentachlorophenol, sometimes to levels as high as 150 times
greater than that in the medium (Fig. 5). Another crustacean,
Daphnia, has also been shown to bioconcentrate pentachlorophenol. In fact, among the several species utilized in a

terrestrial-aquatic model ecosystem, *Daphnia* seemed to bioconcentrate pentachlorophenol to the greatest extent; the bioconcentration factors were 205 for *Daphnia magna,* 132 for *Gambusia affinis,* 26 for *Culex pipiens quinquefasciatus,* 21 for *Physa* spp., and 5 for *Oedogonium cardiacum* (Lu et al., 1978). Goldfish seem to bioconcentrate pentachlorophenol to a much greater extent: up to 1000X (Kobayashi, 1978).

By the end of a 12-hour exposure to 1.0 ppm 2,4,5-trichlorophenol, grass shrimp bioconcentrated this compound by a factor of 32. When goldfish were exposed to 1.8 ppm 2,4,5-trichlorophenol, they died within 24 hours; the trichlorophenol level in the dead fish was 62 times higher than that in the medium (Kobayashi, 1979).

When goldfish were exposed to toxic concentrations of several other chlorophenols the bioconcentration factors derived from the residues in dead fish were: 6.4 for o-chlorophenol, 10.1 for p-chlorophenol, 34 for 2,4-dichlorophenol, 20 for 2,4,6-trichlorophenol, 93 for 2,3,4,6-tetrachlorophenol, and 475 for pentachlorophenol (Kobayashi, 1979).

Under laboratory conditions, various animals - fishes (Kobayashi, 1978; Leach et al., 1978), clams (Kobayashi, 1978), oysters (Schimmel et al., 1978), and grass shrimp (this study) - bioaccumulate chlorophenols very rapidly, within a matter of minutes or a few hours. In most of these animals, the rate of depuration of chlorophenols is also fairly rapid; upon transfer to clean media, they seem to eliminate most of the accumulated chlorophenol within 12 to 96 hours. Some species, however, seem to require a longer period of depuration. For example, eels seem to retain one-third of the accumulated pentachlorophenol even after an 8-day depuration period (Holmberg et al., 1972). The biological half-life of pentachlorophenol in the guppy fish, *Libistes reticulatus* (Ahling and Jernelov, 1969), has been estimated to be about 30 days.

Under field conditions, in a lake contaminated by two major spills of pentachlorophenol-containing wastes from a wood-treatment plant, fishes such as bass, catfish, and sunfish were found to rapidly accumulate pentachlorophenol and its degradation products (2,3,4,5-tetrachlorophenol, 2,3,5,6-tetrachlorophenol, and pentachloranisole) from the water. Although the levels of chlorophenols in the fish decreased as the concentrations of these compounds in the water decreased, they returned to background levels only after 6 to 10 months. The sediment and leaf litter, which retained high concentrations of pentachlorophenol and its degradation products throughout the study, seemed to be a source of chronic pollution, although chronic influx of pentachlorophenol from the contaminated watershed area and periodic release of

wastes from the wood-preservation plant's holding pond may
also have occurred during this period (Pierce and Victor,
1978). Lake trout from Lake Michigan (Lech *et al*., 1978),
and fish raised in disinfected municipal effluents (Kopperman,
et al., 1976) seem to have pentachlorophenol and pentachloranisole in their tissues. In this light, it would be
worthwhile to determine the chlorophenol residues in animals
from the estuarine and marine environments.

Metabolism of Pentachlorophenol

In laboratory model ecosystem studies, *Daphnia* seem to
detoxify pentachlorophenol via conjugate formation (Lu *et al*.,
1978); the conjugates were not identified, however. In blue
crabs injected with pentachlorophenol, much of the pentachlorophenol in the hepatopancreas appeared to be covalently
bound in the ester form (Bose and Fujiwara, 1978); again, the
nature of this conjugation has not been ascertained. Besides
the conjugated pentachlorophenol, pentachloroanisole (methy-
lated metabolite of pentachlorophenol) was also found in the
hepatopancreas of blue crabs.

In our studies with grass shrimp, we found not only pentachloroanisole and conjugated pentachlorophenol, but also
dechlorination products such as 2,3,4,5-tetrachlorophenol,
2,3,4,6-tetrachlorophenol, 2,4,6-trichlorophenol, tetrachlorohydroquinone, and three unknown metabolites. Some of
the conjugated pentachlorophenol seemed to include a glucuronide conjugate. We did not determine whether the grass shrimp
has the ability to eliminate pentachlorophenol as an acetate-
or sulfate-conjugate. But studies of Kobayashi *et al*. (1976)
show that the sulfate conjugation of phenol and pentachlorophenol by the hepatopancreas of the spiny lobster, *Panulirus
japonicus*, is lower than that of the liver of goldfish.

Detoxification of pentachlorophenol in fishes involves
methylation (anisole formation, see Lech *et al*., 1978) as well
as sulfate conjugation and glucuronic acid conjugation
(Kobayashi, 1978, 1979). Algae, insects, snails (Lu *et al*.,
1978) and clams (Kobayashi, 1978) also have the ability to
detoxify pentachlorophenol by conjugation.

Although some of the earlier studies on mammals indicated
the formation of dechlorination products such as tetrachlorophenols and trichlorophenols, more recent studies show that
the dechlorination products of pentachlorophenol in rats
include tetrachlorohydroquinone and trichlorohydroquinone
(Ahlborg, 1978). These animals excrete some of the pentachlorophenol as a conjugate of glucuronic acid.

Microbial degradation of pentachlorophenol involves either reductive or oxidative dehalogenation either *ortho* or *para* to the hydroxy group so as to yield either 2,3,5,6- or 2,3,4,5-tetrachlorophenols, or tetrachlorocatechol or tetrachlorohydroquinone, respectively. Further degradation leads to ring cleavage, $^{14}CO_2$ evolution, and additional chloride ion liberation (Kaufman, 1978).

Monitoring of aquatic organisms for chlorophenol pollution should include the determination of the residues of not only chlorophenols but also of their metabolites and photodegradation products. In this regard, it is important to note that the metabolism of xenobiotics such as hexachlorobenzene and pentachlorobenzene can also lead to the formation of pentachlorophenol, tetrachlorohydroquinone, and certain sulfur-containing compounds (Lu *et al.*, 1978; Koss and Koransky, 1978).

SUMMARY

Studies on the acute toxicity of 2,4-dichlorophenol, 2,4,5-trichlorophenol, 2,4,6-trichlorophenol, 2,3,4,5-tetrachlorophenol, 2,3,4,6-tetrachlorophenol, 2,3,5,6-tetrachlorophenol, and pentachlorophenol to grass shrimp (*Palaemonetes pugio*) at known stages of the molt cycle revealed that, with the exception of 2,4-dichlorophenol, the various chlorophenols were more toxic to molting shrimp than to non-molting, intermolt shrimp. 2,4-Dichlorophenol was, however, equally toxic to molting and non-molting shrimp. Radiotracer studies with ^{14}C-2,4,5-trichlorophenol and ^{14}C-pentachlorophenol indicated that the higher toxicity of these chlorophenols to molting shrimp is due to an increased bioaccumulation of these compounds during the period shortly after molting.

In general, the toxicity of chlorophenols could not be correlated to either the number of Cl-atoms in these compounds or to the pK_a values (ionization constants). For example, lower chlorinated phenols such as 2,4,5-trichlorophenol and 2,3,4,5-tetrachlorophenol were more toxic to intermolt shrimp than was pentachlorophenol. Although 2,4,5-trichlorophenol and 2,4,6-trichlorophenol have essentially the same pK_a, the former chlorophenol was more toxic than the latter. Among tetrachlorophenols, however, there was a positive correlation between the pK_a and the observed toxicity.

Pentachlorophenol, 2,3,4,5-tetrachlorophenol, and 2,3,4,6-tetrachlorophenol caused dose-dependent inhibition of limb regeneration in the grass shrimp. The median effective concentrations causing this inhibition were lower than the

median lethal concentrations of these chlorophenols for intermolt shrimp. These chlorophenols did not alter the duration
of the intermolt cycle. The two trichlorophenols tested
neither caused dose-dependent inhibition of limb regeneration
nor altered the duration of the intermolt cycle.

When exposed to media containing pentachlorophenol or
2,4,5-trichlorphenol, grass shrimp bioaccumulated these
chlorophenols very rapidly. When such chlorophenol-loaded
shrimp were transferred to clean seawater, most of the
accumulated chlorophenol was lost from the shrimp within 12
to 24 hours. The eggs attached to the pleopods of berried
shrimp have also accumulated pentachlorophenol rapidly, but
they were unable to eliminate this chlorophenol as rapidly as
did the shrimp.

Qualitative and quantitative analysis of the metabolites
of ^{14}C-pentachlorophenol in the grass shrimp indicated that
methylation (anisole formation), glucuronide conjugation, and
dechlorination are among the pathways of pentachlorophenol
disposition in this shrimp. These results are discussed in
light of the previous work on other organisms and available
information on chlorophenolic contamination of the aquatic
ecosystem.

ACKNOWLEDGMENTS

This investigation was supported by Grant R-804541 from
the U.S. Environmental Protection Agency. We are thankful to
Dr. Norman L. Richards, Associate Director for Extramural
Activities, Gulf Breeze Environmental Research Laboratory, for
his suggestions and encouragement during this investigation.

LITERATURE CITED

Ahlborg, U. G. 1978. Declorination of pehtachlorophenol *in
    vivo* and *in vitro*. In: Pentachlorophenol: Chemistry,
    Pharmacology, and Environmental Toxicology (K. R. Rao,
    ed.), pp. 115-130, Plenum Press, New York.

Ahlborg, U. G., and K. Larsson. 1978. Metabolism of tetra-
    chlorophenols in the rat. Arch. Toxicol., 4Q: 63-74.

Ahling, B., and A. Jernelov. 1969. IVL Stockholm Internal
    Report 269/69, 17.10.69.

Anderson, B. G. 1948. The apparent thresholds of toxicity to *Daphnia magna* for chlorides of various metals when added to Lake Erie water. Trans. Amer. Fish. Soc., 78: 96-113.

Barr, A. J., J. H. Goodnight, J. P. Sall, and J. T. Helwig. 1976. A User's Guide to SAS-76, 329 p., SAS Institute, Raleigh, North Carolina.

Bevenue, A. and H. Beckman. 1967. Pentachlorophenol: A discussion of its properties and its occurrence as a residue in human and animal tissues. Residue Rev., 19: 83-134.

Blackman, G. E., M. H. Parke, and G. Graton. 1955. The physiological activity of substituted phenols II. Relationships between physical properties and physiological activity. Arch. Biochem. Biophys., 54: 55-71.

Bliss, D. E. 1956. Neurosecretion and the control of growth in a decapod crustacean. In: Bertil Hanström, Zoological Papers in Honor of his Birthday, November 20, 1956 (K. G. Wingstrand, ed.), pp. 56-75, Zoological Institute, Lund, Sweden.

Borthwick, P. W., and S. C. Schimmel. 1978. Toxicity of pentachlorophenol and related compounds to early life stages of selected estuarine animals. In: Pentachlorophenol: Chemistry, Pharmacology, and Environmental Toxicology (K. R. Rao, ed.), pp. 141-146, Plenum Press, New York.

Bose, A. K., and H. Fujiwara. 1978. Fate of pentachlorophenol in the blue crab, *Callinectes sapidus*. In: Pentachlorophenol: Chemistry, Pharmacology, and Environmental Toxicology (K. R. Rao, ed), pp. 83-88, Plenum Press, New York.

Bringmann, G., and R. Kuhn. 1977. Befunde der Schadwirkung wasser gefardender Stoffe gegen *Daphnia magna*. Z. fur wasser und Abwasser forschung, 10: 161-166.

Buikema, A. L., Jr., M. J. McGinnis, and J. Cairns, Jr. 1979. Phenolics in aquatic ecosystems: a selected review of recent literature. Marine Environ. Res., 2: 87-181.

Buser, H. -R., and H. P. Bosshardt. 1976. Determination of polychlorinated dibenzo-p-dioxins and dibenzofurans in commercial pentachlorophenols by combined gas chromatography-mass spectrometry. J. Ass. Offic. Anal. Chem., 59: 562-569.

Conklin, P. J., and K. R. Rao. 1978. Toxicity of sodium pentachlorophenate to the grass shrimp, *Palaemonetes pugio*, in relation to the molt cycle. In: Pentachlorophenol: Chemistry, Pharmacology, and Environmental Toxicology (K. R. Rao, ed.), pp. 181-192, Plenum Press, New York.

Crandall, C. A., and C. J. Goodnight. 1959. The effects of various factors on the toxicity of sodium pentachlorophenate to fish. Limnol. Oceanogr., 7: 233-239.

Doedens, J. D. 1964. Chlorophenols. In: Kirk-Othmer Encyclopedia of Chemical Technology, 2nd ed., Vol. 5, 325-338, John Wiley & Sons, New York.

Dougherty, R. C. 1978. Human exposure to pentachlorophenol. In: Pentachlorophenol: Chemistry, Pharacology, and Environmental Toxicology (K. R. Rao, ed.), pp. 351-362, Plenum Press, New York.

Farquharson, M. E., J. C. Gage, and J. Northover. 1958. The biological actions of chlorophenols. British J. Pharmacol., 13: 20-24.

Finney, D. J. 1971. Probit Analysis, 3rd edition, 333 pp., Cambridge University Press, London.

Fogels, A., and J. B. Sprague. 1977. Comparative short-term tolerance of zebrafish, flagfish, and rainbow trout to five poisons including potential reference toxicants. Water Research, 11: 811-817.

Holmberg, B. S. Jensen; A. Larsson, K. Lewander, and M. Olsson. 1972. Metabolic effects of technical pentachlorophenol (PCP) on the eel *Anguilla anguilla* L. Comp. Biochem. Physiol., 43B: 171-183.

Ingolls, R. S., P. E. Gaffney, and P. C. Stevenson. 1966. Biological activity of halophenols. J. Water Poll. Control Fed., 38: 629-635.

Jolley, R. L., G. Jones, W. W. Pitt, and J. E. Thompson. 1975. Chlorination of organics in cooling waters and process effluents. In: Proceedings of the Conference on the Environmental Impact of Water Chlorination (R. D. Jolley, ed.), pp. 115-152, Oak Ridge National Laboratory, Oak Ridge, Tennessee.

Kaila, K. and J. Saarikoski. 1977. Toxicity of pentachloro-phenol and 2,3,6-trichlorophenol to the crayfish (Astacus fluviatilis L.). Environ. Pollut., 12: 119-123.

Kaufman, D. 1978. Degradation of pentachlorophenol in soil, and by soil microorganisms. In: Pentachlorophenol: Chemistry, Pharmacology, and Environmental Toxicology (K. R. Rao, ed.), pp. 27-40, Plenum Press, New York.

Kobayashi, K. 1978. Metabolism of pentachlorophenol in fishes. In: Pentachlorophenol: Chemistry, Pharmacology, and Environmental Toxicology (K. R. Rao, ed.), pp. 89-106, Plenum Press, New York.

Kobayashi, K. 1979. Metabolism of pentachlorophenol in fish. In: Pesticide and Xenobiotic Metabolism in Aquatic Organisms (M. A. Q. Kahn, J. J. Lech, and J. J. Menn, eds,), pp. 131-144, ACS Symposium Series, 99; American Chemical Society, Washington, D.C.

Kobayashi, K., S. Kimura, and H. Akitake. 1976. Studies on the metabolism of chlorophenols in fish. VII. SUlfate conjugation of phenol and PCP by fish livers. Bull. Japan. Soc. Sci. Fish., 42: 171-177.

Kopperman, H. L., D. W. Kuehl, and G. E. Glass. 1976. Impact of water chlorination. Oak Ridge National Lab-oratory Conf. 751096, p. 327.

Koss, G., and W. Koransky. 1978. Pentachlorophenol in different species of vertebrates after administration of hexachlorobenzene and pentachlorobenzene. In: Penta-chlorophenol: Chemistry, Pharmacology, and Environmental Toxicology (K. R. Rao, ed.), pp. 131-137, Plenum Press, New York.

Kutz, F. W., R. S. Murphy, and S. C. Strassman. 1978. Survey of pesticide residues and their metabolites in urine from general population. In: Pentachlorophenol: Chemistry, Pharmacology, and Environmental Toxicology (K. R. Rao, ed), pp. 363-370, Plenum Press, New York.

Lander, L., K. Lindstrom, M. Karlsson, J. Nordin, and L.
    Sorenson. 1977. Bioaccumulation in fish of chlorinated
    phenols from Kraft Mill bleachery effluents. Bull.
    Environ. Contam. Toxicol., 18: 663-673.

Lech, J. J., A. H. Glickman, and G. N. Statham. 1978.
    Studies on the uptake, disposition and metabolism of
    pentachlorophenol and pentachloroanisole in rainbow trout
    (*Salmo gairdneri*). In: Pentachlorphenol: Chemistry,
    Pharmacology, and Environmental Toxicology (K. R. Rao,
    ed.), pp. 107-114, Plenum Press, New York.

Lee, D. R., adn A. L. Buikema, Jr. 1979. Molt-related
    sensitivity of *Daphnia pulex* in toxicity testing. J.
    Fish Res. Board Canada, 36: 1129-1133.

Levin, J. -O., C. Rappe, and C. -A. Nilsson. 1976. Use of
    chlorophenols as fungicides in sawmills. Scandinavian J.
    Work Environ. Hlth., 2: 71-81.

Lu, P. -Y., R. L. Metcalf, and L. K. Cole. 1978. The en-
    vironmental fate of ^{14}C-pentachlorophenol in laboratory
    model ecosystems. In: Pentachlorophenol: Chemistry,
    Pharmacology, and Environmental Toxicology (K. R. Rao,
    ed.), pp. 53-64, Plenum Press, New York.

McLeese, D. W., V. Zitko, and M. R. Peterson. 1979.
    Structure-lethality relationships for phenols and other
    aromatic compounds in shrimp and clams. Chemosphere,
    8: 53-57.

Nilsson, C. -A., and L. Renberg. 1974. Further studies on
    impurities in chlorophenols. J. Chromatogr., 89: 325-333.

Nilsson, C. -A., A. Norstrom,, K. Andersson, and C. Rappe.
    1978. Impurities in commercial products related to
    pentachlorphenol. In: Pentachlorophenol: Chemistry,
    Pharmacology, and Environmental Toxicology (K. R. Rao,
    ed.), pp. 313-324, Plenum Press, New York.

Pierce, R. H., Jr., and D. M. Victor. 1978. The fate of
    pentachlorophenol in an aquatic system. In: Pentachloro-
    phenol: Chemistry, Pharmacology, and Environmental
    Toxicology (K. R. Rao, ed.), pp. 41-52, Plenum Press, New
    York.

Price, R. K. J., and R. F. Uglow. 1979. Some effects of certain metals on development and mortality within the moult cycle of *Crangon crangon* (L.). Marine Environ. Res., 2: 287-299.

Rao, K. R. (ed.). 1978. Pentachlorophenol: Chemistry, Pharmacology and Environmental Toxicology, 416 p., Plenum Press, New York.

Rao, K. R.; P. J. Conklin, and A. C. Brannon. 1978. Inhibition of limb regeneration in the grass shrimp, *Palaemonetes pugio,* by sodium pentachlorophenate. In: Pentachlorophenol: Chemistry, Pharamcology, and Environmental Toxicology (K. R. Rao, ed.), pp. 193-203, Plenum Press, New York.

Rao, K. R., F. R. Fox, P. J. Conklin, A. C. Cantelmo, and A. C. Brannon. 1979. Physiological and biochemical investigations of the toxicity of pentachlorphenol to crustaceans. In: Marine Pollution: Functional Responses (W. B. Vernberg, F. R. Thurberg, A. Calabrese, and F. J. Vernberg, eds.), pp. 307-340, Academic Press, New York.

Rockwell, A. L., and R. A. Larson. 1978. Aqueous chlorination of some phenolic acids. In: Water Chlorination: Environmental Impact and Health Effects (R. L. Jolley, L. H. Gorcher, D. H. Hamilton, Jr., eds.), Ann Arbor Science Publishers Inc., Ann Arbor, Michigan.

Saroglia, M. G., and G. Scarano. 1979. Influence of molting on sensitivity to toxics of the crustacean *Penaeus kerathurus* (Forskal). Ecotoxicol. Environ. Safety, 3: 310-320.

Schimmel, S. C., J. M. Patrick, Jr., and L. F. Faas. 1978. Effects of sodium pentachlorphenate on several estuarine animals: toxicity, uptake and depuration. In: Pentachlorophenol: Chemistry, Pharmacology, and Environmental Toxicology (K. R. Rao, ed.), pp. 147-156, Plenum Press, New York.

Weber, E. 1956. Einwirkung von Pentachlorophenolnatrium auf Fische und Fischnahrtiers. Biol. Zentralbl., 84: 81-93.

Weis, J. S. 1976. Effects of mercury, cadmium, and lead salts on limb regeneration and ecdysis in the fiddler crab, *Uca pugilator*. U.S. Fish Wildl. Serv., Fish Bull., 74: 464-467.

Weis, J. S. 1977. Limb regeneration in fiddler crabs:
    species differences and effects of methyl mercury. Biol.
    Bull., 152: 263-274.

Weis, J. S., and L. H. Mantel. 1977. DDT as an accelerator
    of regeneration and molting in fiddler crabs. Estuarine
    Coastal Mar. Sci., 4: 461-466.

Weis, J. S., and P. Weis. 1979. Effects of mercury, cadmium,
    and lead compounds on regeneration in estuarine fishes
    and crabs. In: Marine Pollution: Functional Responses
    (W. B. Vernberg, F. P. Thurberg, A. Calabrese, and F. J.
    Vernberg, eds.), pp. 151-170, Academic Press, New York.

Wong, A. S., and D. G. Crosby. 1978. Photolysis of penta-
    chlorophenol in water. In: Pentachlorophenol:
    Chemistry, Pharmacology, and Environmental Toxicology
    (K. R. Rao, ed.), pp. 19-26, Plenum Press, New York.

BIOCHEMICAL STRESS RESPONSES
OF MULLET *MUGIL CEPHALUS* AND POLYCHAETE WORMS
NEANTHES VIRENS TO PENTACHLOROPHENOL

Peter Thomas
R. Scott Carr
Jerry M. Neff

Department of Biology
Texas A&M University
College Station, Texas

INTRODUCTION

Pentachlorophenol (PCP) and its sodium salt (Na-PCP) have
many industrial and agricultural applications as herbicides,
insecticides, bactericides and wood preservatives. These
chlorophenols are the second most used pesticides in the
United States and production in 1977 was estimated to be about
80 million pounds (Cirelli, 1978). PCP can occur in the
aquatic environment, particularly in runoff waters from wood
treatment plants (Pierce et al., 1977). Antifouling paints
and oil well drilling muds (Robichaux, 1975) are additional
sources of PCP contamination for estuarine environments. PCP
and NA-PCP are powerful metabolic poisons which uncouple
oxidative phosphorylation (Weinbach, 1957; Bevenue and Beckman, 1967). PCP has been detected in fish tissues (Pierce and
Victor, 1978) and high concentrations of PCP in aquatic environments have been associated with fish kills (Holmberg et
al., 1972; Pierce et al., 1977).
Pentachlorophenol is also toxic to several species of
estuarine and marine crustaceans and in sublethal concentrations affects a variety of morphological, ultrastructural,
physiological and biochemical processes (van Dijk et al.,
1977; Kaila and Saarikoski, 1977; Brannon and Conklin, 1978;
Doughtie and Rao, 1978; Cantelmo et al., 1978; Rao et al.,
1979). However, there is relatively little information on the

toxicity of PCP to non-crustacean estuarine organisms
(Borthwick and Schimmel, 1978). There are no reports on the
biochemical responses of polychaetes to PCP, although PCP has
been shown to affect feeding activity in *Arenicola cristata*
(Rubinstein, 1978). An alteration of energy metabolism
(Holmberg *et al.*, 1972) and enzyme levels (Böstrom and
Johansson, 1972) has been shown in eels exposed to PCP for
four days, but the acute biochemical responses of teleosts to
PCP have not been investigated.

In teleosts stressful stimuli rapidly activate interrenal
and adrenergic systems which mediate the secondary metabolic
responses (Mazeaud *et al.*, 1977). Plasma cortisol has been
proposed as a good single indicator of certain kinds of
environmental stress (Strange and Schreck, 1978). Rapid
activation of the hypothalamo-pituitary-interrenal axis has
been found in teleosts after exposure to several pollutants
(DiMichele and Taylor, 1978; Schreck and Lorz, 1978). Of the
pollutants examined so far, only cadmium in lethal concentrations does not elevate plasma cortisol levels (Schreck and
Lorz, 1978). PCP and other chlorophenols are extremely toxic
to fish (Kobayashi, 1979). It is not known if chlorophenols
activate the hypothalamo-pituitary-interrenal axis. High
circulating cortisol levels in teleosts may provide an early
indication of PCP contamination of the environment.

This paper describes the effects of PCP exposure on
several biochemical parameters in juvenile mullet, *Mugil
cephalus* and the sandworm *Neanthes virens* (=*Nereis virens*
Sars, Pettibone, 1963). Coelomic fluid osmolality and glucose
concentrations and tissue ascorbic acid levels were monitored
in *N. virens* during acute lethal and chronic sublethal exposure to PCP. The effects of acute exposure to lethal levels
of PCP on plasma cortisol concentrations and several secondary
stress responses in *M. cephalus* were examined. Aspects of
carbohydrate, lipid and electrolyte metabolism were investigated in order to determine which homeostatic mechanisms
fail during exposure to lethal levels of the toxicant. Pollutant exposure has also been shown to alter ascorbic acid metabolism in teleosts (Mayer *et al.*, 1978). The effect of PCP
exposure on liver ascorbic acid concentrations was also investigated in the present study and the possible utility of
this parameter in teleosts and polychaetes as an index of
environmental contamination is discussed.

MATERIALS AND METHODS

Animals

Sandworms *Neanthes virens* were obtained from a commercial bait supplier in Maine. Artificial seawater made by mixing artificial sea salts (Instant Ocean Aquarium Systems, Inc., Eastlake, Ohio) with distilled water was used throughout this study. The worms were maintained at 10°C in a cold room in the dark in 35 °/oo ± 2 °/oo Instant Ocean seawater in 100 l fiberglass tanks equipped with Eheim filters. The polychaetes were fed fresh minced clam, *Rangia cuneata* during acclimation. Food was supplied during the 12 day experiment but not during the two short-term exposures (48 and 96 hours).

Juvenile *Mugil cephalus*, length 10-20 cm, were collected with a seine near Port Aransas, Texas in September, 1979. The fish were acclimated for two weeks in the laboratory in large recirculating marine aquaria. Mullet were maintained at 33 °/oo salinity, temperature at 20 ± 0.5°C and lights were on from 0700-1900 hours daily. Fish were fed once daily with Tetramin algal flake.

Experimental Treatments

N. virens were exposed to PCP in a cold room at 10°C in the dark. Replicate coelomic fluid samples were withdrawn with a microsyringe and transferred to non-heparinized capillary tubes. The coelomic fluid was centrifuged at 12,000 g for three minutes and the supernatant subsequently analyzed for glucose and osmolality with a YSI glucose analyzer and a Wescor vapor pressure osmometer, respectively.

Parapodial tissue was removed from the live worms with forceps, blotted dry, weighed, and homogenized in ice cold degassed .25M $HClO_4$ (10% weight/volume). All subsequent procedures were performed at 0-4°C. The homogenate was centrifuged at 27,000 g for 25 minutes. The supernatant was transferred to clean test tubes and respun at the same speed for an additional 15 minutes. The supernatants were analyzed directly for ascorbic acid by high-performance liquid chromatography (HPLC).

Six mullet were transferred to each exposure aquarium containing 40 l of 33 °/oo S. artificial seawater and two charcoal filters. The fish were undisturbed for 16 hours prior to sacrifice between 1000 and 1100 hours on the 14th day after transfer. Fish were removed with a dip net and blood samples were collected in syringes by cardiac puncture. All the fish in each tank were bled within two minutes. Special care was

taken to minimize disturbance of fish during sampling in order
to prevent nonspecific activation of the interrenal gland.
The blood was centrifuged in heparinized tubes and the plasma
was removed for biochemical analysis. Glucose, cholesterol
and osmolality were measured on the same day. The remaining
sample was stored at -60°C until analyzed for cortisol.
Immediately after blood collection the mullet were sacrificed
and placed on ice. Pieces of liver tissue were removed within
40 minutes of death, weighed, and frozen in liquid nitrogen.
Samples were subsequently stored at -60°C for up to two weeks
until analyzed for ascorbic acid and glycogen. Mullet
carcasses were stored at -60°C for up to two months until
analyzed for PCP accumulation.

PCP Exposure

Concentrated aqueous solutions of pentachlorophenol were
made using a sand column. The column was charged with 25 gm
of PCP (Aldrich, Gold Label 99+% pure) which had been dissolved in a small volume of ether and evaporated with a stream
of nitrogen on approximately one pound of coarse sand. The
charged sand was subsequently packed between layers of clean
sand in a glass column. The sand was kept in place with glass
wool plugs. Aluminum foil was wrapped around the column to
exclude light. Water was forced through the column (rate 1
ml/min.) with a peristaltic pump. The concentration of PCP in
the elute stabilizes after two days operation and is constant
for an additional 20 days. *N. virens* were chronically exposed
to a constant concentration of PCP for 12 days using this
dosing system. PCP solutions for acute exposure of *N. virens*
were prepared by stirring PCP in seawater for 48 hours. The
sand column elute was collected and added in a single dose to
the aquaria for acute exposures of mullet (3-24 hours) to PCP.
Pentachlorophenol concentrations were maintained during the
longer exposures by changing half of the seawater with fresh
seawater of known PCP concentration daily. The PCP concentration in each exposure tank was monitored daily for the
duration of the experiments.

Biochemical Analysis

Plasma cortisol was measured by radioimmunoassay using
antiserum obtained from New England Nuclear. This antiserum
cross reacts 17% with corticosterone and 25.7% with cortisone.
The radioimmunoassay for cortisol determination in teleost

plasma was validated by thin-layer chromatography and high-
performance liquid chromatography (Thomas *et al.*, in preparation). Corticosterone was also found, but in insufficient
concentrations to significantly affect the radioimmunoassay
for cortisol. Preliminary purification on a Sephadex LH20
column was therefore unnecessary. Fifty µl plasma samples
were extracted in five ml methylene dichloride. The aqueous
layer was aspirated and the organic layer was evaporated to
dryness under a stream of air. The sides of the tube were
washed three times with a small volume of methyl alcohol. The
three washings were transferred to a clean tube and the methyl
alcohol was evaporated under a stream of air. This procedure
prevented possible interference in the assay by plasma
proteins when the sample was redissolved in assay buffer. The
concentration of charcoal was increased fourfold to prevent
nonspecific interference by lipids (Rash *et al.*, 1979) which
are present in fish plasma in high concentrations (Larsson
and Fänge, 1977). Three dilutions of each sample were
assayed. Recovery was always above 80%. The assay typically
had a sensitivity of 10 pg, and a within assay variance of
6.76% (mean of ten determinations, 32·5±2·2 ng/ml).
   Glucose was analyzed in mullet plasma with a YSI glucose
analyzer (model 23A, Yellow Springs Instrument Co., Yellow
Springs, Ohio) which measures the H_2O_2 released upon the enzymatic oxidation of glucose. The assay has a sensitivity of
1 mg/dl in a 25 µl plasma sample. Total plasma cholesterol
was measured colorimetrically after enzymatic conversion using
a commercial kit (Pierce Chemical Company). The osmolality
of mullet plasma was determined with a vapor pressure osmometer (model 5100, Wescor Inc., Logan, Utah). Liver glycogen was measured by the technique of Montgomery (1957). The
α,α'-dipyridyl method (Zannoni *et al.*, 1974) was used for the
determination of hepatic ascorbic acid in mullet.
   The α,α'-dipyridyl method was found to be unsatisfactory
for analysis of ascorbic acid in polychaete tissues due to
interferring substances and low tissue concentrations.
Ascorbic acid was measured in 10 µl samples by a sensitive and
specific HPLC technique developed in this laboratory (Carr and
Neff, in preparation). Ascorbic acid was separated on an
anion exchange column (Whatman Partisil 10 SAX, 25 cm x 4.6
mm) and detected amperometrically (model LC-4, Bioanalytical
Systems Inc.). A 0·06M sodium acetate solution adjusted to
pH 4·6 with 1M acetic acid was used as the mobile phase which
was prepared with double distilled, millipore filtered
(0.45µm) vacuum-degassed, deionized water. Ascorbic acid
eluted in approximately six minutes at a flow rate of 2.9 ml/
min. The assay had a detection limit of 2-3 ng ascorbic acid
in a 10 µl sample. Tissue samples in which no ascorbic acid
was detected were considered to have half the detection limit

value for statistical purposes. Ascorbic acid was detected in greater than 85 percent of the tissue samples analyzed.

PCP Analysis

The concentration of PCP in water samples was measured by a spectrophotometric technique (Carr *et al.*, in preparation). PCP was extracted from 100 ml acidified water samples with 10 ml chloroform, the aqueous layer discarded, and the PCP re-extracted into 2 ml of 0.2 M NaOH for spectrophotometric analysis at 320 nm. There is a linear relationship between optical density and PCP concentration from 2 mg PCP/l to the detection limit of the assay at 50 µg PCP/l.

PCP was analyzed in tissue samples by gas chromatography (Giam, personal communication). Tissue was homogenized in acetone/acetonitrile, diluted with aqueous 5% NaCl, subjected to basic and then acidic extraction with hexane, derivatized with diazomethane, chromatographed on acidic alumina and quantified by gas chromatography with electron capture detection.

All fish data and most polychaete data were analyzed by Student's t-test and significant differences were established at the 0.05 level. The Mann-Whitney U-test (1947) was used for analysis of some polychaete data which did not conform to analysis by parametric tests.

RESULTS

Neanthes virens

During an acute lethal exposure to PCP (720 ppb initially) there was a rapid hypoglycemia in the coelomic fluid of *N. virens* which persisted until death (Table I). At a lower but lethal exposure to PCP (365 ppb initially), a marked hyperglycemia was observed after 21 hours (Fig. 1). Glucose concentrations gradually decreased but did not fall to control levels until the worms were moribund. Ascorbic acid concentrations showed an initial gradual decrease to low levels followed by a dramatic increase after 96 hours, a time when the mortality rate was rapidly increasing. During the chronic sublethal exposure to PCP (120 ppb) similar changes in coelomic fluid glucose levels were observed (Fig. 2). A hyperglycemic response occurred during the first few days of exposure followed by a sudden hypoglycemia at 96 hours. After one week of exposure, however, the coelomic fluid glucose levels were significantly higher than control values. Glucose concentrations returned to control levels after two weeks of exposure.

TABLE 1: Time course of coelomic fluid glucose concentrations in sandworms, Neanthes virens during a static PCP exposure. Initial PCP concentration of 720 µg/l dropped to 450 µg/l after 50 hours. Each value represents the mean of 10 replicate samples. Values significantly different (0.05 level of significance) from controls as analyzed by the t-test and Mann-Whitney U-test denoted by a and b, respectively.

Exposure Time (hours)	Glucose (mg/dl) ± S.E.M.
0 (control)	30.6 ± 10.0
2	27.5 ± 5.1
4	20.4 ± 1.9[b]
8	16.9 ± 3.1[b]
24	16.5 ± 3.2[a]
30 (control)	27.9 ± 5.7
48	18.4 ± 3.4
50 (control)	26.2 ± 4.2

Ascorbic acid levels showed different initial responses to the two lower doses of PCP. There was an initial decrease in ascorbic acid levels and a subsequent dramatic increase prior to death in worms exposed to lethal concentrations of PCP (Fig. 1), whereas tissue levels remained elevated for one week upon exposure to the sublethal dose (Fig. 2). After two weeks exposure to PCP, ascorbic acid concentrations had returned to control levels.

Coelomic fluid glucose concentrations increased from 27.8 mg/dl to 58.1 mg/dl during a 48 hour hypoosmotic salinity stress when worms were transferred from 28 °/oo to 15 °/oo salinity. Glucose and ascorbic acid levels also varied in response to experimental starvation (Table II). The hyperglycemic response of hypoosmotically stressed N. virens suggests a hypermetabolic state due to increased metabolic demands by osmoregulating mechanisms. During starvation both coelomic fluid glucose and tissue ascorbic acid levels dropped initially. Prolonged starvation resulted in dramatic increases in coelomic fluid glucose concentrations and particularly in tissue free ascorbic acid concentrations.

Coelomic fluid osmolality was also monitored during PCP exposure. At 35 °/oo salinity N. virens normally maintained their coelomic fluid at ca. 100 mOs/l hypoosmotic to their environment. An inability to maintain coelomic fluid osmolality was observed only in moribund worms, as indicated by lethargic activity, loss of coelomic turgor and reduced epidermal pigmentation.

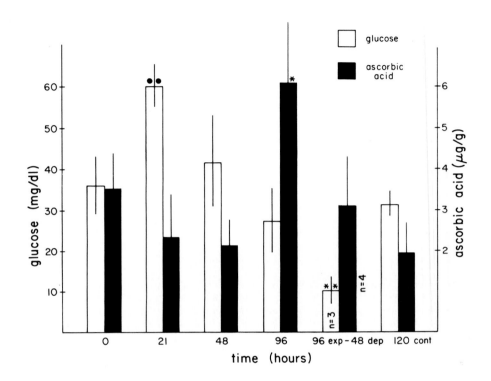

FIGURE 1:   *Time course of coelomic fluid glucose concen-*
trations and tissue free ascorbic acid concentrations in sand-
worms, Neanthes virens, *during a static PCP exposure. Initial*
PCP exposure concentration of 365 μg/l dropped to 100 μg/l
after 20 hours and to below 50 μg/l after 96 hours.   Each
column represents the mean (±S.E.M.) of 12 samples except
where noted otherwise.   Closed circles and asterisks denote
values significantly different from controls as analyzed by
the t-test and Mann-Whitney U-test, respectively (one symbol,
.05 level; two symbols .01 level of significance).

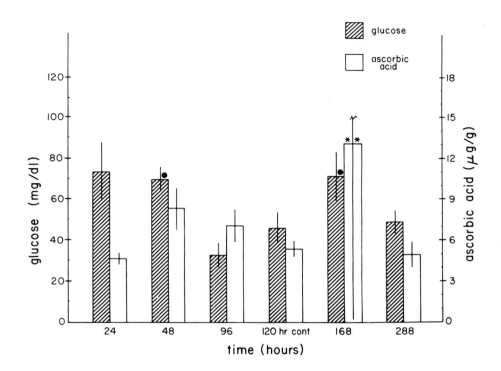

FIGURE 2: *Time course of coelomic fluid glucose concen-*
trations and tissue free ascorbic acid concentrations in sand-
worms, Neanthes virens, *during a 12 day PCP exposure. The PCP*
concentration was maintained at 120 ± 10 μg/l during the ex-
posure period with the aid of a sand column dosing system.
Each column represents the mean (±S.E.M.) of 12 samples.
Closed circles and asterisks denote values significantly
different from controls as analyzed by the t-test and Mann-
Whitney U-test, respectively (one symbol, .05 level; two
symbols, .01 level of significance).

TABLE II: *Time course of coelomic fluid glucose concentrations and free ascorbic acid concentrations in parapodial tissue of sandworms,* Neanthes virens, *during experimental starvation. Each value represents the mean of 12 samples which were taken in replicate for the glucose determinations. Asterisks denote values significantly different from day 0 controls as analyzed by the Mann-Whitney U-test (*, 0.05; **, 0.01 level of significance).*

Days of Starvation	Glucose (mg/dl) ± S.E.M.	Ascorbic Acid (µg/g) ± S.E.M.
0	35.8 ± 6.8	3.46 ± 0.89
5	31.5 ± 2.9	2.53 ± 0.73
16	10.5 ± 1.2**	0.37 ± 0.11*
88	48.0 ± 9.8	9.93 ± 2.04**

Mugil cephalus

Plasma cortisol concentrations rose rapidly in mullet to 300 ng/ml, sixty times resting values, three hours after addition of 200 ppb PCP and declined somewhat after 24 hours exposure (final concentration 164.5 ppb PCP) but were still elevated compared to untreated fish ($P < .002$) (Fig. 3). By this time 33% of the fish were dead. In contrast, plasma cortisol concentrations did not rise until the second day of exposure to 100 ppb (Fig. 3) (final concentration 101 ppb PCP). The magnitude of the corticosteroid stress response was significantly diminished in mullet exposed to the lower dose of PCP, although the mortality in these fish had also reached 33% by 48 hours. Unexposed fish were restrained in a dip net for 30 minutes and exposed to air for one minute every five minutes. Blood was collected for cortisol measurement 30 minutes after the end of this stress. This was considered an extremely stressful procedure which would give an indication of the maximum corticosteroid response in juvenile mullet. Cortisol concentrations rose to 320±60 ng/ml after this treatment. These results suggest that a near maximal corticosteroid response occurs three hours after exposure to 200 ppb PCP.

The dynamics of circulating glucose concentrations upon exposure to the two concentrations of PCP were generally similar to those of plasma cortisol (Table III). Hyperglycemia did not occur until the second day of exposure to 100 ppb PCP, whereas a significant elevation of glucose levels had

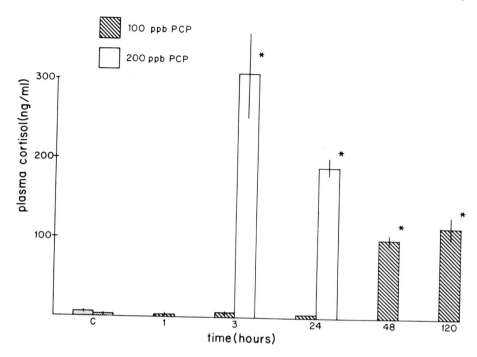

FIGURE 3: *Time course of plasma cortisol concentrations in juvenile striped mullet,* Mugil cephalus, *exposed to 100 ppb (μg/l) and 200 ppb pentachlorophenol. Each column represents the mean (±S.E.M.) of at least six samples. Asterisks denote values significantly different from controls (C).*

occurred after three hours exposure to the high dose. The hyperglycemic response to 200 ppb PCP was not as rapid as the corticosteroid response, however, since plasma glucose had not reached maximum concentrations three hours after addition of the pollutant.

The elevation of plasma glucose was accompanied by a depletion of liver glycogen reserves (Table III). There was a significant depletion of glycogen after three hours exposure to a 200 ppb PCP (P<.01) and by 24 hours reserves had fallen to 7% of normal values. Hepatic glycogen fell to the same levels in the mullet exposed to 100 ppb PCP but depletion occurred at a slower rate and minimum glycogen levels were not found until after 48 hours exposure (Table III). One fish exposed to 200 ppb PCP showed severe disequilibrium after 24 hours and opercular movement had virtually ceased. Glycogen

TABLE 3. Time course of secondary stress responses, bioaccumulation, and mortality in *Mugil cephalus* exposed to pentachlorophenol.

	Dose PCP ppb	Time (hours)					
		0	1	3	24	48	120
Plasma glucose (mg/dl)	100	54.5±6.6	58.2±6.6	53.1±10.1	58.3±4.6	199.0±18.6[c]	192.0±54.8[b]
	200	53.0±9.9		122.4±43.3[c]	253.0±59.0[c]		
Liver Glycogen (mg/gm)	100	60.8±22.6	62.0±17.0	81.0±34.0	49.6±37.0	4.35±1.5[c]	8.69±5.3[b]
	200	86.0±33.0		21.7±1.5[a]	4.34±1.52[c]		
Plasma Osmolality (mOsm/l)	100	300.8±15.4	308.2±25.4	341.2±54.9	314.2±20.5	325.0±10.1	409.3±30.26[c]
	200	315.7±14.2		337.3±38.2	441.5±48.8[a]		
Plasma cholesterol (mg/dl)	100	96.5±17.4	84.7±13.7	81.2±7.8	86.5±10.5	94.0±12.0	99.6±4.5
	200	97.6±13.8		92.2±15.0	82.3±9.8		
PCP accumulation (ppm)	100	.090	NM	2.6	5.4	NM	37.1
	200			8.2	17.8	33	
Mortality %	100	0	0	0	0		50
	200	0		0	33		

Mean (±S.E.M.) of at least six samples

NM - not measured

a - P<.01
b - P<.002
c - P<.001

reserves were 2·38 mg/gm in this fish, approximately half that of other exposed fish at this time, and plasma glucose concentrations had fallen to 19 mg/dl. The results suggest that once hepatic glycogen reserves have fallen to about 5% of the normal value normal glucose levels can no longer be maintained.

Exposure to PCP caused a massive increase in plasma osmolality (Table III), but only after prolonged exposure to the pollutant. A significant elevation of plasma osmolality was shown only with the longest exposures to the two doses of PCP. The highest value (511 mOs/l) was found in the mullet exposed to 200 ppb for 24 hours which had an extremely low glucose level and was on the point of death.

Total plasma cholesterol concentrations were unaltered in mullet exposed to PCP (Table III). Pentachlorophenol was rapidly accumulated in mullet up to approximately 300 times the concentrations in the exposure media (Table III).

Liver ascorbic acid concentrations showed a delayed response to PCP exposure (Fig. 4). No alteration of ascorbic acid was found after 3 hours exposure to 200 ppb PCP, whereas hepatic levels had virtually doubled by 24 hours. In contrast, at this time exposure to the lower dose of PCP caused ascorbic acid concentrations to fall (Fig. 4). Twenty-four hours later hepatic levels had doubled, but after 5 days of exposure ascorbic acid concentrations were no longer significantly elevated.

DISCUSSION

N. virens

Little is known about ascorbic acid metabolism in invertebrates in general, and our knowledge of intermediary metabolism in polychaetes is primarily dependent upon extrapolation from information obtained in studies conducted with other organisms. The pattern of changing glucose concentrations in the coelomic fluid of *Neanthes virens* exposed to various degrees of stress is similar to responses observed in the blood of various vertebrates under similar conditions. In teleosts hyperglycemia is a secondary response to stressful stimuli mediated by the activation of adrenocortical and adrenergic systems (Mazeaud *et al.*, 1977). The glycemic responses observed in *N. virens* may also be under hormonal influence. What systems or hormones are responsible for mediating acute hyperglycemic responses in stressed *N. virens* is not presently known.

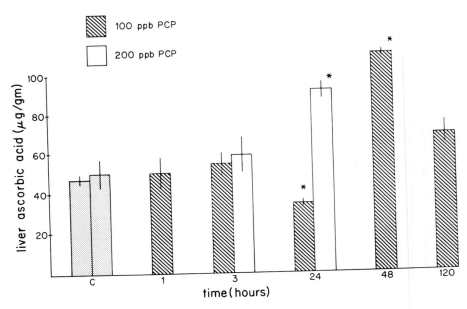

FIGURE 4:   *Time course of liver ascorbic acid concen-*
trations in juvenile striped mullet, Mugil cephalus, *exposed*
to 100 ppb (μg/l) and 200 ppb pentachlorophenol.   Each column
represents the mean (±S.E.M.) of at least six samples.
Asterisks denote values significantly different from controls
(C).

The glycemic responses observed in *N. virens* subjected
to various degrees and durations of PCP-induced stress appear
to follow a reproducible pattern.   Acute toxic PCP exposure
produced prolonged hypoglycemia which suggests a poisoning of
metabolic processes and hence a depressed activity of homeostatic mechanisms resulting in an inability to acclimate to
the altered environment.   During a less severe PCP exposure a
hyperglycemic response was observed suggesting an acceleration
of metabolic processes in an attempt to maintain homeostasis.
The hyperglycemia was a transitory response, however, as
coelomic fluid glucose concentrations gradually returned to
normal or depressed levels depending on the duration and
degree of stress.   During chronic PCP exposure, glucose concentrations again rose initially, followed by hypoglycemia.
Depending on the degree of stress, this hypoglycemic
period may be followed by another rise in glucose levels, as
was also observed in experimentally starved worms.   This bi-
phasic response of glucose concentrations during chronic
stress can be interpreted in several ways.   The second rise in
coelomic fluid glucose concentrations following stress-induced

hypoglycemia may be indicative of a shift from glycogen to lipid reserves. The primary polysaccharide of annelids appears to be glycogen (Scheer, 1969) and this energy reserve would most likely be utilized first during stress. The factors responsible for controlling extracellular glucose concentration might have become exhausted or altered or unable to adequately mediate the release of lipid-derived glucose. Studies designed to determine what mechanisms are involved in the stress-induced glycemic responses of this polychaete are in progress.

The changes observed in tissue free ascorbic acid concentrations in stressed *N. virens* are more difficult to interpret. The majority of studies concerning ascorbic acid requirements in aquatic invertebrates has been conducted with penaeid shrimp (Dehsimaru and Kuroki, 1976; Guary *et al.*, 1976; Lightner 1977; Lightner *et al.*, 1977; Margarelli and Colvin, 1978; Hunter *et al.*, 1979; Margarelli *et al.*, 1979). In several species of penaeid shrimp, starvation and ascorbic acid deficient diets have been observed to result in a disease of the connective tissue called "black death" which is characterized by lesions composed of masses of hemocytes. These experiments demonstrate the dietary requirement for ascorbic acid in these species.

The large increase in free ascorbic acid observed in tissues of chronically starved *N. virens* was an unexpected result. The short-term starvation data (16 days) appeared to support the hypothesis that ascorbic acid is a nutritional requirement for this species and that ascorbate reserves are gradually depleted during starvation. The dramatic increase in ascorbate after prolonged starvation (88 days) might indicate that long-term ascorbate reserves were mobilized during chronic nutritional stress. Likewise *N. virens* exposed to chronic and acute PCP-induced stress were observed to increase the concentration of free ascorbic acid in their parapodial tissues. This evidence might also indicate, however, an ability of this species to synthesize this vital nutrient. Although the information is sparse and inconclusive, it has generally been accepted that the invertebrates lack the capability of synthesizing ascorbic acid (Chatterjee, 1973). This concept may require reevaluation as more information becomes available. From the information obtained in this study, ascorbic acid does not appear to be a limiting factor for survival in this species. The concentration of ascorbic acid in the tissues of *N. virens* may prove to be a good index of stress once the ascorbate metabolism of this animal is better understood.

M. cephalus

   The present study shows that pentachlorophenol is rapidly
accumulated in teleosts, confirming the findings of Kobayashi
and Akitake (1975a). These authors demonstrated that PCP is
absorbed through the gills of goldfish (Kobayashi and Akitake,
1975b) and that the accumulation is dose-dependent (Kobayashi
and Akitake, 1975c). Freshwater fish have been shown to bioconcentrate PCP up to a thousand times the concentration in
the water (Kobayashi and Akitake, 1975b), whereas the maximum
bioconcentration factor for two estuarine teleosts, *Fundulus
similus* and *Mugil cephalus*, was only 41 and 81 respectively
(Schimmel *et al.*, 1978). Our results suggest that the maximum
bioconcentration factor in mullet may be greater than this,
approximately 300 times the PCP levels in the exposure media.
Schimmel *et al.* (1978) reported a 96-hour LC50 of 112 µg PCP/l
for mullet. PCP was found to have a similar toxicity in the
present study.
   In this study a massive rise in plasma cortisol concentrations occurred in juvenile *Mugil cephalus* exposed to 200
ppb pentachlorophenol. Corticosteroid stress responses have
been demonstrated in teleosts exposed to a wide variety of
pollutants (Hill and Fromm, 1968; Lockhart *et al.*, 1972; Grant
and Mehrle, 1973; Donaldson and Dye, 1975; DiMichele and
Taylor, 1978). Cadmium is the only pollutant examined to date
that does not cause interrenal activation in teleosts exposed
to lethal concentrations (Schreck and Lorz, 1978; Thomas and
Neff, in preparation).
   The mechanisms by which pollutants activate the hypothalamo-pituitary-interrenal axis in fish are poorly understood. Schreck and Lorz (1978) suggested from their studies
with cadmium and copper that in teleosts the interrenal may
only be activated by noxious stimuli which cause fright, dis-
comfort, or pain. This hypothesis is supported by studies
showing a rapid rise in cortisol concentrations in fish exposed to oil (Thomas *et al.*, in press b), naphthalene
(DiMichele and Taylor, 1978), and copper (Schreck and Lorz,
1978). However, in the present study plasma cortisol concentrations did not rise until the second day of exposure to
lethal levels (100 ppb) of PCP. The results suggest that the
hypothalamo-pituitary-interrenal axis of mullet is not activated by 100 ppb PCP until a certain amount has been
accumulated in the tissues. Therefore pollutants can activate
the teleost interrenal by more than one mechanism. Circulating cortisol concentrations rose only when the body had
accumulated more than 5 ppm PCP. Increased mortality was
associated with tissue concentrations approximately three
times this level. Kobayashi and Akitake (1975c) found that

goldfish die when tissue concentrations rise above a critical
level. It is likely that homeostatic imbalances occur when
tissue concentrations rise above 5 ppm PCP, causing interrenal
activation, and within a few days, death. High plasma cortisol levels have been associated with imminent death of sockeye salmon after spawning (Fagerlund, 1967).

The degree of activation of the hypothalamo-pituitary-
interrenal axis has been shown to be related to the severity
of the stress (Strange and Schreck, 1978). But after an
initial rise to near maximum levels, plasma cortisol concentrations declined somewhat in mullet exposed to 200 ppb PCP
at a time when mortality had increased. Similarly, a sub-
maximal corticosteroid response was found in mullet exposed
to the lower dose of PCP. Since PCP is a powerful uncoupler
of oxidative phosphorylation (Weinbach, 1957; Bevenue & Beckman, 1967), the steroidogenic capacity of the interrenal may
be reduced by this chlorophenol. The failure of teleosts
exposed to PCP to mature sexually (Crandall & Goodnight, 1962)
suggests that steroidogenic tissues may be sensitive to this
toxicant. However, the failure of mullet to maintain maximum
circulating levels of cortisol could be caused by a variety of
mechanisms including increased clearance due to induction of
the hepatic cytochrome P-450 system (Vizethum and Goerz,
1979). Increased metabolism of steroids has been correlated
with induction of the hepatic cytochrome P-450 system in rainbow trout exposed to polychlorinated biphenyls (Sivarajah et
al., 1978). The functional integrity of the hypothalamo-
pituitary-interrenal axis has been shown to be necessary for
survival of environmental stresses (Chan et al., 1967).
Partial failure of the corticosteroid stress response upon
exposure to PCP may contribute to the toxicity of this
compound.

Plasma cortisol had reached maximum levels after three
hours exposure to the higher dose of PCP, whereas glucose
concentrations were still rising. Different dynamics for
primary and secondary stress responses have been demonstrated
in other studies (Mazeaud et al., 1977; Thomas et al., in
press b).

The osmotic stress response in teleosts is predominantly
under catecholamine control whereas the hyperglycemic response
is also mediated by elevations in circulating corticosteroids
(Mazeaud et al., 1977), so it is not surprising that in the
present study plasma glucose levels were more closely correlated than osmolality to interrenal activation. The plasma
concentration of cholesterol, the third secondary stress
response parameter investigated, was unaltered by any dose of
PCP.

The rise in glucose levels was accompanied by a rapid depletion of hepatic glycogen. Extremely low liver glycogen and
plasma glucose levels were found in the moribund fish. The
rapid depletion of glycogen reserves is probably due to the
increased energy demands following PCP treatment. PCP treatment causes a hypermetabolic state in teleosts characterized
by an increase in oxygen consumption (Crandall and Goodnight,
1962) and energy utilization (Boström and Johansson, 1972;
Holmberg *et al.*, 1972) including glucose (Liu, 1969). The
results suggest that one of the causes of death may be ex-
haustion of hepatic glycogen reserves and subsequent failure
to maintain plasma glucose levels. However, hepatic lipids
were not measured in this study, although they are an important energy store in teleosts (Cowey and Sargent, 1977), so
that the importance of hepatic glycogen stores in maintenance
of blood glucose levels in mullet during an acute hypermetabolic stress is unclear. At later stages failure of
electrolyte and water homeostatic mechanisms may also contribute to the increased mortality. Severe stress causes a
similar elevation of plasma osmolality but the rise is more
rapid (Wydoski *et al.*, 1976). The results suggest that the
increase in serum osmolality may be partly due to a direct
pharmacological effect of pentachlorophenol on electrolyte
regulatory mechanisms in the gills and intestine, since this
chlorophenol is known to inhibit Na^+, K^+ ATPases (Desaiah,
1978).

It was surprising that plasma total cholesterol levels of
mullet were unaltered by PCP treatment, since hypercholes-
terolemia has been observed in this species in response to
other pollutants (Thomas, *et al.*, in press b). A hyper-
cholesterolemic response to PCP was found in fresh water
adapted eels, but not in seawater eels (Holmberg *et al.*,
1972). In contrast, no depletion of hepatic glycogen was
observed in these fish despite a persistent hyperglycemia
(Holmberg *et al.*, 1972). The liver somatic index increased in
eels exposed to PCP (Holmberg *et al.*, 1972), but decreased in
mullet (data not shown). There are considerable interspecific
differences in carbohydrate metabolism in teleosts (Chavin and
Young, 1970; Holmberg *et al.*, 1972; Umminger and Benzinger,
1975; Dave *et al.*, 1975) and also in lipid metabolism
(Bilinski and Lau, 1969; Larsson, 1973; Larsson and Fänge,
1977), so that PCP may have different metabolic effects on
these two species. Alternatively, the differences observed
may be related to the different metabolic and nutritional
conditions of the two species at the time of exposure, since
the eels had been starved while the mullet were feeding.
Starvation has been shown to initially alter lipid metabolism
in eels (Dave *et al.*, 1975). It appears that lipids are pre-

ferentially utilized during chronic stress and carbohydrate
stores are reserved for acute stress (Chavin and Young, 1970).
A better understanding of the effects of nutritional status
and season on the metabolic responses to pollutants is needed
in order to interpret the results of laboratory and field
studies.

Ascorbic Acid Requirements in Teleosts

It is extremely difficult to predict the ecological impact
of pollutant discharges from observed changes in hormonal and
metabolic parameters in organisms tested under laboratory
conditions. Consequently, research has concentrated on
examining the effects of pollutants on the functional integrity of those homeostatic systems which are known to be
vital for resistance to stress and maintenance of growth and
reproductive potential. The precise role of ascorbic acid in
homeostatic mechanisms is still unclear and there is scant information on the effects of pollutant exposure on the concentration of vitamin C in marine organisms. Interpretations of
alterations of ascorbic acid concentrations are complicated by
the widespread distribution of this vitamin in vertebrate
tissues (Hughes *et al.*, 1971; Chatterjee *et al.*, 1975; Hilton
et al., 1979) and its involvement in a wide range of biological processes (Baker, 1967, Lewin, 1976). However, alteration
of ascorbic acid levels has been observed in certain tissues
of teleosts after a mild stress (Wedemeyer, 1969) or after
exposure to a pollutant (Mayer *et al.*, 1978). Ascorbic acid
is necessary for collagen formation during wound healing
(Halver *et al.*, 1969). Also, vitamin C has been implicated in
resistance to stress (Goldsmith, 1961; Baker, 1967), biogenic
amine production (Subramanian, 1977), immune responses
(Leibovitz and Siegel, 1978) and detoxification (Kamm *et al.*,
1973; Zannoni, 1977) and elimination (Mayer *et al.*, 1978) of
drugs and insecticides. These reports suggest that this
micronutrient may play an important role in resistance to
xenobiotics and their metabolism. Pathological lesions
develop after exposure to toxaphene in juvenile channel catfish fed ascorbic acid deficient diets and elimination of this
compound is impaired (Mayer *et al.*, 1978). However, the
dependence on a dietary intake of vitamin C for maximum
resistance to pollutants in other species is uncertain.
Fingerling salmonids and channel catfish require a dietary intake of vitamin C (Kitamura *et al.*, 1965; Halver *et al.*, 1969;
Lim and Lovell, 1978), but this requirement probably decreases
as the fish mature (Sato *et al.*, 1978). On the other hand L-
gulono-lactone oxidase, the enzyme which catalyzes the conversion of L-gulono-γ-lactone to L-ascorbic acid, is present

in the hepatopancreas of Cypriniformes (Yamamoto *et al.*,
1978), which suggests that some teleosts may be able to
synthesize enough vitamin C for their normal requirments.
Even Cypriniformes may need supplemental vitamin C in their
diet during pollutant exposure, since an increased requirement
for this vitamin during biochemical stress (Goldsmith 1961;
Baker, 1967; Hughes *et al.*, 1971; Lewin 1976) and detoxification (Zannoni, 1977; Mayer *et al.*, 1978), has been suggested.
These studies emphasize the need for a better understanding of
the role of ascorbic acid in marine organisms and their
dietary vitamin C requirements during pollutant exposure.

The hepatic concentration of ascorbic acid has been
proposed as a useful indicator of the ascorbic acid status in
teleosts (Hilton *et al.*, 1977; Lim and Lovell, 1978).
Ascorbic acid is rapidly accumulated by the liver (Baker 1967;
Hornig *et al.*, 1972) and is found in high concentration in
this tissue (Halver *et al.*, 1975). An increase in ascorbic
acid concentration in the liver is associated with elevation
of hepatic cytochrome P-450 levels (Zannoni, 1977), while certain enzyme activities decrease as hepatic vitamin C concentration falls (Street and Chadwick, 1975). Hepatic
ascorbic acid concentration may be a useful sublethal indicator of environmental pollution.

Liver ascorbic acid levels in mullet were elevated 24
hours after interrenal activation in response to PCP exposure
which suggests that hepatic ascorbic ·acid accumulation may be
under interrenal control. Cortisol administration to adrenal-
ectomized rats increased liver vitamin C levels (Majumder *et
al.*, 1973), whereas hypophysectomy had an opposite effect
(Hornig *et al.*, 1972). Adrenocorticotrophin administration to
teleosts causes cortisol release and ascorbic acid depletion
from interrenal tissue (Wedemeyer, 1969). The increased
circulating levels of ascorbic acid may subsequently be
accumulated by the liver. The high circulating levels of
cortisol may help this process. Such a mechanism could
explain the results in this study. However, in contrast to
mammals, no increase in plasma ascorbic acid levels was found
in salmonids upon interrenal activation (Wedemeyer, 1969).
Alternatively, corticotrophin could inhibit catabolism of
ascorbic acid (Lahiri and Lloyd, 1962), or cortisol may
stimulate biosynthesis of the vitamin by stimulating L-
gulono-lactone oxidase activity (Majumder *et al.*, 1973). Together these studies suggest that there may be a close
association between interrenal activation and the subsequent
increase in hepatic ascorbic acid, but the actual mechanism is
not known. The initial fall in hepatic ascorbic acid levels
after 24 hours exposure to the low dose of PCP may be due to
increased utilization of hepatic reserves at this time for

detoxification and elimination of PCP (Kobayashi, 1979). Half of the free PCP was converted to PCP-glucuronide in goldfish (Kobayashi, 1979). A requirement for ascorbic acid in mammals for induction of the glucuronic acid pathway in the presence of organochlorine pesticides has been demonstrated (Street and Chadwick, 1975). The alteration of ascorbic acid concentrations upon PCP exposure confirms the results in channel catfish exposed to toxaphene (Mayer *et al.*, 1978). However it cannot be determined from these studies whether the observed changes are a direct result of pollutant exposure or are secondary to the metabolic disturbances produced. The latter interpretation is suggested from the foregoing discussion, whereas a direct effect of PCP on ascorbic acid biosynthesis is suggested from studies in mammals dosed with lindane (Street and Chadwick, 1975). Both mechanisms may be involved in altering hepatic ascorbic acid levels. The present results do suggest, though, that profound changes in ascorbic acid metabolism do occur during lethal and sublethal exposure to pollutants. Future investigations may therefore profit by including measurement of this parameter when examining the effects of pollutant exposure on marine organisms.

SUMMARY

Lethal and sublethal concentrations of pentachlorophenol alter several homeostatic mechanisms in sandworms (*Neanthes virens*) and juvenile striped mullet (*Mugil cephalus*).

Pentachlorophenol concentrations which were acutely lethal to *N. virens* (above 365 µg/l) caused marked coelomic fluid hypoglycemia and ascorbic acid depletion from parapodial tissue within 48 hours. Different biochemical responses were observed when worms were exposed to sublethal levels (120 µg/l PCP). Coelomic fluid glucose rose to approximately two times the control level by 24 hours and subsequently declined to control levels by 96 hours and decreased thereafter. The concentration of tissue free ascorbic acid rose steadily during exposure to the lower concentrations of pentachlorophenol, suggesting synthesis or mobilization of bound ascorbate reserves in response to sublethal stress.

Pentachlorophenol was considerably more toxic to mullet. Mortality reached 33% after 48 hours exposure to 100 µg/l PCP and after 24 hours exposure to 200 µg/l PCP. Plasma cortisol concentrations did not rise until the second day of exposure to the lower dose, whereas a rapid elevation of cortisol occurred in mullet exposed to 200 µg/l PCP. The rise in cortisol concentrations was accompanied by a marked hyperglycemia and a depletion of hepatic glycogen reserves, whereas hepatic ascorbic acid concentration and serum osmolality rose more slowly. Changes in ascorbic acid concentrations were evident after 24 hours exposure to either dose, while serum osmolality was unaltered after 48 hours exposure to 100 μg/l PCP. Total plasma cholesterol levels were not affected by either dose of PCP. The magnitude and the time course of the biochemical responses of mullet were directly related to exposure concentration and accumulation of PCP.

ACKNOWLEDGMENTS

This study was supported by Grant No. OCE 77-24551 from the National Science Foundation. The authors thank M. Weirich for technical assistance, Dr. C. S. Giam and D. Trujillo for tissue analyses, and B. Schult and J. Rierson for clerical assistance.

LITERATURE CITED

Baker, E. M. 1967. Vitamin C requirements in stress. Am. J. Clin. Nutr., 20: 583-590.

Bevenue, A. and H. Beckman. 1967. Pentachlorophenol: A discussion of its properties and its occurrence as a residue in human and animal tissues. Residue Rev., 19: 83-134.

Bilinski, E. and Y. C. Lau. 1969. Lipolytic activity toward long-chain triglycerides in lateral line muscle of rainbow trout (*Salmo gairdneri*) J. Fish. Res. Bd. Can., 26: 1857-1866.

Borthwick, P. W. and S. C. Schimmel. 1978. Toxicity of pentachlorophenol and related compounds to early life stages of selected estuarine animals. *In*: Pentachlorophenol: chemistry, pharmacology and environmental toxicology. pp. 141-146. Ed. by K. R. Rao. New York. Plenum Press.

Boström, S. L. and R. G. Johansson. 1972. Effects of pentachlorophenol on enzymes involved in energy metabolism in the liver of the eels. Comp. Biochem. Physiol., 41B: 359-369.

Brannon, A. C., and P. J. Conklin. 1978. Effect of sodium pentachlorophenate on exoskeletal calcium in the grass shrimp, *Palaemonetes pugio*. *In:* Pentachlorophenol: chemistry, pharmacology and environmental toxicology. pp. 205-211. Ed. by K. R. Rao. New York: Plenum Press.

Cantelmo, A. C., P. J. Conklin, F. R. Fox and K. R. Rao. 1978 Effects of sodium pentachlorophenate and 2,4-dinitrophenol on respiration in crustaceans. In: Pentachlorophenol: chemistry, pharmacology and environmental toxicology. pp. 251-263. Ed. by K. R. Rao. New York: Plenum Press.

Carr, R. S. and J. M. Neff. (Submitted). Determination of ascorbic acid in tissues of marine invertebrates by high performance liquid chromatography with electro-chemical detection.

Carr, R. S., P. Thomas and J. M. Neff. (in preparation). A simple spectrophotometric technique for the analysis of pentachlorophenol in water.

Chan, D. K. O., I. Chester-Jones, I. W. Henderson and J. C. Ranklin. 1967. Studies on the experimental alteration of water and electrolyte composition in the eel *(Anguilla anguilla* L.). J. Endocr., 37: 297-317.

Chatterjee, I. B. 1973. Vitamin C synthesis in animals: evolutionary trends. Sci. Cult. 5: 210-212.

Chatterjee, I. B., A. K. Majumder, B. K. Nandi, and N. Subramanian. 1975. Synthesis and some major functions of vitamin C in animals. Ann. N.Y. Acad. Sci., 258: 24-47.

Chavin, W. and J. E. Young. 1970. Factors in the determination of normal serum glucose levels of goldfish, *Carassium autatus* L. Comp. Biochem. Physiol., 33: 629-653.

Cirelli, D. P. 1978. Patterns of pentachlorophenol usage in the United States of America - an overview. *In:* Penta-chlorophenol: chemistry, pharmacology and environmental toxicology. pp. 13-18. Ed. by K. R. Rao. New York: Plenum Press.

Cowey, C. B. and J. R. Sargent. 1977. Lipid nutrition in fish. Comp. Biochem. Physiol., 57B: 269-273.

Crandall, C. A. and C. J. Goodnight. 1962. Effects of sub-
    lethal concentrations of several toxicants on growth of
    the common guppy, *Lebistes reticulatus*. Limnol.
    Oceanogr., 7: 233-239.

Dave, G., M. L. Johansson-Sjöbeck, A. Larsson, K. Lewander
    and U. Lidman. 1975. Metabolic and hematological effects
    of starvation in the European eel, *Anguilla anguilla* L. I.
    Carbohydrate, lipid, protein and inorganic ion metabolism.
    Comp. Biochem. Physiol., 52A: 423-430.

Dehsimaru, O. and K. Kuroki. 1976. Studies on a purified
    diet for prawn VII. Adequate dietary levels of ascorbic
    acid and inositol. Bull. Japan Soc. Scient. Fish. 42:
    571-576.

Desaiah, D. 1978. Effect of pentachlorophenol on the ATPases
    in rat tissues. In: Pentachlorophenol: chemistry,
    pharmacology and environmental toxicology. pp. 277-283.
    Ed. by K. R. Rao. New York: Plenum Press.

DiMichele, L. and M. H. Taylor. 1978. Histopathological and
    physiological responses of *Fundulus heteroclitus* to
    naphthalene exposure. J. Fish. Res. Bd. Can., 35: 1060-
    1066.

Donaldson, E. M. and H. M. Dye. 1975. Corticosteroid concen-
    trations in sockeye salmon (*Oncorhynchus nerka*) exposed to
    low concentrations of copper. J. Fish. Res. Bd. Can., 32:
    533-539.

Doughtie, D. G. and K. R. Rao. 1978. Ultrastructural changes
    induced by sodium pentachlorophenate in the grass shrimp,
    Palaemonetes pugio, in relation to the molt cycle. In:
    Pentachlorophenol: chemistry, pharmacology and environ-
    mental toxicology. pp. 213-250. Ed. by K. R. Rao. New
    York: Plenum Press.

Fagerlund, U. H. M. 1967. Plasma cortisol concentration in
    relation to stress in adult sockeye salmon during the
    freshwater stage of their life cycle. Gen. Comp.
    Endocrinol., 8: 197-207.

Goldsmith, G. A. 1961. Human requirements for vitamin C and
    its use in clinical medicine. Ann. N.Y. Acad. Sci. 92:
    230-245.

Grant, B. F. and P. M. Mehrle. 1973. Endrin toxicosis in
    rainbow trout (*Salmo gairdneri*). J. Fish. Res. Bd. Can.,
    30: 31-40.

Guary, M., A. Kanazawa, N. Tanaka and J. H. Ceccaldi. 1976.
    Nutrition requirements of prawn-VI. requirement for
    ascorbic acid. Mem. Fac. Fish. Kagoshima Univ. 25:
    53-57.

Halver, J. E., L. M. Ashley and R. R. Smith. 1969. Ascorbic
    acid requirements of coho salmon and rainbow trout.
    Trans. Am. Fish. Soc., 98: 762-771.

Halver, J. E., R. R. Smith, B. M. Tolbert and E. M. Baker.
    1975. Utilization of ascorbic acid in fish. Ann. N.Y.
    Acad. Sci., 258: 81-102.

Hill, C. W. and P. O. Fromm. 1968. Response of the inter-
    renal gland of rainbow trout (*Salmo gairdneri*) to stress.
    Gen. Comp. Endocrinol., 11: 69-77.

Hilton, J. W., R. G. Brown and S. J. Slinger. 1979. The
    half-life and uptake of ^{14}C-1-L-ascorbic acid in selected
    organs of rainbow trout (*Salmo gairdneri*). Comp. Biochem.
    Physiol., 62A: 427-432.

Hilton, J. W., C. Y. Cho and S. J. Slinger. 1977. Evaluation
    of the ascorbic acid status of rainbow trout (*Salmo
    gairdneri*). J. Fish. Res. Bd. Can., 34: 2207-2210.

Holmberg, B., S. Jensen, A. Larsson, K. Lewander and M.
    Olsson. 1972. Metabolic effects of technical penta-
    chlorophenol (PCP) on the eel *Anguilla anguilla* L. Comp.
    Biochem. Physiol., 43B: 171-183.

Hornig, D., H. E. Gallo-Torres and H. Weiser. 1972. Tissue
    distribution of labelled ascorbic acid in normal and
    hypophysectomized rats. Internat. J. Vit. Nutr. Res.,
    42: 487-496.

Hughes, R. E., P. R. Jones, R. S. Williams and P. F. Wright.
    1971. Effect of prolonged swimming on the distribution of
    ascorbic acid and cholesterol in the tissues of the
    guinea pig. Life Sci., 10: 661-668.

Hunter, B., P. C. Margarelli Jr., D. V. Lightner and L. B.
    Colvin. 1979. Ascorbic acid-dependent collagen formation
    in penaeid shrimp. Comp. Biochem. Physiol. 64B: 381-385.

Kaila, K. and J. Saarikoski. 1977. Toxicity of penta-
    chlorophenol and 2, 3, 6-trichlorophenol to the crayfish
    (*Astacus fluviatilis* L.). Environ. Pollut., 12: 119-123.

Kamm, J. J., T. Dashman, A. H. Conney and J. J. Burns. 1973.
    Protective effect of ascorbic acid on hepatotoxicity
    caused by sodium nitrite plus amino-pyrine. Proc. Nat.
    Acad. Sci. USA, 70: 747-749.

Kitamura, S., S. Ohara, T. Suwa and K. Nakagawa. 1965.
    Studies on vitamin requirements of rainbow trout, *Salmo
    gairdneri*. 11. On the ascorbic acid. Bull. Japan. Soc.
    Scient. Fish., 31: 818-826.

Kobayashi, K. 1979. Metabolism of pentachlorophenol in
    fish. In: Pesticide and xenobiotic metabolism in
    aquatic organisms. ACS Symposium Series, No. 99. pp.
    131-143. Ed. by M. A. Q. Kahn, J. J. Lech and J. J. Menn.
    Washington, D.C.: American Chemical Society.

Kobayashi, K. and H. Akitake. 1975a. Studies on the metabo-
    lism of chlorophenols in fish-II. Turnover of absorbed
    PCP in goldfish. Bull. Japan. Soc. Scient. Fish., 41:
    93-99.

Kobayashi, K. and H. Akitake. 1975b. Studies on the metabo-
    lism of chlorophenols in fish.-IV. Absorption and ex-
    cretion of phenol by fish. Bull. Japan. Soc. Scient.
    Fish., 41: 1271-1276.

Kobayashi, K. and H. Akitake. 1975c. Studies on the metabo-
    lism of chlorophenols in fish.-I. Absorption and ex-
    cretion of PCP by goldfish. Bull. Japan. Soc. Scient.
    Fish., 41: 87-92.

Lahiri, S. and B. B. Lloyd. 1962. The effect of stress and
    corticotrophin on the concentrations of vitamin C in
    blood and tissues of the rat. Biochem. J., 84: 478-483.

Larsson, A. 1973. Metabolic effects of epinephrine and
    norepinephrine in the eel *Anguilla anguilla* L. Gen.
    Comp. Endocrinol. 20: 155-167.

Larsson, A. and R. Fänge. 1977. Cholesterol and free fatty
    acids (FFA) in the blood of marine fish. Comp. Biochem.
    Physiol., 57B: 191-196.

Leibovitz, B. and B. V. Siegel. 1978. Ascorbic acid, neutrophil function, and the immune response. Internat. J. Vit. Nutr. Res., 48: 159-164.

Lewin, S. 1976. Vitamin C: its molecular biology and medical potential. 231 pp. London: Academic Press.

Lightner, D. V. 1977. "Black Death" disease of shrimps. In: Disease diagnosis and control in North American marine aquaculture and fisheries. Vol. 6 pp. 65-66. Ed. by C. J. Sindermann. Oxford: Elsevier.

Lightner, D. V., L. B. Colvin, C. Brand and D. A. Danald. 1977. Black death a disease syndrome of penaeid shrimp related to a dietary deficiency of ascorbic acid. Proc. World Maricul. Soc. 8: 611-623.

Lim, C. and R. T. Lovell. 1978. Pathology of vitamin C deficiency syndrome in channel catfish (*Ictalurus punctatus*). J. Nutr., 108: 1137-1146.

Liu, D. H. W. 1969. Alterations of carbohydrate metabolism by pentachlorophenol in cichlid fish. Ph.D. Thesis. Corvallis: Oregon State University.

Lockhart, W. L., J. F. Uthe, A. R. Kenney and P. M. Mehrle. 1972. Methylmercury in northern pike (*Esox lucius*): distribution, elimination, and some biochemical characteristics of contaminated fish. J. Fish. Res. Bd. Can. 29: 1519-1523.

Majumder, P. K., S. K. Banerjee, R. K. Roy and G. C. Chatterjee. 1973. Effect of hydrocortisone on the metabolism of L-ascorbic acid in rats. Biochem. Pharmac. 22: 1829-1833.

Mann, H. B. and D. R. Whitney. 1947. On a test of whether one of two variables is stochastically larger than the other. Ann. Math. Stat. 18: 50-60.

Margarelli, P. C., Jr., and L. B. Colvin. 1978. Depletion/repletion of ascorbic acid in two species of penaeid: *Penaeus californiensis* and *Penaeus stylirostris*. Proc. World Maricul. Soc. 9: 235-241.

Margarelli, P. C., Jr., B. Hunter, D. V. Lightner and L. B. Colvin. 1979. Black death: an ascorbic acid deficiency disease in penaeid shrimp. Comp. Biochem. Physiol. 63A: 103-108.

Mayer, F. L., P. M. Mehrle and P. L. Crutcher. 1978.
    Interactions of toxaphene and vitamin C in channel cat-
    fish. Trans. Am. Fish. Soc. 107: 326-333.

Mazeaud, M. M., F. Mazeaud and E. M. Donaldson. 1977. Pri-
    mary and secondary effects of stress in fish: some new
    data with a general review. Trans. Am. Fish. Soc. 106:
    201-212.

Montgomery, R. 1957. Determination of glycogen. Arch.
    Biochem. Biophys. 67: 378-386.

Pierce, R. H., Jr., C. R. Brent, H. P. Williams and S. G.
    Reeves. 1977. Pentachlorophenol distribution in a fresh
    water ecosystem. Bull. Environ. Contam. Toxicol. 18:
    251-258.

Pierce, R. H., Jr., and D. M. Victor. 1978. The fate of
    pentachlorophenol in an aquatic ecosystem. In: Penta-
    chlorophenol: chemistry, pharmacology and environmental
    toxicology. pp. 41-52. Ed. by K. R. Rao. New York:
    Plenum Press.

Pettibone, M. 1963. Marine polychaete worms of the New
    England Region. I. Aphroditidae through Trochochaetidae.
    Smithsonian Inst. U.S. Nat. Mus. Bull. #227, part I.
    356 pp.

Rao, K. R., F. R. Fox, P. J. Conklin, A. C. Cantelmo and A. C.
    Brannon. 1979. Physiological and biochemical invest-
    igations of the toxicity of pentachlorophenol to
    crustaceans. In: Marine pollution: functional res-
    ponses, pp. 307-339. Ed. by W. B. Vernberg, A. Calabrese,
    F. P. Thurberg and F. J. Vernberg. New York: Academic
    Press.

Rash, J. M., I. Jerkunica and D. Sgoutas. 1979. Mechanisms
    of interference of nonesterified fatty acids in radio-
    immunoassays of steroids. Clin. Chem. Acta. 93:
    283-294.

Robichaux, T. J. 1975. Bactericides used in drilling and
    completion operations. In: Conference on environmental
    aspects of chemical use in well-drilling operations.
    Houston, Texas May 21-23, 1975. EPA-560-1-75-004, pp.
    182-198. Washington, D.C.: Office of Toxic Substances,
    United States Environmental Protection Agency.

Rubinstein, N. I. 1978. Effect of sodium pentachlorophenate on the feeding activity of the lugworm, *Arenicola cristata* Stimpson. In: Pentachlorophenol: chemistry, pharmacology and environmental toxicology. pp. 175-179. Ed. by K. R. Rao. New York: Plenum Press.

Sato, M., R. Yoshinaka and S. Ikeda. 1978. Dietary ascorbic acid requirement of rainbow trout for growth and collagen formation. Bull. Japan. Soc. Scient. Fish. 44: 1029-1035.

Scheer, B. T. 1969. Carbohydrates and carbohydrate metabolism: Annelida, Sipunculida, Echiurida. In: Chemical Zoology. Vol IV. Annelida, Echiura and Sipuncula. pp. 135-145. Ed. by M. Florkin and B. T. Scheer. New York: Academic Press.

Schimmel, S. C., J. M. Patrick, Jr., and L. F. Faas. 1978. Effects of sodium pentachlorophenate on several estuarine animals: toxicity, uptake, and depuration. In: Pentachlorophenol: chemistry, pharmacology, and environmental toxicology. pp. 147-155. Ed. by K. R. Rao. New York: Plenum Press.

Schreck, C. B. and H. W. Lorz. 1978. Stress response of coho salmon (*Oncorhynchus kisutch*) elicited by cadmium and copper and potential use of cortisol as an indicator of stress. J. Fish. Res. Bd. Can. 35: 1124-1129.

Sivarajah, K., C. S. Franklin and W. P. Williams. 1978. The effects of polychlorinated biphenyls on plasma steroid levels and hepatic microsomal enzymes in fish. J. Fish. Biol., 13: 401-409.

Strange, R. J. and C. B. Schreck. 1978. Anesthetic and handling stress on survival and cortisol concentration in yearling chinook salmon (*Oncorhynchus tshawytscha*). J. Fish. Res. Bd. Can. 35: 345-349.

Street, J. C. and R. W. Chadwick. 1975. Ascorbic acid requirements and metabolism in relation to organochlorine pesticides. Ann. N.Y. Acad. Sci. 258: 132-143.

Subramanian, N. 1977. On the brain ascorbic acid and its importance in metabolism of biogenic amines. Life Sci. 20: 1479-1484.

Thomas, P. and J. M. Neff. (In press). Applicability of the general adaptation syndrome to teleosts: differential interrenal responses to cadmium in a cichlid (*Sarotheroden mossambicus*) and a mullet (*Mugil cephalus*).

Thomas, P., H. W. Wofford and J. M. Neff. (In press). Preliminary identification of cortisol, cortisone and corticosterone in representatives of the major vertebrate classes by high-performance liquid chromatography.

Thomas, P., B. Woodin and J. M. Neff. (In press). Biochemical responses of the mullet *Mugil cephalus* to oil exposure. 1. Acute responses-interrenal activation and secondary stress responses. Mar. Biol. (In press).

Umminger, B. L. and D. Benziger. 1975. *In vitro* stimulation of hepatic glycogen phosphorylase activity by epinephrine and glucagon in the brown bullhead, *Ictalurus nebulosus*. Gen. Comp. Endocrinol. 25: 96-104.

van Dijk, J. J., C. van der Meer and M. Wijnans. 1977. The toxicity of sodium pentachlorophenolate for three species of decapod crustaceans and their larvae. Bull. Environ. Contam. Toxicol. 17: 622-630.

Vizethum, W. and G. Goerz. 1979. Induction of the hepatic microsomal and nuclear cytochrome P-450 system by hexachlorobenzene, pentachlorophenol and trichlorophenol. Chem.-Biol. Inter., 28: 291-300.

Wedemeyer, G. 1969. Stress-induced ascorbic acid depletion and cortisol production in two salmonid fishes. Comp. Biochem. Physiol. 29: 1247-1251.

Weinbach, E. C. 1957. Biochemical basis for the toxicity of pentachlorophenol. Proc. Nat. Acad. Sci. USA. 43: 393-397.

Wydoski, R. S., G. A. Wedemeyer and N. C. Nelson. 1976. Physiological response to hooking stress in hatchery and wild rainbow trout (*Salmo gairdneri*). Trans. Am. Fish. Soc. 105: 601-606.

Yamamoto, Y., M. Sato and S. Ikeda. 1978. Existence of L-gulonolactone oxidase in some teleosts. Bull. Japan. Soc. Scient. Fish. 44: 775-779.

Zannoni, V. C. 1977. Ascorbic acid and liver microsomal drug metabolism. Acta vitamin. Enzymol. (Milano). 31: 17-29.

Zannoni, V., M. Lynch, S. Goldstein, and P. Sato. 1974. A rapid micromethod for the determination of ascorbic acid in plasma and tissues. Biochem. Med. 11: 41-48.

Section II

HEAVY METALS

EFFECTS OF SILVER ON RESPIRATION
AND ON ION AND WATER BALANCE IN *NEANTHES VIRENS*

J. J. Pereira

Deaprtment of Biology
University of Bridgeport
Bridgeport, Connecticut

K. Kanungo

Department of Biological
and Environmental Sciences
Western Connecticut State College
Danbury, Connecticut

INTRODUCTION

Recent studies on the effects of heavy metals on polychaetes have been concerned with the determination of lethal
levels of the metal or with the accumulation of the metal in
the animal (Bryan and Hummerstone, 1971; Cross *et al.*, 1970;
Reish *et al.*, 1976; Ahsanullah, 1976). Little has been done
on the physiological responses of polychaetes to heavy metals.
   The polychaete *Neanthes virens* was chosen as an experimental animal because it is easy to maintin in the laboratory
and is an important food source for benthic fishes on the
eastern coast of the United States. Its sedentary mode of
life makes it attractive as a pollution indicator species. In
addition, tolerance to heavy metals has been reported in a
closely related species, *Nereis diversicolor* (Bryan, 1974;
Bryan and Hummerstone, 1971, 1973).
   Changes in oxygen consumption, ion concentrations in the
coelomic fluid, and the water content of *N. virens* were measured in response to silver exposure. Silver was used because
of reports of high levels of silver in Long Island Sound

(Schutz and Turekian, 1965) and because silver is one of the
most toxic heavy metals in the aquatic environment (Bryan
1971).

MATERIALS AND METHODS

Source of Animals

The polychaetes used in this study were obtained from a
local bait dealer and had been collected on the coast of
Maine and transported to Connecticut in refrigerated trucks.
The packing boxes contained a layer of damp seaweed to prevent
desication and were purchased on the day of their arrival in
Connectucut. They were screened for surface wounds, missing
segments, and uniform size, and are referred to as "Maine"
animals in this paper.

In addition, specimens of *N. virens* were collected
locally near Charles Island located in Long Island Sound off
the coast of Milford, Connecticut. These animals, which are
referred to as "local" animals, were used for comparison with
the "Maine" animals.

Acclimation and Depuration

Milford Harbor seawater (26±1 ppt) was filtered through
a 1 µm spun orlon filter, added to 80 l aquaria, and constantly recirculated by a standard aquarium filter. The
bottom of the aquarium was covered by a 3-cm layer of fine
sand to assist in eliminating sediment from the gut of the
worms. Bryan and Hummerstone (1971) found that this technique for depuration was preferable to simply starving the
animals in clean seawater with no substrate. Prior to its
use, the sand was acid cleaned using 10% HCl, washed
thoroughly, and autoclaved. This treatment was repeated prior
to each experiment to remove any metal contamination which
might have occurred on contact with sediments from the worms.
Worms (up to six dozen per aquarium) were maintained in
aquaria in an environmental chamber at 15±1°C for five to
seven days prior to each experiment.

Exposure Procedures

A 1000-ppm silver stock solution was prepared by
dissolving 0.39 g of $AgNO_3$ in 250 ml of a 10% HNO_3 solution.
The exposure solution was prepared by pipetting appropriate
amounts of the stock solution into one liter of filtered seawater and then adding 0.5 g of $NaHCO_3$ to neutralize the stock

solution. Exposure solutions for the control animals were
prepared in the same manner except that 10% HNO_3 was substituted for the silver stock solution. The resulting sol-
tuions had a pH of approximately 7.0. Three concentrations
(0.5 ppm, 0.8 ppm, and 1.0 ppm) of silver were used in the
exposure tests.

Groups of up to 12 animals were selected at random from
the depuration tanks and placed in 1500-ml finger bowls with
either silver-containing or control seawater. The seawater
was constantly aerated and changed every 24 hours. All bowls
were maintained in an environmental chamber at 15±1°C.
Twenty-four to seventy-two hour exposure times were used.

For experiments at the 0.5 ppm level of exposure, groups
of two to ten animals were used. For all other levels of
exposure in which "Maine" animals were used, experiments were
conducted using 10 exposed and 10 control animals. The
number of "local" animals tested varied but in no case did a
control or exposed group contain less than five animals. The
exposed and control worms in "Maine" and "local" groups were
divided into three groups. One group was used for analysis
of silver content. The second group was used for measurement
of oxygen consumption and afterwards for collection of
coelomic fluid for ion determination. The third group was
used for the determination of body water content.

Analysis of Silver Content

After exposure, the worms were blotted dry and frozen.
In preparation for silver content determination, the worms
were thawed, weighed, and dried in an oven for 18 hours at
120°C before being weighed again. The dried samples were
digested with quartz-distilled HNO3 and evaporated to dryness
several times until a white residue was obtained. This
residue was taken up in 5% HNO_3 for analysis on a Perkin-
Elmer model 560 atomic absorption spectrophotometer. Silver
content was calculated as ppm on a dry weight basis.

Measurement of Oxygen Consumption

Oxygen consumption of whole animals was measured using
a Gilson Differential Respirometer. Each animal was placed
into a 150-ml reaction flask with 50 ml of the solution to
which it had been exposed. The temperature of the water bath
was kept at 15±1°C. Readings were taken at 15-minute intervals for two hours. Oxygen consumption was expressed as
μl/g/hr.) at STP.

Determination of Inorganic Ions
in the Body Fluids

After measurement of oxygen consumption, worms were removed from the reaction flasks, blotted dry on paper towels, and weighed. They were then transferred to individual, acid-washed watch glasses. The body wall was opened dorsolaterally using iris scissors. Care was taken not to puncture the gut. The dissected animal was allowed to remain on the watch glass until a sufficient amount of coelomic fluid had collected (about five minutes). The coelomic fluid was collected as quickly as possible to prevent loss of water by evaporation, placed in plastic, stoppered vials, and frozen for later ion analysis. Samples were later thawed at room temperature and centrifuged for 30 minutes in an Eppendorf-Brinkman model 5412 centrifuge. Duplicate samples of the supernatant were analysed on a Coleman model 51 flame photometer. The reported values of Ca^{2+}, K^+, and Na^+ represent the average obtained from each set of samples.

Determination of Water Content

The worms were blotted dry and weighed to the nearest 0.01 g, dried to a constant weight ($\propto$ 18 hours) at 120°C, and weighed again. Water content was calculated as a percentage of the original body weight.

Statistical Analysis

Student's "t" test was used to determine the significance of all measured parameters. All lines were plotted using linear regression analysis.

RESULTS

Silver Content

In "Maine" worms the amount of silver retained by the exposed animals varied with the concentration of the exposure medium and with the duration of exposure. Animals exposed to 0.5, 0.8, and 1.0 ppm silver for 24 hours accumulated 18.8± 8.6, 74.6±46.0, and 209.0±48.5 ppm of the metal, respectively. When exposed for 48 hours to 0.5 and 0.8 ppm, the accumulation of silver amounted to 66.8±18.7 and 112.0±52.2 ppm, respectively. At 0.5 ppm, silver accumulation continued to increase for 72 hours, although the largest amount was retained between the 24 and 48-hour exposure period. At 0.8

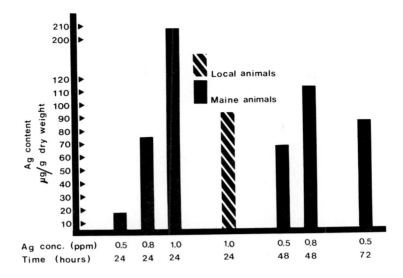

FIGURE 1. *Silver accumulation in* N. virens *after exposure to different concentrations of silver at 24, 48 and 72 hours.*

and 1.0 ppm silver, the rate of metal accumulation was greatest within the first 24 hours. "Local" animals retained only 90.7±53.0 ppm of silver during a 24 hour exposure to 1.0 ppm of the metal. The control "Maine" animals contained silver levels that ranged from undetectable to 10. ppm, while the control "local" animals ranged from 0.5 to 12.3 ppm (Fig. 1 and Table 1).

Oxygen Consumption

Oxygen consumption of "Maine" animals exposed to 0.5 ppm silver for 24, 48 and 72 hours did not differ significantly from that of the controls. Exposure to 0.8 ppm silver for 24 hours did not significantly decrease the rate of oxygen consumption, while exposure to 0.8 ppm of silver for 48 hours significantly (p<0.05) lowered the rate of oxygen consumption. "Maine" animals exposed to 1.0 ppm of silver for 24 hours showed oxygen consumption values which were significantly lower than those of the control animals (p<0.05). The "local" animals showed no such change when similarly exposed.

TABLE 1. Accumulation of Silver in relation to body weight and the rate of oxygen consumption in *N. virens* exposed to silver.

EXPOSURE TO SILVER (Ag Conc/Time) (ppm)/(hr)		SILVER ACCUMULATION (ppm)	WEIGHT (g) WET	DRY	OXYGEN CONSUMPTION (μl/g/hr)
0.5/24	Exposed	18.8 ± 8.62 (N = 3)	4.43 ± 1.09 (N = 3)	0.44 ± 0.13 (N = 3)	20.75 ± 10.20 (N = 10)
	Control	0.19 ± 0.05 (N = 2)	6.89 ± 3.9 (N = 2)	0.80 ± 0.1 (N = 2)	26.37 ± 3.07 (N = 3)
0.5/48	Exposed	66.78 ± 18.85 (N = 3)	5.6 ± 3.35 (N = 3)	0.86 ± 0.44	25.71 ± 12.68 (N = 10)
	Control	**	5.53 ± 1.5 (N = 2)	**	27.56 ± 4.48 (N = 3)
0.5/72	Exposed	87.69 ± 20.03 (N = 2)	4.47 ± 1.77 (N = 2)	0.70 ± 0.20 (N = 2)	24.88 ± 9.58 (N = 10)
	Control	0.53 ± 0.06 (N = 2)	3.61 ± 1.5 (N = 2)	0.51 ± 0.02 (N = 2)	39.11 ± 11.63 (N = 3)
0.8/24	Exposed	74.63 ± 45.97 (N = 10)	2.36 ± 0.57 (N = 10)	0.39 ± 0.19 (N = 10)	50.13 ± 13.19 (N = 7)
	Control	1.06 ± 0.47 (N = 9)	2.71 ± 0.66 (N = 9)	0.40 ± 0.14 (N = 9)	37.75 ± 12.5 (N = 10)
0.8/48	Exposed	112.52 ± 52.19 (N = 10)	1.61 ± 0.36 (N = 10)	0.26 ± 0.09 (N = 10)	45.45 ± 12.41* (N = 8)
	Control	0.79 ± 0.26 (N = 10)	2.08 ± 0.22 (N = 10)	0.28 ± 0.08 (N = 10)	61.75 ± 16.6 (N = 7)
1.0/24	Exposed	209.0 ± 48.48 (N = 10)	2.37 ± 0.37 (N = 10)	0.39 ± 0.14 (N = 10)	24.84 ± 11.28* (N = 9)
	Control	1.26 ± 0.66 (N = 10)	2.27 ± 0.58 (N = 10)	0.41 ± 0.13 (N = 10)	43.76 ± 11.95 (N = 6)
1.0/24 (Local Animals)	Exposed	90.71 ± 52.98 (N = 8)	2.67 ± 1.73 (N = 8)	0.48 ± 0.37 (N = 8)	42.77 ± 16.1 (N = 8)
	Control	3.57 ± 4.61 (N = 6)	3.79 ± 1.53 (N = 6)	0.68 ± 0.46 (N = 6)	46.96 ± 6.97 (N = 7)

* P < 0.05
** Determination Not Done

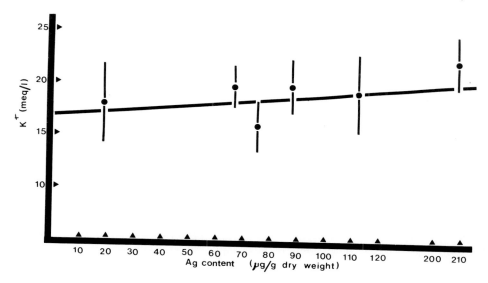

FIGURE 2. [K^+] in the coelomic fluid of N. virens as a function of silver concentration in the animals. The graph includes data from "Maine" animals only. Vertical lines indicate standard deviation. Line determined by linear regression.

Inorganic Ions in the Coelomic Fluids

In "Maine" animals [K+] in the coelomic fluid increased with increasing silver accumulation (Fig. 2). The "Maine" control animals showed a mean [K+] of 15 meq/l (Table 2). Potassium concentrations of the exposed animals are significantly different from the controls for animals containing 88 ppm of silver or more. Calcium concentration decreased in "Maine" worms with silver accumulation (Fig. 3). Sodium showed no significant changes in either exposed "Maine" or "local" worms. In addition, "local" worms exhibited no significant changes in [K^+] when exposed to 1.0 ppm of silver for 24 hours, but did show lower [Ca^{2+}] ($p < 0.01$) than the controls.

Water Content

The water content of exposed "Maine" worms was significantly lower than the controls for those containing 88 ppm silver or more ($p < 0.05$). Exposed "local" worms, although containing over 90 ppm silver, did not differ significantly from controls in their water content.

TABLE 2. Changes in Ion Concentration and Water Content of *N. virens* exposed to silver.

EXPOSURE TO SILVER (Ag Conc/Time) (ppm)/(hr)		ION CONCENTRATION (meq/l) IN THE COELOMIC FLUID			H_2O (% Body Wt.)
		Ca^{2+}	K^+	Na^+	
0.5/24	Exposed	11.83 + 4.37 (N = 10)	17.75 + 3.5 (N = 10)	428.0 + 7.28 (N = 10)	86.17 + 1.53 (N = 3)
	Control	16.42 + 1.23 (N = 3)	13.75 + 0.75 (N = 3)	432.5 + 23.85 (N = 3)	86.35 + 0.64 (N = 2)
0.5/48	Exposed	7.3 + 3.98 (N = 10)	19.65 + 4.7 (N = 10)	424.64 + 16.86 (N = 10)	81.80 + 2.98 (N = 3)
	Control	10.91 + 0.58 (N = 3)	16.25 + 0.75 (N = 3)	429.0 + 1.62 (N = 3)	84.3 + 0.42 (N = 2)
0.5/72	Exposed	7.48 + 3.43* (N = 10)	17.33 + 1.8* (N = 10)	439.3 + 15.22 (N = 10)	82.07 + 0.84* (N = 3)
	Control	12.75 + 2.0 (N = 3)	14.41 + 0.52 (N = 3)	438.23 + 8.8 (N = 3)	86.2 + 0.00 (N = 2)
0.8/24	Exposed	2.75 + 0.85 (N = 9)	15.75 + 2.23 (N = 9)	418.61 + 11.39 (N = 9)	84.95 + 2.03 (N = 10)
	Control	2.91 + 0.82 (N = 9)	14.02 + 1.44 (N = 9)	408.75 + 9.32 (N = 9)	84.56 + 1.44 (N = 10)
0.8/48	Exposed	1.42 + 0.79* (N =10)	18.62 + 3.46* (N = 10)	418.16 + 10.0 (N = 10)	83.90 + 1.31 (N = 10)
	Control	3.75 + 1.06 (N = 10)	15.11 + 1.27 (N = 10)	388.4 + 11.3 (N = 10)	86.20 + 0.71 (N = 10)
1.0/24	Exposed	4.75 + 1.70* (N = 9)	22.42 + 2.39* (N = 9)	423.28 + 13.86 (N = 9)	82.39 + 1.14* (N = 10)
	Control	6.69 + 2.12 (N = 10)	15.8 + 1.03 (N = 10)	414.25 + 17.32 (N = 10)	83.97 + 1.31 (N = 10)
1.0/24 (Local Animals)	Exposed	1.50 + 0.73* (N = 8)	17.25 + 2.06 (N = 8)	428.12 + 11.27 (N = 8)	83.31 + 1.51 (N = 10)
	Control	2.93 + 0.31 (N = 8)	14.56 + 1.21 (N = 8)	428.75 + 46.29 (N = 8)	83.92 + 2.72 (N = 6)

* $P \leq 0.05$

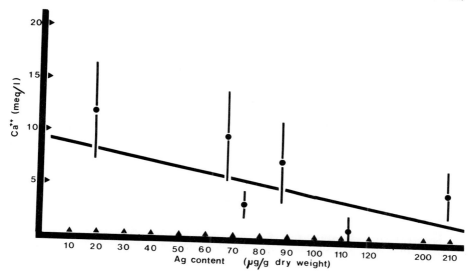

FIGURE 3. $[Ca^{2+}]$ *in the coelomic fluid of* N. virens *as a function of silver concentration in the animals. The graph includes data from "Maine" animals only. The vertical lines indicate standard deviation. The line was determined by linear regression.*

Morphological and Behavioral Changes

During the course of the experiments, the appearance of the worms was observed to change markedly with exposure to silver. Bryan (1976) reported the development of a grayish color on the dorsal surface of *N. diversicolor* which had been exposed to silver. Similar changes were observed in *N. virens*. The exposed worms also tended to lie on their sides with the tail region curled toward the ventral surface (Fig. 5). Exposed worms were also thinner than the controls and the parapodia were swollen with fluid (Figs. 6 and 7). The pharnyx was partially everted in some exposed worms and movements of exposed worms were uncoordinated in comparison to controls. These changes were apparent at all exposure levels and seemed to differ only in degree. The "local" animals appeared to be less affected than the "Maine" animals which were similarly exposed.

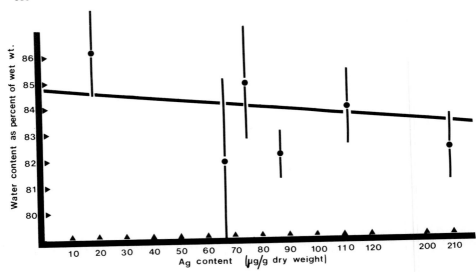

FIGURE 4. *Water content of N. virens as a function of the silver content of the body. The graph includes data from "Maine" animals only. Vertical lines indicate standard deviation. The line was determined by linear regression.*

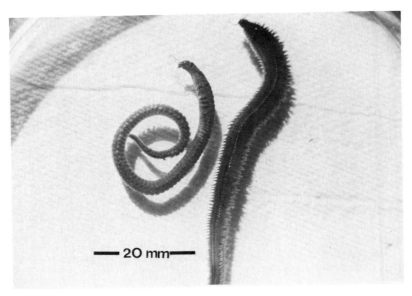

FIGURE 5. *Two worms (exposed on the left, control on the right) showing the "curled" posture of the exposed animals*

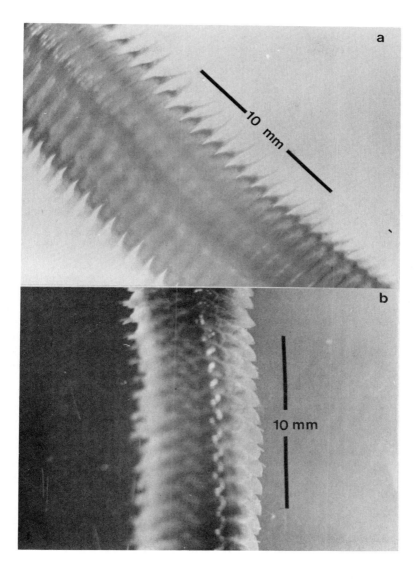

FIGURE 6a.   Normal parapodia of N. virens.
FIGURE 6b.   Edema of parapodia in N. virens *exposed to silver.*

DISCUSSION

The data on silver exposed "Maine" worms clearly show
that they can accumulate high levels of silver within a short
exposure period (Fig. 1). Since the rate of uptake of dissolved materials from solution depends not only on the concentration of the materials but also on the size (weight) of
the animals, the smaller animals with more surface-to-volume
ratio usually show a higher rate of uptake than the larger
aniamls. In our experiments, although efforts were made to
keep the size variation to a minimum between various groups
of "Maine" animals, the average wet weight in the groups
varied from about 2.0 to 7.0 g (Table 1). This size difference alone, however, is not enough to account for the
observed differences in the amount of silver accumulated by
the exposed groups. The two groups of approximately the same
average wet weight, when exposed to 0.8 and 1.0 ppm silver,
accumulated 97.5 ppm and 209 ppm of silver, respectively,
during the 24-hour exposure period. Further, smaller worms
(average wet weight 1.93 g) exposed to 0.8 ppm silver for 48
hours retained a lesser amount of the metal than larger worms
(average wet weight 2.89 g) exposed to 1.0 ppm silver for only
24 hours. Thus, it is reasonable to conclude that the difference in the amounts of silver accumulated by the exposed
groups of "Maine" animals were not size related but were due
to the differences in the concentration of the metal in the
medium.

Bryan and Hummerstone (1971) reported levels of silver
of up to 10 ppm in *N. diversicolor* which was roughly proportional to silver levels in the sediments from which they were
collected. By comparison, *N. virens* in our experiments
accumulated over 200 ppm of silver when the exposure medium
contained only 1.0 ppm of silver. This high level of metal
retention by *N. virens* was most probably due to the type of
medium (seawater rather than sediment) we used. Sediment
restricts the mobility of the worms, may bind metal ions
(Pesch, 1979) and, in a static system, the concentration of
metal in the immediate environment of the animal would
decrease as the animal takes it up. This would explain the
small amount of silver accumulated by *N. diversicolor* as
reported by Bryan and Hummerstone (1971) even when the sediment contained 10 ppm of the metal. Once the animals are
free in a liquid medium where the metal is evenly distributed
by aeration of the medium and the solution changed to
replenish the lost amount, the uptake and retention of metal
by the animals proceed to high levels in a rate-dependent
fashion presumably until death occurs or until some silver-

binding system in the animal is saturated.

Another observation with regard to silver accumulation in *N. virens* was the tendency of "local" animals to retain relatively less of the metal than the "Maine" animals when both groups were exposed to 1.0 ppm of silver for 24 hours (Fig. 1). The "local" animals, whose mean background level of silver was 3.6 ppm, retained 90.7 ± 53 ppm of the metal, whereas the "Maine" animals, with a mean background level of 1.3 ppm, accumulated 209 ± 48.5 ppm silver under identical exposure conditions. The sediment (Long Island Sound) from which the "local" animals were collected contained about 1.7 ppm silver in the sediment (Greig *et al.*, 1977). Since the background level of silver in the "Maine" worms was lower than the "local" animals, we may assume that the silver levels in the "Maine" environment were lower than those in the "local" environment.

The sizes of the worms in the "Maine" and "local" groups were not significantly different to account for the observed difference in metal retention. Thus, it is plausible that the "local" animals may have built up a "tolerance" to silver as a result of their exposure to it in the environment. It is not clear, however, how this "tolerance" allows lower levels of silver accumulation in the "local" animals. Short-term adaptations to a persistent and low-level stress are usually manifested by alterations in physiological and/or behavioral components in the organism (Prosser and Brown, 1961). It is likely that tolerant animals have developed mechanisms that diminish the rate of uptake and/or elevate the rate of loss of the metal from the body so as to effectively lower the net accumulation of the metal in the worm.

One parameter commonly used to assess physiological adjustment of an organism to stress is oxygen consumption. Both a decrease (MacInnes and Thurberg, 1973) and an increase (Thurberg *et al.*, 1974) in oxygen consumption have been reported for various aquatic organisms after exposure to silver. At our experimental conditions, the "Maine" animals which had accumulated 113 ppm of silver or more showed a lower rate of oxygen consumption than that of the corresponding control worms (Table 1). Those that had less than 113 ppm of silver did not show a decrease in respiration rate. Whether this low oxygen utilization by the experimental "Maine" animals was due to the metal acting directly on cellular respiratory systems, such as mitochondria and their associated enzymes, or indirectly by creating ion imbalances in the body fluid, is not clear at this time.

The exposed "local" worms did not show a significant difference in the rate of oxygen consumption from that of the corresponding control group. However, when oxygen uptake of

the experimental "local" group was compared to that of the
"Maine" group exposed to 1.0 ppm silver for 24 hours, the
former had a significantly higher rate of oxygen consumption
than the latter (Table 1). These results indicate that the
rate of oxygen uptake was affected by silver accumulation
after the metal reached a concentration of 113 ppm or more
in the body. By comparison, ion and water balances were
affected at a silver concentration of 88 ppm (Tables 1, 2).
Perhaps different physiological processes in *N. virens* have
different thresholds of tolerance for silver.

    That the exposure to silver adversely affects transport
mechanisms in worms is evident from the observed ion (Figs.
2 and 3) and water (Fig. 4) imbalances. In "Maine" experimental animals, the concentration of K^+ in the coelomic fluid
either stayed the same or increased significantly while the
concentration of Ca^{2+} in the coelomic fluid of exposed animals
either did not change or decreased significantly from those of
the control groups. The water content of the exposed "Maine"
animals also showed a decrease in comparison to that of the
control animals. Increase in $[K^+]$ in coelomic fluid of polychaetes after experimental manipulations has been interpreted
as an artifact seemingly due to trauma inflicted during
sampling which causes breakage of cells and release of
intracellular fluid high in K^+ (Oglesby, 1970). However, this
interpretation has not been accepted by Freel *et al.* (1973)
who, in their study on *Nereis succinea,* did not find a
parallel increase in extracellular concentration of ninhydrin-
positive substances with increases in extracellular $[K^+]$.
While Oglesby's explanation is a reasonable one, it is less
likely that trauma produced the observed extracellular increase in $[K^+]$ in our experiments. This assertion is
supported by the results obtained with the "Maine" and "local"
worms exposed to 1.0 ppm of silver for 24 hours and with the
corresponding controls: (a) that the exposed "Maine" animals
had $[K^+]$ about 7 meq/l higher than the controls while the
exposed "local" animals were only 3 meq/l higher, (b) that
there was a significant difference in $[K^+]$ between the two
experimental sets above when the ion concentration between
the two corresponding control sets was not significantly
different, and (c) $[K^+]$ in the two experimental sets showed a
change parallel with the change in their respective silver
contents. Further, the increase in extracellular $[K^+]$ is not
due to the loss of body water in the exposed worms since the
percent loss of water was minimal while the percent increase
in $[K^+]$ was higher in a given set of worms (Table 1 and 2).
It is reasonable, to conclude therefore, that the observed
increase in $[K^+]$ in the coelomic fluid of worms used in our
experiments is not an experimental artifact but a real change

caused by exposure to silver.

Although biochemical processes regulating K^+ transport in *N. virens* have not been studies in detail, some probable explanation as to the underlying mechanism for the observed K^+ imbalance can be offered by way of extrapolation from other biological systems. It is known that many types of cells maintain a higher $[K^+]$ inside their mitochondria and other cytoplasmic compartments and that such intracellular accumulation of K^+ requires expenditure of cellular energy (Tyler, 1973). With the loss of cellular regulation because of the rise in silver accumulation, the ion (K^+) will eventually enter the extracellular fluid, moving down a diffusion gradient. The observations of Jackim *et al.* (1970) and Gould and Karolus (1974) indicating disruption of enzyme systems due to heavy metals support the above contention.

The concentration of Na^+ in the coelomic fluid did not show any significant change between the exposed and the control animals, perhaps due to the fact that extracellular Na^+ is high in normal animals and any small change caused by exposure to silver was masked by the larger amount present in the coelomic fluid.

Exposure to silver also produced a lowering of $[Ca^{2+}]$ in coelomic fluid of the worms tested (Fig. 3). Avenues for the removal of Ca^{2+} from the extracellular fluid of *N. virens* are not definitely known. Ca^{2+} may be concentrated intracellularly by forming concretions as has been reported by Fowler *et al.* (in press) or it may simply move into the intracellular fluid (Alohan and Huddart, 1979). Alohan and Huddart (1979) have shown that excess K^+ in extracellular fluid induces Ca^{2+} influx into the cells which causes muscle contracture in the pharyngeal muscle of *N. virens*. Hypocalcemia has also been shown to produce muscle tetany in vertebrate systems by increasing the permeability of cells to ions (Guyton, 1966). It is possible, therefore, that in our experiments exposure of worms to silver produced the observed hypercalcemia which, in turn, induced influx of Ca^{2+} into the cells, thus resulting in hypocalcemia. This induced hypocalcemia may have contributed to the muscle contracture and curled posture which was observed in the exposed worms. However, this proposed relationship between low calcium and muscle contracture is unclear since some control animals also had low $[Ca^{2+}]$ but did not exhibit contracture.

Since in polychaetes, and in annelids in general, the movement of coelomic fluid is brought about by muscle contraction, a sustained contraction of body wall muscles would force fluid from the coelomic spaces around the gut into the parapodia. Parapodia, which are devoid of muscles, would thus swell with the accumulated fluid, thereby producing the parapodial edema which was observed in the exposed worms.

SUMMARY

1. *Neanthes virens* was used to study the effects of
exposure to silver on respiration, on the concentration of
Na^+, K^+, and Ca^{2+} in the coelomic fluid, and on the water
balance of the worms. These parameters were correlated with
the levels of silver accumulated by the worms.

2. Animals were obtained from two geographical
locations: from Maine where the levels of silver in the
environment are presumed to be lower, and from Long Island
Sound near Milford, Connecticut where silver has been demonstrated to exist in the sediments.

3. Worms from Maine accumulated levels of silver as
high as 209 ppm (dry weight), whereas Long Island Sound worms
accumulated levels only as high as 90 ppm.

4. In "Maine" worms, exposure to silver resulted in a
decrease in the rate of oxygen consumption only after worms
had accumulated 113 ppm silver. The oxygen consumption rates
of exposed "local" animals were not significantly different
from that of the controls, but were significantly higher than
"Maine" worms which were similarly exposed.

5. In "Maine" worms, exposure to silver produced an
increase in $[K^+]$ and a decrease in $[Ca^{2+}]$ in coelomic fluid
after a body burden of 88 ppm silver was reached. The concentration of Na^+ did not show any significant change.
"Local" worms showed no change in $[Na^+]$ or $[K^+]$ but showed a
decrease in $[Ca^{2+}]$ in their coelomic fluid.

6. Water content of exposed "Maine" worms showed a
decrease, while "local" worms did not exhibit any such losses.

7. Exposure to silver also resulted in localized edema
in the parapodia and in curled posture in both the "Maine"
and the "local" worms, although the "local" worms were
affected to a lesser degree than the "Maine" worms.

8. It is postulated that the "local" worms, as a
result of their chronic exposure to sublethal concentrations
of silver in their natural environment, developed a tolerance
to silver. This tolerance could account for the low level
of silver in the body and the relatively smaller alteration
of their physiological state as compared to that of the
"Maine" animals.

ACKNOWLEDGEMENTS

     We wish to express our gratitude to the following
persons at the National Marine Fisheries Service Milford,
Connecticut Laboratory: to Dr. Frederick Thurberg for
providing facilities in his laboratory, for frequent discussions regarding the project and for reading the manuscript
critically; to Margaret Dawson for her helpful suggestions;
to Richard Greig for determining silver concentrations by
atomic absorption spectroscopy and to David Nelson for providing the silver stock solution.

LITERATURE CITED

Ahsanulla, M. 1976. Acute toxicity of cadmium and zinc to
     seven invertebrate species from Western Port, Victoria.
     Aust. J. Mar. Freshwater Res. 27: 187-196.

Alohan, F. L. and H. Huddart. 1979. Spontaneous activity of
     annelid visceral muscle and related calcium movements.
     The effect of KCl depolarization, caffeine, acetycholine
     and adrenaline. Comp. Biochem. Physiol. 63C, 161-171.

Bryan, G. W. 1971. The effects of heavy metals (other than
     mercury) on marine and estuarine organisms. Proc. Roy.
     Soc. Lond. B Series 177: 389-410.

Bryan, G. W. 1974. Adaptation of an estuarine polychaete to
     sediments containing high concentrations of heavy
     metals. In "Pollution and Physiology of Marine
     Organisms." (F. J. Vernberg, and W. B. Vernberg, eds.)
     Academic Press. New York. pp. 123-137.

Bryan, G. W. 1976. Some aspects of heavy metal tolerance in
     aquatic organisms. In "Effects of Pollutants on Marine
     Organisms", Vol. 2 (A. P. M. Lockwood, Ed.) pp. 7-34.
     Society for Experimental BIology Seminar Series. Cam-
     bridge University Press.

Bryan, G. W. and L. G. Hummerstone. 1971. Adaptation of the
     polychaete Nereis diversicolor to estuarine sediments
     containing high concentrations of heavy metals. J. Mar.
     Bio. Assoc. of U.K. 51: 845-863.

Bryan, G. W. and L. G. Hummerstone. 1973. Adaptation of the polychaete *Nereis diversicolor* to manganese in estuarine sediments. J. of Marine Biol. Assoc. of U.K. 53: 859-872.

Cross, F. A., T. W. Duke and J. N. Willis. 1970. Biogeochemistry of trace elements in a coastal plain estuary: Distribution of manganese, iron, and zinc in sediments, water and polychaetous worms. Chesapeake Science 11(4): 221-234.

Fowler, B. A., N. G. Carmichael, K. S. Squibb, and D. W. Engel In press. Factors affecting trace metal toxicity to marine organisms. II. Cellular mechanisms. In: Biological Monitoring of Marine Pollutants. (A. Calabrese, F. P. Thurberg, W. B. Vernberg and F. J. Vernberg, eds.). Academic Press, New York.

Freel, R. W., S. G. Medler, and M. E. Clark. 1973. Solute adjustments in the coelomic fluid and muscle fibers of a euryhaline polychaete, *Neanthes succinea*, adapted to various salinities. Biol. Bul. Mar. Biol. Lab. Woods Hole, Mass. 144: 289-303.

Gould, E. and J. J. Karolus. 1974. Physiological response of the cunner, *Tautogolabrus adspersus,* to cadmium: V. Observations on the biochemistry. In Physiological Response of the Cunner, *Tautogolabrus adspersus,* to cadmium. NOAA Tech. Rep. NMFS SSRF-681. pp. 21-25.

Greig, R. A., R. Reid and D. R. Wenzloff. 1977. Trace metal concentration in sediments from Long Island Sound. Mar. Pollution Bull. 8: 183-188.

Guyton, A. C. 1966. Test Book of Medical Physiology 3rd edition. W. B. Saunders Co., Philadelphia.

Jackim, E., J. M. Hamlin and S. Sonis. 1970. Effects of metal poisoning on five liver enzymes in the killifish (*Fundulus heteroclitus*) J. Fish. Res. Bd. Can. 27: 283-390.

MacInnes, J. and F. Thurberg. 1973. Effects of metals on behaviour and oxygen consumption of the mud snail. Mar. Poll. Bull. 4: 185-186.

Oglesby, L. C. 1970. Studies on the salt and water balance of *Nereis diversicolor:* I. Steady state parameters. Comp. Biochem. and Physiol. 36: 446-449.

Pesch, C. E. 1979. Influence of three sediment types on copper toxicity of the polychaete *Neanthes arenaceodentata.* Mar. Biol., 52: 237-245.

Prosser, C. L. and F. A. Brown. 1961. Comparative Animal Physiology. W. B. Saunders Co., Philadelphia.

Reish, D. J., J. M. Martin, F. M. Piltz, and J. Q. Word. 1976. The effect of heavy metals on laboratory populations of two polychaetes with comparisons to the water quality conditions and standards in Southern California waters. Water Res. 10: 299-302.

Schutz, D. and K. Turekian. 1965. The investigation of geographical and vertical distribution of several trace elements in seawater using neutron activation analysis. Biochem. Cosmochim Acta 20: 259-313.

Thurberg, F. P., A. Calabrese and M. A. Dawson. 1974. Effects of silver on oxygen consumption of bivalves at various salinities. In: Pollution and Physiology of Marine Organisms. (F. J. Vernberg and W. B. Vernberg, eds.). Academic Press, New York pp. 67-78.

Tyler, D. D. 1973. The mitochondrion. In: Cell Biology in Medicine. (E. E. Bittar, ed.). pp. 107-149. John Wiley & Sons, New York.

FACTORS AFFECTING TRACE METAL UPTAKE AND TOXICITY TO ESTUARINE ORGANISMS. I. ENVIRONMENTAL PARAMETERS

David W. Engel
William G. Sunda

National Marine Fisheries Service, NOAA
Southeast Fisheries Center
Beaufort Laboratory
Beaufort, North Carolina

Bruce A. Fowler

Environmental Toxicology Branch
National Institute of Environmental Health Sciences
Research Triangle Park, North Carolina

INTRODUCTION

Trace metals are of biological interest because of their role as micronutrients (Fe, Zn, Cu, Mn, Co, Mo) and toxins (Cu, Hg, Ag, Cr, Cd, Zn, Ni). Some metals, such as Cu and Zn, can act in either a stimulatory or inhibitory mode depending on their level of availability. Investigations into the interaction between trace metals and marine organisms have been intensified recently because of increased anthropogenic inputs of these metals into aquatic systems. The question of how much metal contamination can occur before damaging an ecosystem is quite pertinent, especially since under appropriate conditions natural environmental levels of trace metals such as copper can be toxic to some organisms (Anderson and Morel, 1978). In these instances any increase above natural concentrations may have deleterious effects.

Trace metals can exist in a variety of different chemical forms in natural waters including free ions, inorganic complexes, organic complexes, and metal adsorbed on or incorporated into particulate matter (Stumm and Brauner, 1975). The chemical forms of a metal in aqueous media will depend on both the chemical properties of the individual metal and the chemical composition of the natural water. Recent investigations with copper (Sunda and Guillard, 1976; Andrews *et al.*, 1977; Anderson and Morel, 1978; Waiwood and Beamish, 1978), cadmium (Sunda *et al.*, 1978) and zinc (Anderson *et al.*, 1978) have demonstrated that the toxicity and bioavailability of trace metals is highly dependent on their chemical form. These investigations have shown that biological response (i.e., toxicity and accumulation) to dissolved trace metals is a function of the free metal ion concentration which is determined not only by the total dissolved concentration, but also by the extent of metal complexation to both organic and inorganic ligands.

Such relationships between chemical form of trace metals and their bioavailability is shown schematically in Figure 1. In the figure we point out that food (i.e. ingested material)

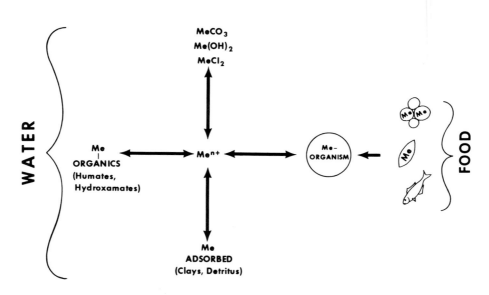

FIGURE 1. A conceptual model of the different forms of metal (Me) present following its addition to fresh, estuarine, or marine water. The double ended arrows indicate that the metal is in a dynamic equilibrium with the various components of the water column, and that a change in any one aspect of the system can shift the inter-relationships. For simplicity the model omits some possible interactions such as the ad-sorption of metal complexes on surfaces.

as well as water can be a source of trace metals to marine
fauna. In this paper we will deal only with water as a pathway and, therefore, with processes initiated by the passage of
a trace metal across external membranes. The double-ended
arrows in the figure indicate that trace metal reactions can
occur in both directions and that all components of the system
are interrelated. Therefore, any shift in one portion of the
system will affect other components as well. For example, an
increase in organic complexation of a trace metal will tend to
decrease inorganic complexation and metal adsorption as well
as the metal's reaction with components of organisms (e.g.
with proteins and membranes). As discussed above, the ionic
metal concentration (Me^{n+}) (or more correctly it's activity
or chemical potential) is an extremely important variable in
determining trace metal-biota interactions, but this variable
itself is determined by the chemical poise of the entire
system (i.e. the amount of total metal and the extent to which
the metal has combined with different components of the
system, including organisms).

   With this simplified model in mind, we will first examine
the aquatic chemistry of copper, cadmium, and silver with
emphasis on organic and inorganic complexation in estuarine
seawater. We will then attempt to relate the toxicity and
bioaccumulation of these metals to their chemical speciation.

CHEMICAL SPECIATION OF COPPER, CADMIUM AND SILVER

   Trace metals can form complexes with inorganic ligands
(primarily Cl^-, CO_3^{2-}, OH^-) and organic ligands (e.g.
humates). Copper, cadmium and silver all have different
affinities for binding to these various ligands. Measurements
with ion-selective electrodes in filtered, organic-rich river
water ($\sim$ 20 mg DOC/ℓ) containing 10^{-6}M trace metal additions
show that copper is much more highly complexed than cadmium
or silver (Fig. 2). For example, at pH 7.5 the ratios of
free metal ion to total metal concentration are $10^{-4 \cdot 1}$,
$10^{-1 \cdot 0}$, $10^{-0 \cdot 6}$ (0.008%, 10%, 25%), respectively, for copper,
cadmium and silver which indicates that copper is at least
three orders of magnitude more tightly complexed than the
other two metals. The trace metals are complexed primarily to
organic ligands as evidenced by the complete loss of the river
water's capacity to bind cadmium after UV photo-oxidation of
organic matter (Fig. 2). Similar loss of complexing capacity
after photooxidation has also been observed for copper (Sunda
and Hanson, 1979). Our results agree favorably with those
of Mantoura *et al.* (1978) who used gel filtration to measure

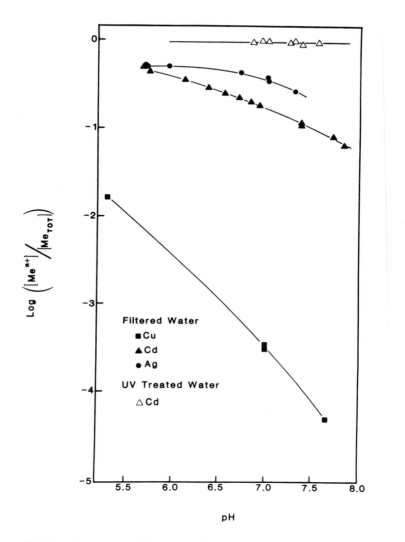

FIGURE 2. Logarithm of the fraction of the total dissolved metal (copper, cadmium, or silver) present as the free
ion as a function of pH for filtered Newport River water containing $10^{-6}M$ additions of $CuSO_4$, $CdCl_2$ or $AgNO_3$. Free metal
ion concentrations, $[Me^{n+}]$, were measured at 25°C using Orion
ion-selective electrodes. pH was varied by alteration of
P_{CO_2} or addition of $NaHCO_3$ and was measured with a glass
electorde. $[Me_{TOT}]$ refers to the total dissolved concentration of copper, cadmium or silver and can be assumed to be
equal to the concentration of added metal.

stability constants for organic ligands isolated from
different natural marine and fresh waters. Their conditional
stability constants (pH 8) for copper were 3 to 5 orders of
magnitude higher than those for cadmium, indicating that
copper has a much greater affinity for organic complexation
than cadmium. Of the three metals we examined, silver has
the lowest tendency to form organic complexes.

As has been noted in previous studies (Sunda and Hanson,
1979), the complexation of copper by natural organic ligands
increases noticeably with pH due to the weak acidic nature of
the natural organics (Fig. 2). This pH dependence also
occurs with cadmium and silver, but to a lesser extent
(e.g. at 10^{-6}M total metal concentrations and pH 7.0 to 7.7,
$\Delta pCu/\Delta pH$ is ~ 1.2, whereas $\Delta pCd/\Delta pH$ is ~ 0.4). Thus, the complexation of copper increases relative to that of cadmium and
silver with increasing pH. Because of copper's strong
affinity for organic ligands, we can expect its chemical
speciation in most estuarine waters to be dominated by
organic complexes (Mantoura et al., 1978; Gillespie and
Vacarro, 1978; Sunda and Gillespie, 1979). The extent of
organic binding in turn will be determined by the concentration and chemical composition of dissolved organic matter,
pH and salinity (Sunda and Hanson, 1979). Inorganic complexes (primarily $CuCO_3$), however, may be the dominant
chemical species of copper in seawater of low organic content
such as found in areas of low productivity or newly upwelled
seawater.

Unlike copper, we can expect the chemistry of cadmium
and silver to be dominated by inorganic complexes in most
estuarine waters. This is due both to their relatively weak
affinity toward organic complexation and to a strong tendency
toward complexation with chloride (Fig. 3). Increased
salinity also will tend to decrease organic complexation due
to increased concentrations of calcium and magnesium which
compete with trace metals for available chelation sites
(Mantoura et al., 1978). Finally, the fact that rivers
usually have higher concentrations of dissolved organic matter
than seawater further favors the dominance of inorganic
complexes with increased salinity. Our measurements of
cadmium and silver and complexation in natural estuarine
waters confirms the importance of chloro-complexes in controlling the speciation of these metals. The measurements of
cadmium complexation in river and estuarine water were made
at different salinities using a cadmium ion-selective
electrode. Portions of the water samples were either untreated, filtered (glass fiber) or filtered plus UV-photo-
oxidized. As before, all samples contained 10^{-6}M additions of
$CdCl_2$ in order to have a sufficient cadmium concentration for

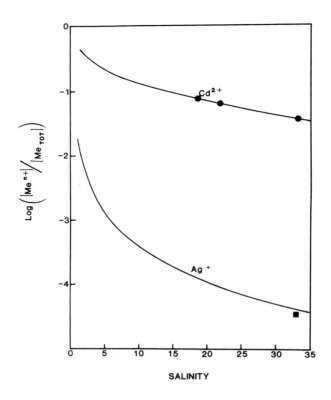

SALINITY

FIGURE 3. Logarithm of the fraction of total dissolved metal (cadmium and silver) present as the free ion as a function of salinity as determined by the formation of chloro complexes. Relationship for cadmium (22°C) is taken from Sunda et al. (1978) and that for silver is computed from equlibrium relationships for the formation of chloro complexes ($AgCl°$ and $AgCl_2{}^{2-}$). Stability constants ($K_1 = 10^{3.31}$ and $K_2 = 10^{1.94}$ at zero ionic strength) used in these calculations were selected from Sillen and Martell (1964). Activity coefficients used were computed from the Davies modification of Debeye Huckel equation (Stumm and Morgan, 1970). Solid symbols denote values from ion-selective electrode measurements at 25°C in seawater from the Newport River estuary. Water for silver ion measurements was filtered and contained 10^{-6} and $10^{-7}M$ $AgNO_3$. Water for Cd measurements contained $10^{-6}M$ $CdCl_2$ and was not filtered, however, filtration and filtration with subsequent UV-photooxidation had no effect on measured free cadmium ion concentrations.

analysis with the ion-selective electrode. The measurements
indicated that cadmium in the zero salinity river samples was
about equally partitioned between free ion, organic complexes,
and metal associated with particles. In higher salinity
samples (19, 22 and 33 $°/_{oo}$) there was an increase in total
complexation of cadmium, but there was no change in cadmium
binding with sample treatment. Thus, there was no apparent
binding of cadmium by particulate matter or by organic
ligands, and the complexation of cadmium was determined solely
by the formation of chloro-complexes (Fig. 3).

Similar measurements were conducted with a silver ion-
selective electrode in filtered, 33 $°/_{oo}$ seawater collected
near the mouth of the Newport River estuary. The fraction of
the total silver measured as free silver ion ($10^{-4.48}$) in
samples containing either 10^{-7} of 10^{-6}M AgNO$_3$ was essentially
the same as that ($10^{-4.40}$) which we computed from thermo-
dynamic equations based on the formation of chloro-complexes
(Fig. 3). As can be seen from our measurements and calculations, silver is much more strongly complexed by chloride
ion than is cadmium. In seawater at 35 $°/_{oo}$, 3% of the
cadmium is present as free cadmium ion with the remainder
present primarily as the neutrally charged complex $CdCl_2{}^0$. By
contrast, free silver ion accounts for only 0.003% of the
total metal at this same salinity with 99% present as $AgCl_2-$.

BIOLOGICAL RESPONSE

Our investigation of the chemistries of copper, cadmium
and silver in natural waters has shown that speciation of
trace metals can be highly variable. The dominant chemical
forms under a given set of conditions depends upon the metal,
the concentration of total dissolved metal and environmental
parameters, such as the concentration of dissolved organic
matter, salinity and pH. The question which arises is, "What
are the implications of differences in chemical speciation of
these three metals on the biological response of organisms
exposed to these metals?" This question will be discussed
for the toxicity and biological accumulation of copper,
cadmium and silver. We will both briefly review results of
previous investigations (particularly for copper) as well as
reporting on recent experiments we have conducted dealing
with the effects of salinity and chloride complexation on the
toxicity and accumulation of cadmium and silver.

Copper has long been known to be an extremely toxic
metal in aquatic systems with its main pollution sources
being domestic effluents, industrial wastes and antifouling

paints. The free cupric ion has been shown to be the dissolved species of copper which determines its toxicity to
aquatic organisms. Factors which control cupric ion activity
such as the level of organic and inorganic complexation will,
therefore, also control copper toxicity (Sunda and Lewis,
1978).

By use of chemically defined media employing cupric ion
buffers, it has been possible to quantify the toxicity of
dissolved copper as functions of free cupric ion concentration within environmentally realistic ranges. These
studies have shown that concentrations of free cupric ion
that are toxic to a number of marine organisms, including
phytoplankton (Sunda and Guillard, 1976; Anderson and Morel,
1978; Jackson and Morgan, 1978; Reuter *et al.*, 1979), bacteria
(Sunda and Gillespie, 1979), and eggs of fish and sea urchins
(Guida *et al.*, 1978; Engel and Sunda, 1979), approach or are
within the estimated range for seawater ($10^{-9.3}$ - $10^{-10.3}$) in
which there is little or no organic complexation.[1] In
previous work with fish eggs and larvae it has been noted that
copper toxicity is highly dependent upon the stage of development (Fig. 4). Egg hatching of the Atlantic silverside
Menidia menidia, for example, is affected at cupric ion concentrations that are two orders of magnitude below those
which affect survival of larvae. This implies that low levels
of environmental copper contamination could have a marked
effect on egg survival, but no direct effect on the survival
of larvae. The life cycle stage at which exposure to the
contaminant occurs would therefore be important in determining
possible effects of copper toxicity on fish populations.

Accumulation of copper by marine organisms also appears
to be related to the free cupric ion concentration in the
water as demonstrated recently in our laboratory for the
accumulation of dissolved copper by the oyster, *Crassostrea
virginica* (Zamuda and Sunda, 1979). Such free ion dependence
for accumulation has also been demonstrated for phytoplankton
(Sunda and Guillard, 1976). In both cases the use of cupric
ion buffer systems employing different concentrations of
synthetic chelators allowed the maintenance of cupric ion concentrations which were representative of estimated environmental levels.

As with copper, the toxicity of cadmium also has been
shown to be related to free cadmium ion concentration, rather

[1]*This range is computed for total copper concentration of
10^{-9} - 10^{-8} molar (i.e. 0.06 - 0.6 ppb) and is based on free
cupric ion being 5% of the total dissolved copper.*

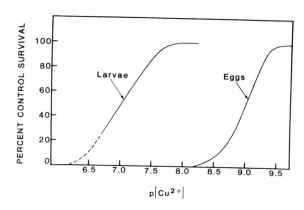

FIGURE 4.   The relationship between 24 hour survival of larvae and egg hatch for silverside, M. menidia and $p[Cu^{2+}]$ (-log of cupric ion concentration).   The curve for larval survival is replotted from Simon et al. (1979) and the curve for egg hatch represents the combined results of Engel and Sunda (1979) and Simon et al. (1979).

than total cadmium concentration (Sunda *et al.*, 1978).   Since cadmium in seawater should be complexed primarily by chloride ion the level of complexation will be an increasing function of salinity (Fig. 3).   In previous experiments with grass shrimp, *Palaemonetes pugio,* increased cadmium complexation with increased salinity was accompanied by a decrease in the toxicity of dissolved cadmium which, in turn, could be related quantitatively to a decrease in free cadmium ion concentration (Sunda *et al.*, 1978).

Recently, we also have found similar salinity relationships in experiments testing the toxicity of cadmium to eggs and larvae of the silverside exposed to a range of salinities, 5 to 35 $^o/_{oo}$, and to a range of cadmium concentrations, 0.9 to 9.0 µM.   Larval survival and egg hatch increased with salinity at all cadmium additions (Fig. 5A) and were found to relate inversely to the computed free cadmium ion concentration (Fig. 5B)[2].   The relationship between survival and

[2]*In fact, larval survival and egg hatch with different combinations of cadmium concentration and salinity showed an even closer correlation with computed free cadmium ion activity.   Activities were computed from cadmium ion concentration using the Davies equation to calculate activity coefficients.*

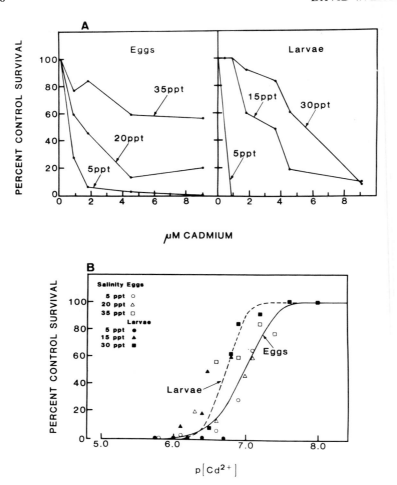

FIGURE 5.   Egg hatch and four day larval survival of
silversides at three salinities, pH 8.0-8.2 and temperature
20°C; (A) as a function of total dissolved cadmium concentration (verified by atomic absorption analysis) and (B) as a
function of p[Cd²⁺], -log of free cadmium ion concentraiton.
Experimental details concerning laboratory spawning of fish
and larval fish rearing are as described previously (Simon et
al., 1979).   There were 15 larvae and ca. 60 eggs per treatment.   Different salinity media were prepared by diluting
filtered estuarine seawater with distilled water.   Media of
salinity <15°/₀₀ contaned 0.5 m moles/liter NaHCO₃ to adjust
pH to 8.

free cadmium ion was similar for eggs and larvae, and both sets of data probably could be explained on the basis of a single curve. This contrasts with the response to copper where the eggs were two orders of magnitude more sensitive than larvae (Figs. 4 and 5B). Although on a free ion basis, copper was far more toxic to the eggs than cadmium, the toxicity of the two metals to the larvae was similar. As with grass shrimp (Sunda et al., 1978), the levels of free cadmium ion required to cause larval fish mortality or egg hatch inhibition (> $10^{-7.5}$M) were about three orders of magnitude above our estimate of free cadmium ion concentration in seawater ($\leq 10^{-10.5}$M). This estimate was computed from recent measurements of cadmium concentrations ($\leq 10^{-9}$M; Boyle et al., 1977) and the total level of cadmium complexation by chloride ion (Fig. 3).

As further test of the effects of the chemical speciation of cadmium on its vioavailability, its accumulation by juvenile oysters, Crassostrea virginica, also was studied. Oysters were exposed to 0.89 and 8.9 μM concentrations of total dissolved cadmium (^{115}Cd + stable cadmium) for 7 days at salinities of 10, 20 and 30 °/oo (Fig. 6). Cadmium accumulation by the oysters was followed using the radionuclide ^{115}Cd. The rate of cadmium uptake[3] by the oysters decreased with increasing salinity, at both dissolved cadmium concentrations. When the level of accumulation at seven days is replotted as a function of free cadmium ion concentration, there is a linear log-log relationship between cadmium ion concentration and cadmium accumulation (Fig. 7). Thus, accumulation as well as toxicity of cadmium is a function of free cadmium ion rather than the total dissolved cadmium concentrations.

These results of both the cadmium accumulation experiments with oysters and the toxicity experiments with silverside eggs and larvae are in agreement with our previously discussed chemical model (Fig. 1). Also, our results are in agreement with the data of von Westernhagen and co-workers (von Westernhagen et al., 1974; von Westernhagen et al., 1975; and von Westernhagen and Dethlefsen, 1975) where they showed similar effects of salinity on the toxicity of cadmium to the eggs of fish from the Baltic Sea. Their data probably also can be explained on the basis of a free cadmium ion model.

As discussed previously (Fig. 3), the complexation of

[3]Total accumulation cadmium and silver concentrations in oysters and grass shrimp were determined by dividing accumulated ^{115}Cd and ^{110}Ag concentrations by the measured specific activities of the added isotopes.

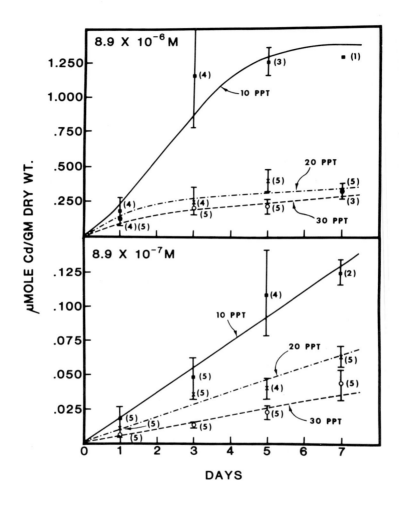

FIGURE 6. *Accumulation of cadmium by juvenile American oysters, Crassostrea virginica, as function of salinity, time of exposure, and total dissolved cadmium concentration at 20°C and pH 8.1-8.2. Numbers in parentheses refer to the numbers of individuals per point ± SE. Cadmium accumulation was followed using ^{115m}Cd, and the data at $8.9x10^{-6}M$ cadmium was replotted from Engel and Fowler (1979). Oysters were supplied by Frank M. Flowers and Sons, Inc., Bayville, N.Y. Preparation of media is the same as described in the legend of Figure 5.*

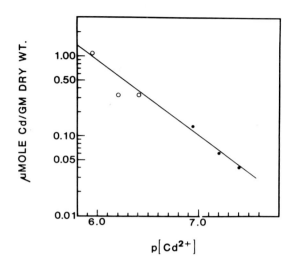

FIGURE 7.   Concentration of cadmium accumulated by oysters at four days plotted as a function of computed p[Cd²⁺] (-log free cadmium ion concentration).   Computations of p[Cd²⁺] are based on cadmium complexation relationship in Figure 3.   (Sunda et al., 1978).

silver, like cadmium, also varies with salinity due to the formation of chloro-complexes. Silver, however, differs from cadmium in two important respects: (1) silver is complexed by chloride much more strongly than cadmium (Fig. 2) and (2) un-like cadmium, changes in the bioavailability of dissolved silver with salinity cannot be explained simply on the basis of a free silver ion model. This latter observation was made in silver accumulation experiments we recently completed with grass shrimp. The total added silver concentrations (^{110}Ag + stable silver) were 2.5×10^{-8}M to 2.1×10^{-7}M and the test salinities ranged from 4 to 32 °/₀₀ at 20°C. The ^{110}Ag was used as a tracer of the total dissolved silver in the ex-posure media and was used to determine uptake of Ag by the shrimp[3]. As noted with cadmium, the accumulation of silver by the grass shrimp decreased with increasing salinity, however,

[3]*Total accumulation cadmium and silver concentrations in oysters and grass shrimp were determined by dividing accumu-lated ^{115}Cd and ^{110}Ag concentrations by the measured specific activities of the added isotopes.*

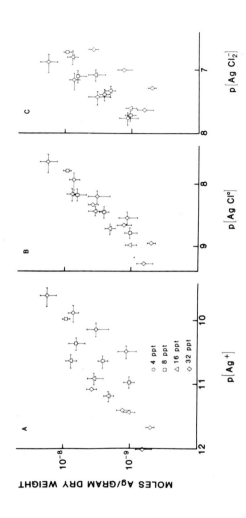

FIGURE 8. Accumulation of silver by grass shrimp, Palaemonetes pugio, after exposure to silver for four days at 20°C at four different salinities. Accumulation was measured using ^{110}Ag as a radiotracer. Data are plotted as functions of the negative log of concentrations of Ag^+, $AgCl^\circ$ and $AgCl_2^-$ computed from total dissolved silver concentrations and stability constants listed in the legend of Figure 3. Vertical error bars represent $SE \pm 1$ with $n = 10$. Horizontal bars represents ranges based on initial added concentrations and final measured ^{110}Ag values. pH values measured initially and finally were in the range 7.8–8.1.

unlike cadmium the accumulation of silver after four days was not a function of free silver ion concentration. Instead, silver accumulation was most closely related to the computed concentration of the mono-chloro complex ($AgCl^0$) (Fig. 8). Linear log-log regressions of silver concentration in the shrimp vs. the concentrations of different chemical species of silver had the following R^2 values: $[Ag+]$, 0.72; $[AgCl^0]$, 0.90; and $[AgCl_2-]$, 0.51.

It is possible that silver accumulation is determined by a combination of both free silver ion concentration and total silver concentration and that the close correlation with concentration of monochloro complex is merely fortuitous. However, it is also possible that the correlation with $AgCl^0$ results directly from a greater permeability of the neutrally charged complex across cell membranes. In support of this hypothesis is the fact that neutrally charged species are orders of magnitude more permeable through lipid membranes than charged species (Finkelstein and Cass, 1968; Gutknecht and Tosteson, 1973).

SUMMARY

The data presented demonstrate that the chemical speciation of dissolved copper, cadmium and silver affect the biological responses of organisms exposed to these metals, and from these data several conclusions can be drawn:
1. The aquatic chemistry of the three metals are complex and broad generalizations among metals concerning their chemistries and biological effects cannot be made.
2. The toxicity and biological accumulation of copper and cadmium appears to be determined primarily by free ion concentration (or activity); for silver, on the other hand, chlorocomplexes may also play an important chemical role in the accumulation of this metal by marine organisms. Thus, although our free ion model (Fig. 1) is found to be valid in many, if not most instances, it does not appear to hold true in all cases.
3. The toxicity and bioaccumulation of trace metals are manifestations of chemical processes occurring both in the water and in the organism, and it is therefore imperative in designing and interpreting metal response experiments with aquatic organisms that chemical speciation of the metal be considered.

ACKNOWLEDGMENTS

This research is supported in part by the U.S. Environmental Protection Agency under an interagency agreement relating to the Federal Interagency/Environmental Research and Development Programs, and in part through an interagency agreement between the Department of Energy and the National Marine Fisheries Service, NOAA. S.E.F.C. Contribution No. 80-52B.

LITERATURE CITED

Anderson, D. M., and F. M. M. Morel. 1978. Copper sensitivity of *Gonyaulax tamarensis*. Limnol. Oceanogr. 23: 283-295.

Anderson, M. A., F. M. M. Morel, and R. R. L. Guillard. 1978. Growth limitation of a coastal diatom by low zinc ion activity. Nature, Lond. 276: 70-71.

Andrew, R. W., K. E. Biesinger and G. E. Glass. 1977. Effects of inorganic complexing on the toxicity of copper to *Daphnia magna*. Water Res. 11: 309-315.

Boyle, E. A., F. R. Sclater, and J. M. Edmond. 1977. The distribution of dissolved copper in the Pacific. Earth Planet. Sci. Letters. 37: 38-54.

Engel, D. W. and B. A. Fowler. 1979. Factors influencing cadmium accumulation and its toxicity to marine organisms. Environ. Health Persp. 28: 81-88.

Engel, D. W., and W. G. Sunda. 1979. Toxicity of cupric ion to eggs of the spot *Leiostomus xanthurus* and the Atlantic silverside *Menidia menidia*. Mar. Biol. 50: 121-126.

Finkelstein, A., and A. Cass. 1968. Permeability and electrical properties of thin lipid membranes. J. Gen. Physiol. 52: 145s-173s.

Gillespie, P. A., and R. F. Vaccaro. 1978. A bacterial bioassay for measuring the copper-chelation capacity of seawater. Limnol. Oceanogr. 23: 543-548.

Guida, V., D. W. Engel, and W. G. Sunda. 1978. The effect of cupric ion on the embryonic development of the purple sea urchin, *Arbacia punctulata*. Annual Report of the Beaufort Laboratory to the U.S. Department of Energy. 451-467.

Gutknecht, J., and D. C. Tosteson. 1973. Diffusion of weak acids across lipid bilayer membranes: Effects of chemical reactions in the unstirred layers. Science 182: 1258-1261.

Jackson, G. A., and J. J. Morgan. 1978. Trace metal-chelator interactions and phytoplankton growth in seawater media: Theoretical analysis and comparison with reported observations. Limnol. Oceanogr. 24: 268-282.

Mantoura, R. F. C., A. Dickson, and J. P. Riley. 1978. The complexation of metals with humic materials in natural waters. Estuar. Coast. Mar. Sci. 6: 387-408.

Reuter, J. G., J. J. McCarthy, and E. J. Carpenter. 1979. The toxic effect of copper on *Oscillatoria (Trichodesmium) theibautii*. Limnol. Oceanogr. 24: 558-562.

Sillen, L. G. and A. E. Martell. 1964. Stability constants, Special Publ. no. 17. The Chem. Soc. of London.

Simon, B. J., D. W. Engel and W. G. Sunda. 1979. Effect of cupric ion on eggs and larvae of the Atlantic silverside, *Menidia menidia*. Annual Report of the Beaufort Laboratory to the Department of Energy.. 632-640.

Stumm, W., and D. A. Brauner. 1975. Chemical speciation. In: J. P. Riley and G. Skirrow (eds.). Chemical oceanography. Vol. 1. pp. 173-234. New York: Academic Press.

Stumm, W., and J. J. Morgan. 1970. Aquatic Chemistry. Wiley Interscience, New York. 1-583.

Sunda, W., D. W. Engel, and R. M. Thuotte. 1978. Effect of chemical speciation on toxicity of cadmium to grass shrimp, *Palaemonetes pugio:* Importance of free cadmium ion. Environ. Sci. Technol. 12: 409-413.

Sunda, W. G., and P. A. Gillespie. 1979. The responses of a marine bacterium to cupric ion and its use to estimate cupric ion activity. J. Mar. Res. 37: 761-777.

Sunda, W. G., and R. R. L. Guillard. 1976. The relationship between cupric ion activity and the toxicity of copper to phytoplankton. J. Mar. Res. 34: 511-529.

Sunda, W. G., and P. J. Hanson. 1979. Chemical speciation
    of copper in river water. Effect of total copper, pH
    carbonate and dissolved organic matter. pp. 147-180.
    In: E. A. Jenne., Chemical modeling in aqueous systems.
    ACS Symposium Series. No. 93.

Sunda, W. G., and J. M. Lewis. 1978. Effect of complexation
    by natural organic ligands on the toxicity of copper to
    a unicellular alga, *Monochrysis lutheri*. Limnol.
    Oceanogr. 23: 870-876.

von Westernhagen, H., and V. Dethlefsen. 1975. Combined
    effects of cadmium and salinity on development and
    survival of flounder eggs. J. Mar. BIol. Assoc. U.K. 55:
    945-957.

von Westernhagen, H., V. Dethlefsen, and H. Rosenthal. 1975.
    Combined effects of cadmium and salinity on development
    and survival of garpike eggs. Helgoländer wiss.
    Meeresunters. 27: 268-282.

von Westernhagen, H., H. Rosenthal, and K. R. Sperling. 1974.
    Combined effects of cadmium and salinity on development
    and survival of herring eggs. Helgoländer wiss.
    Meeresunters. 26: 415-433.

Waiwood, K. G., and F. W. H. Beamish. 1978. Effects of
    copper, pH and hardness on the critical swimming per-
    formance of rainbow trout *Salmo gairdneri* Richardson.
    Wat. Res. 12: 611-619.

Zamuda, C. D., and W. G. Sunda. 1979. The effect of physico-
    chemical speciation of copper upon its bioavailability to
    the American oyster, *Crassostrea virginica*. Annual
    Report of the Beaufort Laboratory to the U.S. Department
    of Energy. 592-610.

FACTORS AFFECTING TRACE METAL UPTAKE AND
TOXICITY TO ESTUARINE ORGANISMS
II. CELLULAR MECHANISMS

Bruce A. Fowler
Neil G. Carmichael
Katherine S. Squibb

National Institute of Environmental Health Sciences
Laboratory of Organ Function and Toxicology
Research Triangle Park, North Carolina

David W. Engel

National Marine Fisheries Service, NOAA
Southeast Fisheries Center
Beaufort Laboratory
Beaufort, North Carolina

INTRODUCTION

There are many factors which must be considered in
evaluating the uptake and toxicity of trace metals in marine
organisms. The chemical and nonbiological factors influencing
these processes have been reviewed in the preceding paper
(Engel *et al.*, 1980). This report will examine some of the
known biological mechanisms for trace metal uptake by commercially important marine species and present data from some
recent studies from our laboratory concerning the role of
metal-binding proteins and intracellular concretions in the
accumulation of cadmium by marine shellfish. Information of
this type is essential to understanding both the relationship
between metal uptake and toxicity in the marine species themselves and how these processes may cause marine organisms to
act as vectors for human exposure. This discussion should

0-12-718450-3

also permit some comparison with the mechanisms of metal
accumulation known in mammals.

Uptake of Metals by Lysosomes
and Membrane-Bound Cytoplasmic Bodies

A number of authors have described cellular uptake and/
or deposition of metals in aquatic or marine invertebrates
following water-borne exposure to these elements in seawater.
X-ray microanalytical studies have demonstrated mercury
(Fowler *et al.*, 1975), lead (Brown, 1977, 1978; Marshall and
Talbot, 1979), iron (Fowler *et al.*, 1975; Lowe and Moore,
1979) and copper (Walker, 1977) in lysosomes or morphologi-
cally similar membrane-bound bodies within cells from various
tissues from a variety of marine invertebrates. Figure 1
shows the morphological appearance of these structures in
mantle tentacle epithelial cells of the quahog clam (*Mercenaria mercenaria*) following exposure to 10 ppm mercury (as
mercuric chloride) in seawater. These structures were found
to possess detectable concentrations of both mercury and iron
by X-ray microanalysis (Fowler *et al.*, 1975).

These membrane-bound cytoplasmic bodies have been demonstrated to play an important role in the uptake of metals
into cells (George *et al.*, 1975; Schulz-Baldes, 1977) in
addition to serving as an intracellular storage site for
metals. For example, Schulz-Baldes (1977) has suggested that
lead is accumulated in kidney cells of molluscs by pinocytotic
uptake of soluble lead from the kidney lumen with subsequent
storage of the lead in the cytoplasmic vesicles. Such a
process appears to serve as a detoxification mechanism for
the cell due to the chemically inert form of the lead within
the vesicles (Walker, 1977).

Studies in mammals exposed to metals have shown similar
cytoplasmic bodies in various tissues and have yielded similar
results. Mercury (Fowler *et al.*, 1974), lead (Fowler *et al.*,
1980), iron (Fowler *et al.*, 1975), copper (Goldfischer and
Moskal, 1966), gold (Stuve and Galle, 1970), zinc (Brun and
Brunk, 1970), cadmium (Fowler and Nordberg, 1978; Squibb *et
al.*, 1979) and silver (Carmichael *et al.*, unpublished observations) have been demonstrated in lysosomes of liver or
kidney cells following exposure. The main point to be
derived from the above studies is that deposition of metals in
membrane-bound cytoplasmic bodies or lysosomes appears to be
a general biological process which occurs in a variety of
diverse phyla.

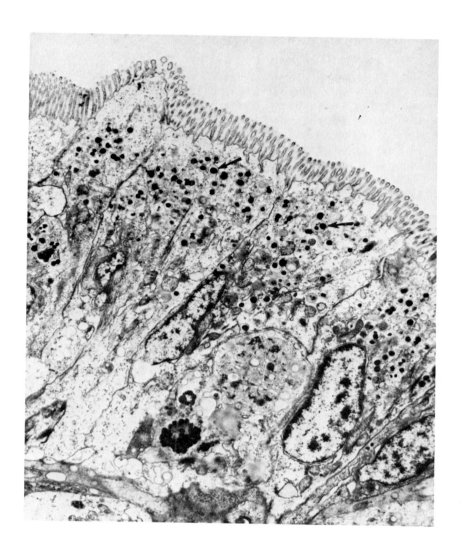

FIGURE 1. Epithelial cell from the mantle tentacles of a quahog clam (Mercenaria mercenaria) exposed to Hg^{+2} in seawater at a concentration of 10 ppm for seven days showing numerous electron dense cytoplasmic bodies (arrows) x9,576.

Calcium Phosphate Granules
and Concretions

Many invertebrate phyla produce intracellular calcified
structures and recently information has become available to
show that in some species these calcified structures play an
important role in metal accumulation. In particular the
kidney of bivalve molluscs appears to have a role in accumulating and possibly excreting metals via calcium phosphate
concretions (Ghiretti *et al.*, 1971; Doyle *et al.*, 1978;
Carmichael *et al.*, 1979a). Recent studies in our laboratories
have shown that the kidney of clams (*Mercenaria mercenaria*)
can concentrate isotopes of cadmium, zinc and manganese from
solution and sequester these isotopes, particularly manganese,
in intracellular calcium phosphate concretions. These concretions (Fig. 2) can be purified by centrifugation and are
found to contain a third of all kidney ^{54}Mn after only one
week of exposure (Carmichael *et al.*, 1979b).

The hepatopancreas of the blue crab is also associated
with metal accumulation and is well known to contain intracellular calcium phosphate granules which are thought to act
as a calcium reserve for recalcifying the exoskeleten
following molting (Becker *et al.*, 1974). In a recent preliminary study, blue crabs (*Callinectes sapidus*) were exposed
to 700 ppb cadmium in seawater for 5 days. After the crabs
had been flushed for 24 hours in clean seawater, the hepatopancreas was removed and frozen. A homogenate was made from
this tissue and prepared in the same way as in the previous
clam studies to give a pellet of granules and a clear supernatant fraction. These were analyzed for cadmium and, while
the supernatant had a low cadmium content (approx. 1 ppm)
and no evidence of low molecular weight metal binding
proteins, the granule pellet contained 810 ppm cadmium (dry
weight).

It appears from these studies that some invertebrates
may have a mechanism for handling trace elements which is
unique to their particular species. These intracellular
calcium phosphate granules may serve a variety of functions
in invertebrate systems but their capacity to sequester trace
metals indicates that they might play an important role in
trace element metabolism.

Metal-Binding Proteins

In recent years a number of studies have been conducted
concerning the role of metal-binding proteins in the uptake
and metabolism of metals in marine organisms. Studies on
specific metal-binding proteins have been reported for cadmium

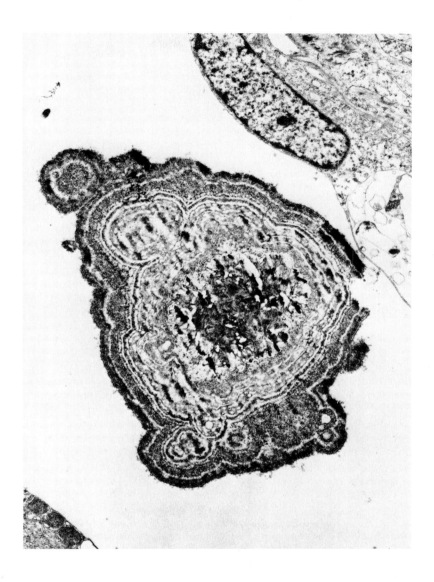

FIGURE 2. *Electronmicrograph showing appearance of a concretion in cross section illustrating the concentric bands of differing electron opacity. x18,705.*

(Casterline and Yip, 1975; Frazier, 1976, 1979; Zaroogian and Cheer, 1976; Nöel-Lambot, 1976; Nöel-Lambot *et al.*, 1978; Howard and Nickless, 1977a, b, 1978; Jennings *et al.*, 1979; Talbot and Magee, 1978; Olafson *et al.*, 1979a, b; Ridlington and Fowler, 1979; Carmichael *et al.*, 1979b), zinc (Olafson, *et al.*, 1979a, b; Nöel-Lambot, 1976; Carmichael *et al.*, 1979b), copper (Roesijadi, 1980) and manganese (Carmichael *et al.*, 1979b). Considerable work has been done on the characterization of a low molecular weight cadmium binding protein (CdBP) present in marine bivalves exposed to cadmium (Ridlington and Fowler, 1977, 1979; Talbot and Magee, 1978; Marshall and Talbot, 1979). This soluble CdBP has a molecular weight of 6-10,000 daltons, but an ultraviolet absorbance spectrum different from that observed with metallothionein, a 6-10,000 molecular weight mammalian cadmium-binding protein. Amino acid analysis of this protein from oysters (Ridlington and Fowler, 1977; 1979) disclosed an amino acid composition dominated by aspartic and glutamic amino acids rather than cysteine as found in mammalian metallothionein. More recently, Marshall and Talbot (1979) have reported a virtually identical amino acid composition for a similar low molecular weight cadmium binding protein isolated from mussels. From the studies by Ridlington and Fowler (1979) the cadmium binding capacity of the molluscan protein appears to be lower (1 g atom Cd/mole protein) than that observed in mammals (7g atom Cd/mole protein).

The function of this protein is at present unknown, however, studies from our laboratory have shown that the binding of cadmium to this protein increases with time in oysters (*Crassostrea virginica*) exposed to 100 ppb cadmium in seawater for up to 7 days (Fig. 3). The binding of cadmium to the low molecular weight protein peak (Fractions 30-45) increases at each succeeding time point. Previous studies (Engel and Fowler, 1979a) have shown that cadmium binding to this protein plays a role in determining the onset of cellular toxicity in oysters as judged by electron microscopy and gill tissue respiration.

In order to further evaluate the relationship between cellular toxicity and CdBP levels and to determine the effects of concomitant selenium exposure on the tissue accumulation and toxicity of cadmium to oysters, a series of experiments were performed. The potential interactive effects of selenium on cadmium toxicity to oysters was studied because selenium, which has also been shown to be readily accumulated by marine bivalves (Fowler and Benayoun, 1976), is known to alter the biological responses of mammals to cadmium (Gasiewicz and Smith, 1978). It is also important to note that selenium may be released into the environment as a result of increased fossil fuel utilization (Ketchum, 1975)

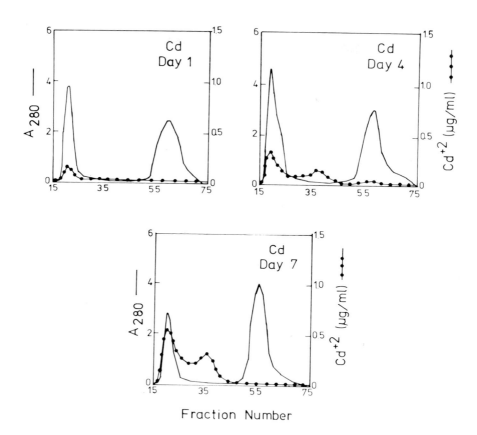

FIGURE 3.   Oysters were exposed to 100 ppb Cd in their water for 1, 4 or 7 days. Cytosolic fractions of whole oyster homogenates were chromatographed on G-75 Sephadex columns eluted with 0.01 M Phosphate buffer (7.8).

and, hence, future decisions concerning the predicted impact of these elements on the environment must take elemental interactions into account.

For these experiments randomly selected oysters *(Crassostrea virginica)* were collected in a marsh near the Southeast Fisheries Center of the National Marine Fisheries Service in Beaufort, North Carolina. The oysters were acclimated in flowing seawater for 72 hours at a salinity of

$34°/_o$ and a temperature of $20°C$. After acclimation, the oysters were divided into four groups and exposed in a flowing seawater system to cadmium at 400 ppb for 1, 4, 7, 11 and 14 days as previously described (Engel and Fowler, 1979b) or cadmium + selenium (Se^{+6}) at 400 ppb for each. Non-exposed control and selenium (400 ppb Se^{+6}) oysters were also sampled at each time point. Tissue respiration values and cadmium determinations were made as previously described (Engel and Fowler, 1979a, b). Isolation of cadmium-binding proteins was performed according to the procedure of Ridlington and Fowler (1979) with the exception that the isolated 27,000 x g supernatant fraction was not lyophilized prior to G-75 Sephadex column chromatography. Data were statistically analyzed by ANOVA procedures (Snedecor and Cochran, 1967).

Results from these experiments indicated that overall accumulation of cadmium in whole oysters exposed to cadmium was not significantly affected by concurrent exposure to selenium (Fig. 4). Although total accumulation of cadmium on individual days showed some differences between the two cadmium treatment groups, the results showed no consistent difference in whole oyster cadmium accumulation from day to day and overall accumulations were not significantly different. Gill concentrations of cadmium (Fig. 5), however, showed statistically significant differences between the cadmium and cadmium + selenium groups from days 7-14 of exposure. Gills from oysters exposed to cadmium + selenium contained statistically lower (p < .05) levels of cadmium relative to those oysters exposed to cadmium alone.

Data showing the effects of exposure to cadmium, selenium or cadmium + selenium on gill-tissue respiration, expressed as percent of non-exposed control oysters, are given in Figure 6. As has been previously reported (Engel and Fowler, 1979a, b), cadmium exposure increased gill-tissue respiration. This increased respiration was, however, partially but not statistically decreased by concomitant exposure to selenium from days 4-14. Selenium exposure alone was not found to significantly alter gill-tissue respiration relative to controls.

The intracellular binding of cadmium was studied by gel filtration of the 27,100 x g supernatant fraction of whole oysters exposed to cadmium or cadmium + selenium for 7 days (Fig. 7). These data indicate that there were no qualitative, nor apparent quantitative, changes in the binding of cadmium to the high or low molecular weight protein species found in the supernatant fraction after exposure to cadmium or cadmium + selenium at the dose levels used.

The results of these studies indicate that exposure of oysters to cadmium plus selenium at equal dose levels

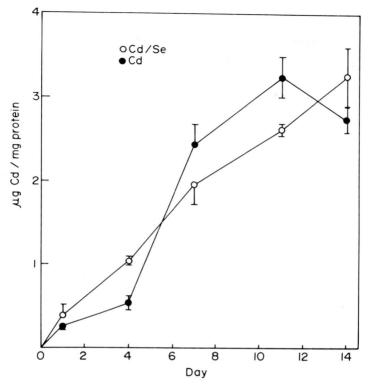

FIGURE 4. Mean ± standard deviation concentrations of cadmium in whole oysters exposed to 400 ppb cadmium or 400 ppb cadmium + 400 ppb selenium for 1-14 days in seawater.

produces little change in whole oyster cadmium accumulation or in the binding of cadmium to the previously reported low molecular weight oyster cadmium-binding protein. Gill-tissue accumulation of cadmium was reduced by concomitant selenium exposure, however, and there is evidence to suggest that the increase in respiration caused by cadmium may be diminished by selenium. Further studies are needed to elucidate the exact mechanism of cadmium toxicity to oyster gill tissue and also to study the impact of other environmental agents such as selenium on cadmium toxicity and specific binding of cadmium to low molecular weight proteins.

These findings do indicate that certain essential organ systems such as the gill may be more susceptible to interactive effects than others and serve to illustrate the potential problems associated with attempting to relate whole animal metal burdens to specific toxic effects. These

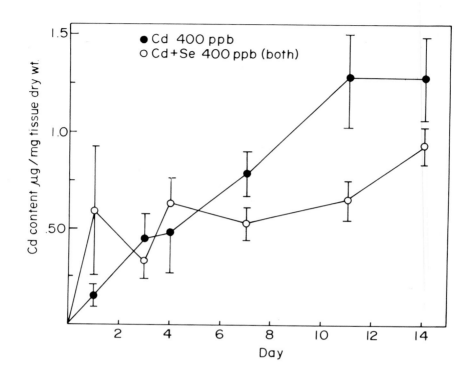

FIGURE 5. Mean ± standard deviation cadmium concentration in gill tissue of oysters exposed to 400 ppb cadmium or 400 ppb cadmium + 400 ppb selenium for 1-14 days in seawater.

differences could well be due to organ differences in metal ion uptake and storage mechanisms. As previously discussed, the formation of concretions, cytoplasmic membrane-bound vesicles and metal-binding proteins varies in different organs as well as in different species of marine animals.

It is important to emphasize that none of the mechanisms of accumulation described in this report are static or exclusive of one another. Uptake and storage of metals may well include all or most of these systems. Thus, metal uptake by the gill may involve pinocytosis and transfer to the lysosomal system as shown by George *et al* (1975) for iron. Uptake via the lysosomal system may induce formation of metal-

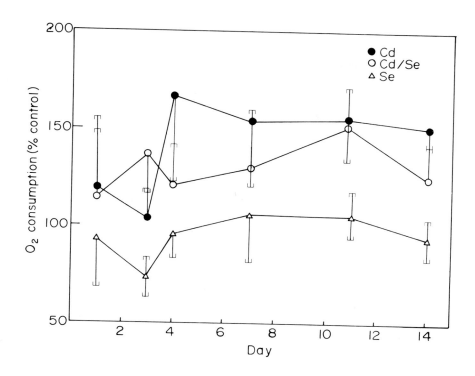

FIGURE 6. Gill-tissue respiration for oysters exposed to cadmium (400 ppb), selenium (400 ppb) or cadmium (400 ppb) + selenium (400 ppb) for 1-14 days expressed as percent of control. (mean percent ± standard error). Although the increase in respiratory rate observed with cadmium is reduced with the combination of cadmium + selenium, the data are not statistically different.

binding proteins resulting in accumulation of the metal within the same organ or in transport to storage and excretory organs such as the kidney and digestive gland. Within these organs excretory products such as concretions may form. Although these organs may show high levels of an accumulated metal, the metal may now be in its most stable and therefore least toxic form. All of these mechanisms of handling toxic trace elements must be considered when studying the toxicity of metals to marine organisms.

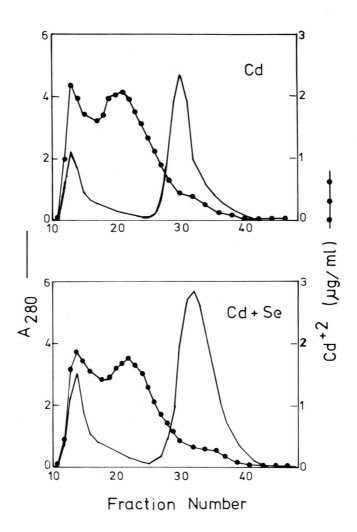

FIGURE 7. Oysters were exposed to 400 ppb Cd or 400 ppb Cd + 400 ppb Se in their water for 7 days. Cytosolic fractions of whole oyster homogenates were chromatographed on G-75 Sephadex columns eluted with 0.01 M Phosphate buffer (pH 7.8).

Environmental Significance

The effects of intracellular binding of toxic metals within marine shellfish on the absorption, distribution, excretion and toxicity of these elements in mammals following consumption of these organisms has received relatively little study, but such data are of potentially great importance. Mammalian studies with cadmium indicate that the absorption of dietary cadmium is affected by the dietary levels of other trace elements such as calcium, iron and zinc (Spivey-Fox *et al.*, 1979; Washko and Cousins, 1977). All of these elements are potentially present in high levels in marine shellfish. Also, the mammalian metabolism and toxicity of cadmium has been shown to differ substantially from that of the free cadmium ion when it is bound to the low molecular weight protein, metallothionein (Cherian, 1979). In light of these studies and others, data on the intracellular binding of metals in marine organisms can greatly aid in predicting the bioavailability and potential toxicity to humans of metal exposure via consumption of commercially important marine species.

SUMMARY

Many marine organisms consumed as food by man are known to concentrate potentially toxic trace metals from the environment but data on the basic biological processes involved in this phenomenon are relatively limited. Such data are of extreme importance both with respect to understanding how marine shellfish survive in metal-polluted areas and assessing the role of these organisms as vectors for human exposure to metals. Combined ultrastructural, analytical and biochemical studies from our laboratories and those of others have identified lysosomes, low molecular weight metal-binding proteins and concretions as major intracellular compartments responsible for trace metal accumulation in several species of marine shellfish. Lysosomes have been found to play a role in the intracellular compartmentation of iron, mercury, lead, copper and zinc in marine invertebrates. Low molecular weight proteins (6-10,000 MW) have recently been found to bind cadmium, copper, manganese and zinc in various organs of commercially important marine species under both laboratory and field exposure conditions. Intracellular calcium-phosphorous concretions in kidneys of marine bivalves and the hepatopancreas of crustaceans have also been shown to be primarily responsible for the high levels of several

potentially toxic trace metals found in these organs. Recent
studies in oysters have indicated that the accumulation of
cadmium by gills can be altered by concomitant exposure of
oysters to cadmium and selenium, although total cadmium
accumulation by this marine species is not changed by the
elemental interaction. This suggests that mechanisms of
metal uptake and storage differ within different organ
systems. A major point to be derived from these studies is
that marine species appear to possess a variety of mechanisms
for metabolizing and accumulating toxic metals from the
environment. The significance of this information rests with
an understanding of how these mechanisms are related to
survival of the organisms and how they might influence the
toxicity of metals to humans when the marine organisms are
consumed as food.

ACKNOWLEDGMENT

     This work was supported in part by the U.S. Environmental
Protection Agency through an interagency agreement related to
the Federal Interagency Energy/Environmental Research and
Development Program. S.E.F.C. Contribution No. 80-53B.

LITERATURE CITED

Becker, G. L., Chen, C-H., Greenawalt, J. W. and Lehninger,
     A. L. 1974. Caclium phosphate granules in the hepato-
     pancreas of the blue crab *Calinectes sapidus*. J. Cell.
     Biol. 61: 316-326.

Brown, B. E. 1977. Uptake of copper and lead by a metal-
     tolerant isopod *Asellus meridianus*. Freshwater Biol.
     7: 235-244.

Brown, B. E. 1978. Lead detoxification by a copper tolerant
     isopod. Nature 276: 388-440.

Brun A. and Brunk, U. 1970. Histochemical indications for
     lysosomal localization of heavy metals in normal rat
     brain and liver. J. Histochem. Cytochem. 18: 820-827.

Carmichael, N. G., Squibb, K. S. and Fowler, B. A. 1979a. Metals in the molluscan kidney: A comparison of two closely related bivalve species (Argopecten), using X-ray microanalysis and atomic absorption spectroscopy. J. Fish. Res. Bd. Canada. 36: 1149-1155.

Carmichael, N. G., Squibb, K. S., Engel, D. W. and Fowler, B. A. 1979b. Metals in the molluscan kidney: Uptake and subcellular distribution of ^{109}Cd, ^{54}Mn and ^{65}Zn by the clam, Mercenaria mercenaria. Comp. Biochem. Physiol. 65A: 203-206.

Casterline, Jr., J. L. and Yip, G. 1975. The distribution and binding of cadmium in oyster, soybean, and rat liver kidney. Arch. Environ. Contam. Toxicol. 3: 319-329.

Cherian, M. G. 1979. Metabolism of orally administered cadmium-metallothionein in mice. Environ. Hlth. Perspec. 28: 127-130.

Doyle, L. J., Blake, N. J., Woo, C. C. and Yevich, P. 1978. Recent biogenic phosphorite: Concretions in mollusc kidneys. Science 199: 1431-1433.

Engel, D. W. and Fowler, B. A. 1979a. Factors influencing cadmium accumulation and toxicity to marine organisms. Environ. Hlth. Perspec. 28: 81-88.

Engel, D. W. and Fowler, B. A. 1979b. Copper and cadmium induced changes in the metabolism and structure of molluscan gill tissue. In: Marine Pollution: Functional Responses. W. B. Vernberg, F. P. Thurberg, A. Calabrese and F. J. Vernberg (eds). Academic Press, New York, pp 239-255.

Engel, D. W., Sunda, W. G. and Fowler, B. A. In Press. Factors affecting trace metal uptake and toxicity to estuarine organisms I. Environmental parameters. In: Biological Monitoring of Marine Pollutants. F. J. Vernberg, A. Calabrese, F. P. Thurberg and W. B. Vernberg (eds). Academic Press, New York.

Fowler, B. A., Brown, H. W., Lucier, G. W. and Beard, M. E. 1974. Mercury uptake by renal lysosomes of rats ingesting methyl mercury hydroxide: Ultrastructural observations and energy dispersing X-ray microanalysis Arch. Path. 98: 297-301.

Fowler, B. A. and Nordberg, G. F. 1978. The renal toxicity
    of cadmium metallothionein: Morphometric and X-ray
    microanalytical studies. Toxicol. Appl. Pharmacol.
    46: 609-624.

Fowler, B. A., Kimmel, C. A., Woods, J. S., McConnell, E. E.,
    and Grant, L. D. In Press. Chronic low level lead tox-
    icity in the rat. III. An integrated toxicological
    assessment with special reference to the kidney. Toxicol.
    Appl. Pharmacol.

Fowler, B. A., Wolfe, D. A. and Hettler, W. F. 1975. Mercury
    and iron uptake by cytosomes in mantle epithelial cells
    of quahog clams (*Mercenaria mercenaria*) exposed to
    mercury. J. Fish. Res. Bd. Canada. 32: 1767-1775.

Fowler, S. W. and Benayoun, 1976. Influence of environmental
    factors on selenium flux in the marine invertebrates.
    Marine Biol. 37: 59-68.

Frazier, J. M. 1976. The dynamics of metals in the American
    oyster *Crassostrea virginica*. II. Environmental
    effects. Chesapeake Sci. 17: 188-197.

Frazier, J. M. 1979. Bioaccumulation of cadmium in marine
    organisms. Environ. Hlth. Perspec. 28: 75-79.

Gasiewicz, T. A. and Smith, J. C. 1978. Interaction between
    cadmium and selenium in rat plasma. Environ. Hlth.
    Perspect. 25: 133-136.

George, S. G., Pirie, B. J. S. and Coombs, T. L. 1975.
    Absorption, accumulation and excretion of iron-protein
    complexes by *Mytilus edulis*. Internat. Conf. Heavy
    Metals in the Environment, Toronto, Ontairo Canada,
    October 27-31, 1975. pp 887-900.

Ghiretti, F., Salvato, B., Carlucci, S. and DePieri, R.
    1971. Manganese in *Pinna nobilis*. Experientia 28:
    232-233.

Goldfischer, S. and Moskal, J. 1966. Electron probe micro-
    analysis of liver in Wilson's disease: Simultaneous
    assay for copper and for lead deposited by acid phos-
    phatase activity in lysosomes. Amer. J. Pathol. 48:
    305-316.

Howard, A. G. and Nickless, G. 1977a. Heavy metal complexation in polluted molluscs I. Limpets (*Patella vulgata* and *Patella intermedia*). Chem. Biol. Interact. 16: 107-114.

Howard, A. G. and Nickless, G. 1977b. Heavy metal complexation in polluted molluscs II. Oysters *(Ostrea edulis* and *Crassostrea gigas)*. Chem. Biol. Interact. 17: 257-263.

Howard, A. G. and Nickless, G. 1978. Heavy metal complexation in polluted molluscs III. Periwinkles *(Littorina littorea)*, cockles *(Cardium edule)* and scallops *(Chlamys opercularis)*. Chem. Biol. Interact. 23: 227-231.

Jennings, J. R., Rainbow, P. S. and Scott, A. G. 1979. Studies on the uptake of cadmium by the crab *Carcinus maenas* in the laboratory. II. Preliminary investigation of cadmium-binding proteins. Marine Biol. 50: 141-149.

Ketchum, B. H. 1975. Biological implications of global marine pollution. In: The changing global environment. S. F. Singer (ed.). D. Reidel Co., Dordrecht, Holland, pp 311-328.

Lowe, D. M. and Moore, M. N. In Press. The cytochemical distributions of zinc (Zn II) and iron (Fe III) in the common mussel, *Mytilus edulis,* and their relationship with lysosomes. J. Mar. Biol. Assoc. U.K.

Marshall, A. T. and Talbot, V. 1979. Accumulation of cadmium and lead in the gills of *Mytilus edulis:* X-ray microanalysis and chemical analysis. Chem. Biol. Interact. 27: 111-123.

Nöel-Lambot, F. 1976. Distribution of cadmium, zinc and copper in the mussel *Mytilus edulis*. Existence of cadmium-binding proteins similar to metallothioneins. Experientia 32: 324-326.

Nöel-Lambot, F., Gerday, C. and Disteche, A. 1978. Distribution of Cd, Zn, and Cu in liver and gills of the eel, *Anguilla anguilla* with special reference to metallothioneins. Comp. Biochem. Physiol. 61C: 177-187.

Olafson, R. W., Kearns, A. and Sim, R. G. 1979a. Heavy metal induction of metallothionein synthesis in the hepato-pancreas of the crab. *Scylla serrata.* Comp. Biochem. Physiol. 62B: 417-424.

Olafson, R. W., Sim, R. G. and Boto, K. G. 1979b. Isolation and chemical characterization of the heavy metal-binding protein metallothionein from marine invertebrates. Comp. Biochem. Physiol. 62B: 407-416.

Ridlington, J. W. and Fowler, B. A. 1977. Isolation and partial characterization of an inducible cadmium binding protein from American oyster. Pharmacologist 19: 780. Abstract.

Ridlington, J. W. and Fowler, B. A. 1979. Isolation and partial characterization of a cadmium-binding protein from the American oyster (*Crassostrea virginica*). Chem. Biol. Interact. 25: 127-138.

Roesijadi, G. 1980. The influence of copper on the clam *Protothaca staminea:* Effects on gills and occurrence of copper-binding proteins. (Submitted).

Snedecor, G. W. and Cochran, W. G. 1967. Statistical Methods. Iowa State Press, Ames, Iowa. pp. 339-380.

Schulz-Baldes, M. 1977. Lead transport in the common mussel, *Mytilus edulis.* In: D. S. McLusky and A. J. Berry (eds)., Physiology and behavior of marine organisms. Pergamon Press, New York. pp 211-218.

Spivey-Fox, M. R., Jacobs, R. M., Lee Jones, A. O., and Fry, B. E. 1979. Effects of nutritional factors on the metabolism of dietary cadmium at levels similar to those of man. Environ. Hlth. Perspec. 28: 107-114.

Squibb, K. S., Ridlington, J. W., Carmichael, N. G. and Fowler, B. A. 1979. Early cellular effects of circulating cadmium-thionein on kidney proximal tubules. Environ. Health Perspect. 28: 287-296.

Stuve, J. and Galle, P. 1970. Role of mitochondria in handling of gold by the kidney. A study by electron microscopy and electron-probe microanalysis. J. Cell. Biol. 44: 667-676.

Talbot, V. and Magee, R. J. 1978. Naturally-occurring heavy metal-binding proteins in invertebrates. Arch. Environ. Contam. Toxicol. 7: 73-81.

Walker, G. 1977. "Copper" granules in the barnacle *Balanus balanoides*. Marine Biol. 39: 343-349.

Washko, P. W. and Cousins, R. J. 1977. Role of dietary calcium and calcium binding protein in cadmium toxicity in rats. J. Nutr. 107: 920-928.

Zaroogian, G. E. and Cheer, S. 1976. Accumulation of cadmium by the American oyster, *Crassostrea virginica*. Nature 261: 408-409.

THE EFFECTS OF COPPER AND CADMIUM ON THE BEHAVIOR
AND DEVELOPMENT OF BARNACLE LARVAE

W. H. Lang
D. C. Miller
P. J. Ritacco
M. Marcy

Environmental Protection Agency
Environmental Research Laboratory
South Ferry Road
Narragansett, RI

INTRODUCTION

     Chemicals contribute significantly to American living
standards, however, it is apparent that some chemicals are a
threat to both human and environmental health. Increasing
public awareness of environmental issues coupled with major
disasters involving toxic chemicals (Heritage, 1978) have
highlighted the need to identify and regulate toxic substances. The Toxic Substance Control Act of 1976 is the first
law to attempt comprehensive regulation of the chemical
industry. Implementation is not simple, however, with many
problems existing in obtaining and assessing toxicological
information (Walsh, 1978).
     Because of the sheer number of existing chemicals and
rate of production of new products, comprehensive testing of
all substances is impossible. "Tiered" or sequential toxicity
testing methods are often favored (Duthie, 1977; Walsh, 1978)
in which relatively quick and inexpensive "first level
screening tests" are used in an effort to rapidly rank
chemicals as potential hazards (Draggan and Giddings, 1978).
Those chemicals proving to be especially dangerous because of
their high level of toxicity and/or large volumes produced
will then be subjected to increasingly rigorous testing and
evaluation.

For studies of aquatic environmental effects, complete life-cycle or multiple generation chronic toxicity tests, experimental ecosystems, and field studies undoubtedly provide a more satisfactory toxicity assessment, yet time and cost factors limit their use. Short-term acute toxicity tests still represent the starting point for essentially all toxicological studies (Draggan and Giddings, 1978). Environmental realism is sacrificed, with the main goals being to determine, rapidly and at realistic cost, relative toxicity and to provide an initial "range finding step" for possible sequential testing (Duthie, 1977). Historically, the end point for acute toxicity tests of aquatic organisms has been 50% mortality using, more often than not, the adult life stage.

A need exists to increase the sensitivity and scope of information derived from initial screening tests. Two approaches to these goals under study by our group have been to conduct exposure tests on "critical life stages" (Macek and Sleight, 1977) and to incorporate multiple sublethal stress indices (Sprague, 1971; Davis, 1976; Waldichuk, 1979).

The objective of this study is to examine possible early sublethal indicators of stress using invertebrate larval stages. Nauplii of the common estuarine barnacle *Balanus improvisus* were exposed for 48 to 96 hrs to two well documented toxicants of the aquatic environment, copper and cadmium. Five stress indices were monitored at 24 hr intervals: mortality, rate of development, normality of molting, spontaneous swimming speeds, and photobehavior. Of particular interest in this study is the evaluation of behavioral responses as stress indices. The use of a newly developed video-computer analysis system (Greaves and Wilson, 1980) now allows a relatively rapid and accurate assessment of multiple movement parameters.

METHODS AND MATERIALS

Experimental Animals

Adult *Balanus improvisus* were collected with their rock substrate from the Pettaquamscutt River estuary, Narragansett, Rhode Island, and held at 15°C in filtered 15°/oo natural seawater under constant illumination. Adults were fed newly hatched *Artemia salina* (Aquafauna Brand[1], Macau, Brazil) three times weekly.

[1]*Mention of commercial products does not constitute or imply endorsement by the Environmental Protection Agency.*

Newly released stage I-II barnacle nauplii were collected from adult culture dishes by concentrating the larvae with a light beam and removing them by pipette into 0.45 μm filtered seawater at 15 or 30 °/$_{oo}$. For all experiments, larvae were obtained from adults maintained in the laboratory 4-6 weeks. Larvae used for the 96-hr Cu experiment were released during November, 1978; larvae used for the 48-hr Cu and 96-hr Cd experiments were released June-July, 1979.

Bioassay Procedures

Stage II nauplii were exposed to Cu or Cd under static conditions at constant temperature (15°C) and 15 or 30 °/$_{oo}$ salinity. Metal solutions were renewed at 24-hr intervals.

Nauplii were fed a mixed algal diet of *Tetraselmis suecia* and *Skeletonema costatum* grown axenically in a modified f/2 medium (Guillard and Ryther, 1962). Trace metals used in the culture medium included only $FeCl_2$, $ZnSO_4$, $MnCl_2$, with Na·EDTA. Seawater (30°/$_{oo}$) for the algal medium was obtained from the Pettaquamscutt River.

The algae were mixed to approximately equal volumes, filtered through a 44 μm Nitex screen, and diluted with 30 or 15 °/$_{oo}$ seawater to a density of 2-4 x 10^4 cells/ml to feed the larvae.

For metal exposures, newly released nauplii were transferred into 7 cm Carolina culture dishes (40 nauplii/ dish) each containing a 60 ml mixture of seawater, algae and metal toxicant. Nauplii were maintained under constant illumination during the 96 hr test period. At 24 hr intervals dead nauplii were counted and removed while the live nauplii were transferred to chambers for video-taping. Following video-taping, nauplii were returned to fresh exposure medium.

After the final video-taping (48 or 96 hr, depending on the experiment), surviving nauplii were preserved with formalin in seawater. The percentage of stage II and III nauplii was determined and stage III nauplii were examined for abnormal external morphology.

Primary stock solutions of 10,000 ppm Cu in dilute nitric acid and 100 ppm Cd ($CdCl_2$) in deionized water were diluted to secondary working solutions with deionized water (1-4 ppm Cu, 10-100 ppm Cd). Final test solutions were obtained by serial dilution with filtered seawater at 15 or 30°/$_{oo}$.

Total Cu or Cd was analyzed at each 24 hr interval using a Perkin Elmer model 360 atomic absorption spectrophotometer equipped with a HGA 2100 heated graphite furnace. Absorption of samples was used to quantify Cu or Cd as measured against

standards in a similar matrix. No attempt was made to
determine partitioning of the test metals within the exposure
system.

Behavioral Analysis Procedures

     Beginning with the 24 hr exposure interval, nauplii were
transferred to 3.2 x 3.2 cm lucite chambers for video-taping
of swimming and photobehavior. Taping was conducted at 1300-
1500 each day in a 20° temperature controlled darkened room.
Each sample was placed on a Wild dark-field base with substage
illumination filtered to 830 nm (15 nm halfwidth). A Cohu
4400 television camera mounted above the chamber recorded
larval movement in a horizontal plane (Lang *et al.*, 1979a, b).
     Chambers were held in a 15°C temperature box adjacent to
the camera and microscope. Between actual taping, a container
of ice water was used to cool the microscope stage. The
temperature of chamber water increased by 0.4 to 1.0°C during
taping. A horizontal light stimulus provided by a 100 W, 12 V
high intensity tungsten light source (Oriel corporation),
fitted with a 480 nm bandpass filter (7 nm halfwidth), was
used to test the photoresponse of each test group. Light
intensity was regulated with neutral density filters and
stimulus duration was controlled by an electromagnetic
shutter. Each sample of nauplii was tested for photoresponse
at a high intensity ($4 W/m^2$) and low intensity ($0.04 W/m^2$).
Previous studies have shown light-adapted stage II nauplii are
most responsive to 480 nm light and that the intensities
tested normally invoke a strong positive phototaxis at the low
intensities and marked positive photokinesis at the high
intensity (Lang *et al.*, 1979a, 1980).
     Each replicate sample of up to 40 nauplii was placed on
the microscope stage and room lights extinguished. Following
a 15 to 30 sec delay, video-taping was initiated. Nauplii
were taped for approximately 10 sec with substage (830 nm)
illumination only, followed by a 6-6.5 sec stimulus at low
intensity, a 15-30 sec interval without the stimulus, and
finally a 6-6.5 sec stimulus at high intensity.
     Video tapes of naupliar swimming behavior were played
back through a video-to-digital processor (Greaves, 1975;
Wilson and Greaves, 1979; Greaves and Wilson, 1980) for
computer analysis of mean linear velocity (MLV) and direction
of travel (DOT) of nauplii within the camera's field of view.
Analysis of variance was used to test for significant
differences in linear velocity and the chi-square test was
used as a measure of randomness for directional distributions.
Details of computer analysis and statistics are described in
Lang *et al.* (1979a).

Three behavioral parameters were investigated for the possible effects of exposure to Cu or Cd: MLV of control nauplii (those exposed only to 830 nm light), the phototactic (DOT) response to a 480 nm light stimulus and, the photo-kinetic (change in MLV) response of nauplii to 480 nm light. Figure 1 illustrates computer output quantifying these parameters.

Control MLV was determined for a 6 sec interval while nauplii were exposed to 830 nm illumination. The MLV of nauplii during the 480 nm light stimulus was calculated and compared to the control MLV to determine if a significant change ($P \leq .05$) in MLV (photokinesis) occurred. The mean DOT of each nauplius was calculated during a 2.5 sec segment of the 6 sec 480 nm light stimulus (Fig. 1). The 2.5 sec interval chosen allows for a 1 sec delay following stimulation for orientation of the nauplii (see Lang et al., 1979a) yet restricts our phototactic determination to a brief "immediate" response to light. Positive phototaxis is defined as a mean DOT $\pm$ 30° toward the light stimulus; negative phototaxis as $\pm$ 30° in the opposite direction (Fig. 1).

RESULTS

Three experiments were conducted. Nauplii were exposed to Cu (20-160 ppb) and to Cd (50-200 ppb) for 96 hrs exposure at constant temperature (15°C) and two salinities 15, 30°/oo. In addition, nauplii were exposed to Cu (10-80 ppb) as fed (with algae) and starved (without algae) groups for 48 hr periods at 15°C and 30 °/oo. Since the control responses of the nauplii in the three bioassays varied from each other, the experiments are considered separately.

Mortality

For the 96 hr Cu assay, a nearly uniform mortality of about 20% occurred within 24 hrs for all control and Cu-exposed replicate cultures. This mortality appears to be inherent of the larval hatch used, independent of salinity or Cu concentration. Allowing for this initial loss of larvae, significant mortality (about 40%) occurred only at 160 ppb Cu (30 °/oo) between 72 and 96 hrs (Table 1). For the 48 hr Cu assay (Table 2) there was no initial mortality. Exposure to Cu from 10-70 ppb produced no significantly increased mortality in either fed or starved nauplii.

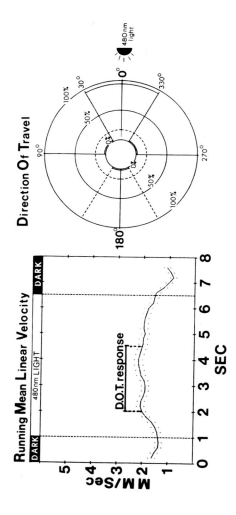

FIGURE 1. Computer determined behavioral parameters. The linear velocity of each nauplius is initially calculated at 0.1 sec intervals. A group mean for each experimental condition is then calculated (Running Mean Linear Velocity). The group mean linear velocities (MLV) during the 6.5 sec light stimulus are presented in this study (Tables 4-8). Direction of travel (DOT) is determined for a 2.5 sec interval as shown in the figure. The mean DOT for each nauplius in the sample group is determined and a percent distribution calculated using 60° directional intervals. Only a summary of percent positive (0° ± 30°) and negative (180° ± 30°) phototaxis is presented in this study (Fig. 5-7).

Cu – ppb Desired	Measured	N	Mortality $\bar{x}$ 96 hr	sd	%Stage III $\bar{x}$ 96hr	sd
30‰ cont.	≤2	120	25	7.6	95	1.2
20	18.5	120	33	8.7	92	5.0
40	37.0	120	29	12.5	87	4.0
80	72.0	119	25	2.9	75	6.1
160	138.5	114	64	20.0	0	
15‰ cont.	≤2	117	30	7.2	89	5.1
20	20.8	120	23	3.5	86	2.1
40	37.1	117	21	5.7	91	2.0
80	74.2	119	31	17.1	84	12.5

Table 1. Mortality (%) and percent development to stage III of Balanus improvisus nauplii exposed to copper for 96 hrs. Total sample number tested (N) is given. The mean ($\bar{x}$) and standard deviation (sd) for mortality and % stage III are based on three replicate cultures at each experimental condition.

For the 96 hr Cd assay (Table 3), exposure to 200 ppb Cd at 15 °/₀₀ salinity resulted in 92% mortality while at 30 °/₀₀ salinity only 18% of the nauplii died. At concentrations of 50- and 100 ppb Cd, no significant mortality occurred at either salinity (Table 3).

Development

Balanus improvisus has a larval development of six naupliar stages and a cyprid (Lang, 1979). Non-feeding stage I nauplii hatch from egg masses incubated within the adult mantle cavity and are then released into the water column. Almost immediately, stage I nauplii molt to stage II planktotrophic larvae. Development to stage III is dependent on adequate diet, and at 15°C normally occurs from 2 to 4 days.

Although adult acclimation and larval culture conditions of temperature, diet and salinity were similar in all cases, differences in developmental rates for controls of each experimental group were observed. Control nauplii for the 96 Cu assay (Table 1) exhibited 89 and 95% development to stage III at 15 and 30°/₀₀ salinity, respectively. Daily

Cu ppb Desired Measured		N	Mortality $\bar{x}$ 48 hr sd		%Stage III $\bar{x}$ 48hr sd	
UNFED	cont.	≤3	78	1.3	1.8		
	10	9.0	79	0	0		
	20	19.1	79	1.3	1.8		
	40	36.0	80	0	0		
	80	68.0	80	0	0		
FED	cont.	≤ 3	75	0	0	65	24.7
	10	9.6	73	2.9	4.2	57	2.9
	20	20.0	67	1.2	1.8	48	6.1
	40	39.0	80	1.2	1.8	28	9.9
	80	69.5	78	2.6	3.7	5	0.3

TABLE 2. Mortality (%) and percent development to stage III of Balanus improvisus *nauplii exposed to copper for 48 hrs. Results are based on two replicate cultures at each experimental condition. No development occurred with unfed larvae.*

qualitative observations indicated most molting occurred between 48 to 72 hrs for nauplii at 30°/oo while relatively more nauplii were molting by 48 hrs at 15 °/oo. Development for the control fed nauplii of the 48 hr Cu assay (Table 2) was evidently faster. At 48 hr from 47 to 82% ($\bar{x}$ ≈ 65%) of the control replicates were stage III. As expected, starved nauplii did not molt beyond stage II.

Nauplii used for the 96 hr Cd assay (Table 3) exhibited a difference in molting rate at the two salinities. At 15 °/oo, results were similar to the 96 hr Cu assay, with 97% of the control nauplii attaining stage III. However, at 30°/oo development to stage III was only 59%.

Both Cu and Cd either delayed development to stage III or totally suppressed molting within the 96 hr exposure interval. At the highest Cu concentration tested (160 ppb, 30 °/oo) nauplii failed to molt to stage III while at 80 ppb the percent development was slightly depressed at 30 °/oo (Table 2).

The above results infer that Cu can delay or suppress the molting process. When molting does occur, Cu-exposure appears to have an additional effect. Virtually all nauplii (>96%) which were exposed to Cu and successfully developed to stage III exhibited morphological abnormalities (Fig. 2). Twisted, short setae and minor deformities of appendages were common at concentrations of 10 ppb (Fig. 2c). With increasing Cu concentrations effects were more pronounced as most setation was absent and appendages were grossly deformed (Fig. 2d).

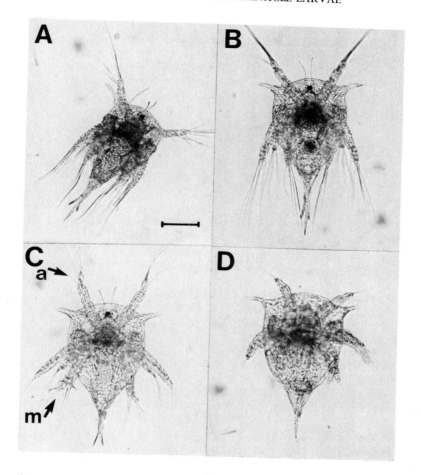

FIGURE 2. Effects of Cu exposure on barnacle naupliar molting: (A) Stage II Balanus improvisus nauplius; (B) normal stage III nauplius; (C) stage III nauplius exposed to 10 ppb Cu, with short, twisted setae on antenna (a) and deformed mandibular endopodite (m): (D) Stage III nauplius exposed to 80 ppb Cu possessing few setae and with grossly deformed appendages. Scale bar = 0.1 mm.

Cd – ppb Desired Measured		N	Mortality $\bar{x}$ 96hr sd		% Stage III $\bar{x}$ 96hr sd	
15‰ cont.	<.2	113	0.8	1.4	97	1.7
50	49.0	115	7.8	0.3	94	0
100	100.5	120	1.7	1.4	31	2.1
200	194.8	118	92.0	8.6	0	0
30‰ cont.	<.2	113	6.2	1.5	59	10.4
50	49.8	107	3.8	2.0	59	10.1
100	99.5	120	2.4	2.4	22	12.1
200	201.8	119	18.0	8.4	0	0

TABLE 3. Mortality (%) and percent development to stage III of Balanus improvisus *nauplii exposed to cadmium for 96 hrs. See Table 1 for further explanation.*

Exposure to Cd halted development to stage III at 200 ppb and reduced percent-molting to less than 50% of control rates at 100 ppb (Table 3). No effect on rate of development was seen at 50 ppb Cd. Those nauplii attaining stage III exhibited normal morphology at all Cd exposure concentrations.

Spontaneous Swimming Speeds

The MLV of 24 hr old control nauplii swimming under "dark" conditions (830 nm illumination) ranged from 1.26-1.51 mm/sec (Tables 4-8). MLV's of about 1.5 mm/sec were recorded for stage II nauplii captured from plankton samples (Lang *et al.*, 1979a). Previous results (Lang *et al.*, 1980, in press) indicate 24 hr old nauplii may swim faster at higher salinities. In contrast, present results indicate no significant difference in MLV between 24 hr controls at 15 $^\circ$/$_\circ\circ$ and 30 $^\circ$/$_\circ\circ$ in the 96 hr Cu assay (Tables 4, 5) and significantly ($P \leq 0.05$) lower 24 hr MLV at 30 $^\circ$/$_\circ\circ$ in the 96 hr Cd assay nauplii (Table 7) relative to MLV at 15 $^\circ$/$_\circ\circ$ (Table 8). Starvation significantly ($p \leq 0.05$) reduced the MLV of 24 hr nauplii relative to fed groups (Table 6), a result also observed in previous studies (Lang *et al.*, 1980).

Stage III nauplii generally swim faster than stage II nauplii (Lang *et al.*, 1979a). An increase in MLV would be expected at 96 hrs relative to 24 hrs as the test population shifts from predominantly stage II to predominantly stage III. However, nauplii in the process of molting essentially lose mobility and MLV may decline during periods of intense molting activity. The spontaneous MLV of control nauplii was greater at 96 hr relative to 24 hr in all cases (Figs. 3, 4), but the increase was only significant ($P \leq 0.05$) in the 96 hr Cd assay (Fig. 4).

TABLE 4. Mean linear velocity of Balanus improvisus nauplii for 96 hr Cu exposure at 30 °/oo, 15°C during 6.5 sec "dark" (840 nm light) and 480 nm light stimuli intervals for two light intensities. MLV value is based on three replicate tests; sample size range indicates number tested per stimulus condition. Plus (+) indicates a significant (P ≤ 0.05) increase in light-induced MLV relative to the "dark" MLV for each treatment group.

| | EXPOSURE LEVEL (ppb Cu) | SAMPLE RANGE | STIMULUS RESPONSE (mm/sec) | | |
			"dark"	$0.02 \ W/m^2$	$2.0 \ W/m^2$
24 HOURS	Control	36–51	1.4	2.0 +	2.5 +
	20	29–35	1.6	2.8 +	2.6 +
	40	27–37	1.8	2.9 +	3.1 +
	80	28–35	1.6	2.5 +	2.6 +
	160	23–29	1.2	1.2	1.3
48 HOURS	Control	33–41	1.7	2.5 +	2.5 +
	20	18–27	1.5	2.8 +	2.8 +
	40	25–35	2.4	2.6	2.9
	80	20–24	2.2	2.1	2.2
	160	22–27	0.5	0.9 +	0.6
72 HOURS	Control	29–39	1.9	2.6 +	2.4 +
	20	18–31	1.2	1.7 +	1.5 +
	40	21–28	1.0	2.0 +	1.9 +
	80	30–35	1.2	1.6	1.4
	160	15–21	0.4	0.3	0.4
96 HOURS	Control	31–35	1.7	2.4 +	2.5 +
	20	16–24	0.6	0.8	1.1
	40	25–29	0.8	0.7	0.9
	80	25–32	0.4	0.4	0.4
	160	11–13	0.1	0.1	0.05

TABLE 5.   Mean linear velocity of Balanus improvisus
nauplii for 96 hr Cu exposure at 15°/∘∘, 15°C.   See also
Table 4.

	EXPOSURE LEVEL (ppb Cu)	SAMPLE RANGE	STIMULUS RESPONSE (mm/sec)		
			"dark"	0.02 W/m^2	2.0 W/m^2
24 HOURS	Control	20-29	1.4	1.9 +	2.2 +
	20	28-39	1.7	2.3 +	3.1 +
	40	34-43	1.9	2.8 +	4.0 +
	80	25-30	1.7	3.0 +	2.8 +
48 HOURS	Control	29-34	1.6	1.8	1.8
	20	21-26	1.3	1.7 +	1.8 +
	40	23-30	1.7	1.9	2.1
	80	25-36	1.9	2.6	3.0 +
72 HOURS	Control	24-31	1.7	2.4 +	2.3 +
	20	19-21	1.1	1.0	0.7
	40	21-28	0.9	1.9	1.0
	80	32-35	1.2	1.5	1.4
96 HOURS	Control	18-25	1.8	2.5	3.0
	20	23-38	0.8	0.9	0.8
	40	31-47	0.6	0.7	0.6
	80	32-41	0.2	0.4	0.3

TABLE 6. Mean linear velocity of Balanus improvisus nauplii for 48 hr Cu exposure at 30 °/oo, 15°C. See also Table 4.

| | EXPOSURE LEVEL (ppb Cu) | SAMPLE RANGE | STIMULUS RESPONSE (mm/sec) | | |
			"dark"	0.02 W/m^2	2.0 W/m^2
24 HR - FED	Control	31-41	1.5	1.9 +	2.1 +
	10	29-43	1.8	2.0	2.2
	20	30-46	1.6	1.8	2.6 +
	40	22-26	1.7	2.2 +	2.2 +
	80	20-26	1.7	1.9	2.1
48 HR - FED	Control	24-34	1.4	1.6	2.3 +
	10	21-31	1.0	1.1	1.4 +
	20	18-26	1.2	1.2	1.6
	40	27-35	1.1	1.3	1.5
	80	26-33	0.9	1.3 +	1.3 +
24 HR-STARVED	Control	25-29	1.2	1.3	1.5
	10	33-42	1.6	1.5	1.5
	20	40-41	1.3	1.2	1.5 +
	40	31-41	1.1	1.2	1.4 +
	80	35-46	0.9	1.1	1.2
48 HR.-STARVED	Control	32-39	1.4	1.2	1.2
	10	32-35	1.3	1.3	1.4
	20	30-38	1.0	1.1	1.1
	40	32-39	0.9	0.8	0.9
	80	36-40	0.4	0.4	0.4

TABLE 7. Mean linear velocity of Balanus improvisus nauplii for 96 hr Cd exposure at 30 °/oo, 15°C. See also Table 4.

	EXPOSURE LEVEL (ppb Cd)	SAMPLE RANGE	STIMULUS RESPONSE (mm/sec)		
			"dark"	0.02 W/m^2	2.0 W/m^2
24 HOURS	Control	46-51	1.3	1.3	1.4 +
	50	34-58	1.6	1.7	1.5
	100	50-55	1.5	1.6	1.5
	200	60-72	1.0	1.0	1.0
48 HOURS	Control	57-68	1.7	1.9	1.4
	50	51-67	1.5	1.7	1.6
	100	51-68	1.4	1.6 +	1.4
	200	51-53	1.2	1.1	1.0
72 HOURS	Control	50-54	1.7	1.7	1.7
	50	52-57	2.0	2.0	1.9
	100	45-59	1.6	1.7	1.6
	200	31-37	0.8	0.8	0.8
96 HOURS	Control	44-49	2.0	1.9	1.6
	50	44-60	1.8	1.9	1.8
	100	40-56	1.5	1.7	1.7
	200	13-15	0.7	0.8	0.9

TABLE 8. *Mean linear velocity of* Balanus improvisus *nauplii for 96 hr Cd exposure at 15 °/oo, 15°C. See also Table 4.*

	EXPOSURE LEVEL (ppb Cd)	SAMPLE RANGE	STIMULUS RESPONSE (mm/sec)		
			"dark"	0.02 W/m²	2.0 W/m²
24 HOURS	Control	42-56	1.5	1.7	1.7
	50	50-63	1.5	1.7 +	1.9 +
	100	49-63	1.2	1.4 +	1.6 +
	200	59-62	0.8	0.9	1.1 +
48 HOURS	Control	56-61	1.4	1.5	1.5
	50	45-63	1.1	1.3	1.3
	100	58-76	1.5	1.7 +	1.4
	200	19-28	0.6	0.7	0.7
72 HOURS	Control	47-65	2.2	2.4	2.0
	50	48-57	1.8	2.1 +	1.8
	100	51-62	1.8	2.0 +	1.6
	200	-	-	-	-
96 HOURS	Control	53-62	2.1	2.5 +	2.1
	50	46-52	2.3	2.4	1.9
	100	57-68	1.6	1.8	1.6
	200	-	-	-	-

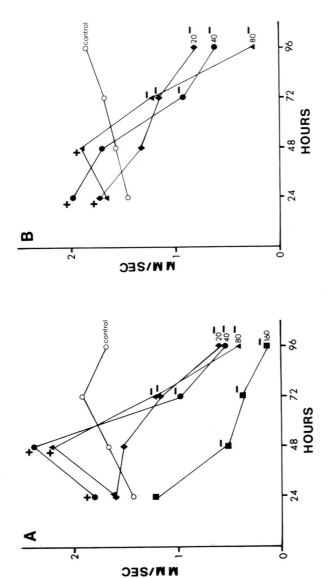

FIGURE 3. Spontaneous mean linear velocity of barnacle nauplii exposed to copper at 30 °/∘∘ (A) and 15 °/∘∘ (B). Exposure levels (ppb Cu) are indicated. A significant ($p \leq 0.05$) increase (+) or decrease (-) in velocity relative to control is noted for each interval (see Tables 4-5 for sample numbers).

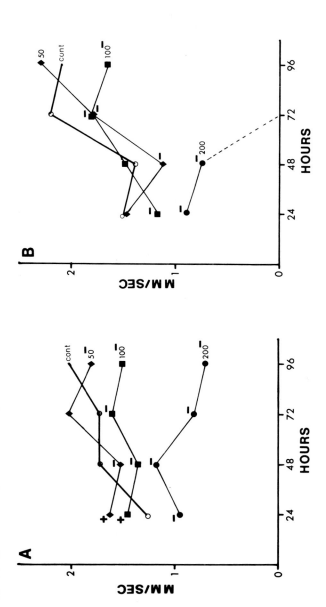

FIGURE 4. Spontaneous mean linear velocity of barnacle nauplii exposed to cadmium at 30 °/oo (A) and 15 °/oo (B). Exposure levels (ppb Cd) are indicated. A significant (p ≤ 0.05) increase (+) or (-) in velocity relative to control is noted for each time interval (see Tables 7-8 for sample numbers).

   Exposure to Cu may initially have a stimulatory effect
on naupliar spontaneous MLV (Lang *et al.*, 1979b), although
this response is not always evident (Lang *et al.*, in press).
In the 96 hr Cu assay, there was an initial (24 hr) stimulatory effect at Cu concentrations from 20 -80 ppb (Fig. 3,
Tables 4,5); by 48 hrs, this effect disappears at 20 ppb Cu.
Also, by 48 hrs hyperactivity is less pronounced at 15 °/oo
than at 30 °/oo. At 72 hrs all nauplii exposed to Cu have
MLV's significantly slower than controls. This probably
reflects both a direct slowing of stage II nauplii and also a
loss of swimming ability in deformed stage III nauplii.
   Although Cu-exposed nauplii initially tended to swim
faster in the 96 hr assay (Fig. 3, Tables 4, 5), no significant (P $\leq$ 0.05) hyperactivity occurred at 24 hrs in fed
nauplii for the 48 hr Cu assay (Table 6). Starved nauplii
exhibited significant hyperactivity at 10 ppb Cu. Starvation
depressed the 24 hr control MLV and at a given exposure level,
starved nauplii had reduced MLV's relative to fed nauplii
(Table 6). At 48 hrs, MLV of starved nauplii was significantly reduced at $\geq$ 20 ppb Cu, indicating a depressing
effect of Cu independent of molting.
   Cadmium induced hyperactivity initially (24 hr) at 50
to 100 ppb exposure for nauplii at 30 °/oo (Fig. 4a), but at
15 °/oo MLV was significantly depressed at 100 ppb Cd (Fig.
4b). Exposure to 200 ppb Cd depressed MLV at the onset for
both salinities and surviving nauplii at 15 °/oo were virtually immobile by 72 hrs. From 48 to 96 hrs, response of
both groups to 50 ppb Cd varied from no significant
difference to depressed spontaneous MLV while exposure to
100 ppb Cd induced a consistently depressed MLV (Fig. 4).

Photobehavior

   The immediate response of light-adapted stage II nauplii
to a 480 nm light stimulus has been characterized (Lang *et
al.*, 1979a, 1980). Laboratory-hatched stage II nauplii
(24 hr old) exhibit a strong positive phototaxis at 480 nm
light intensities from about 10^{-3} W/m^2 to at least 1 W/m^2 and
positive photokinesis occurs from about 10^{-1} W/m^2 to at
least 4.0 W/m^2 (Lang *et al.*, 1980). The percent response
for positive phototaxis tends to decrease at the higher light
intensities with maximal response near 10^{-2} W/m^2. At 24 hrs,
salinity had no consistent effect on photobehavior.
   Cu-control nauplii at 30°/oo showed a consistent
positive phototaxis and photokinesis throughout the 96 hr
(Fig. 5a, Tables 4,5) and 48 hr (Fig. 6, Table 6) test
periods. Starved control nauplii in the 48 hr test exhibited
similar positive phototaxis but photokinesis was suppressed

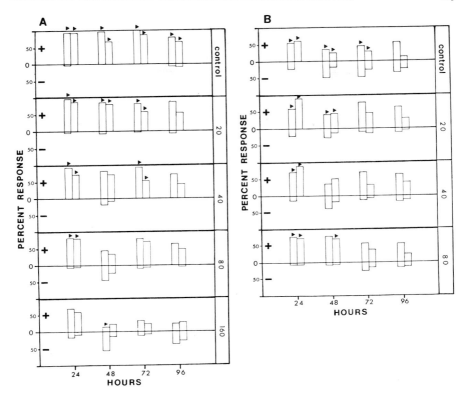

FIGURE 5.   Photobehavior of barnacle nauplii exposed to indicated copper concentrations (ppb) at 30 °/₀₀   (A) and 15 °/₀₀ (B).   Paried bars show percent positive and negative phototaxis at 0.04 (left) and 4.0 (right) W/m² light stimulus. Shaded bars indicate a random response (p ≤ 0.05); triangles above bars indicate a significant (p ≤ 0.05) increase in swimming speed during light stimulation.

(Fig. 6, Table 6). Cu-control nauplii at 15 $°/_{00}$ (Fig. 5b), however, shifted to a mixed phototactic response by 48 hrs, with about one half of the responding nauplii orienting in a negative direction.

At 30$°/_{00}$, exposure to Cu tended to depress phototaxis, and in some cases, induced a reversal to negative phototaxis. For nauplii used in the 96 hr assay, some decline in positive phototaxis and loss of photokinesis occurred at exposure levels of 20 ppb and 40 ppb Cu, particularly at 96 hrs (Fig. 5a). At 80 ppb Cu, a partial reversal to negative phototaxis was evident at 48 hr while at 160 ppb Cu negative

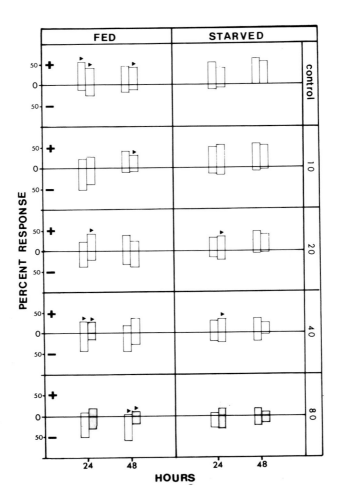

FIGURE 6.   *Photobehavior of fed and starved barnacle nauplii exposed to indicated Cu concentration (ppb) at 30°/oo. See Figure 5 for explanation of figure.*

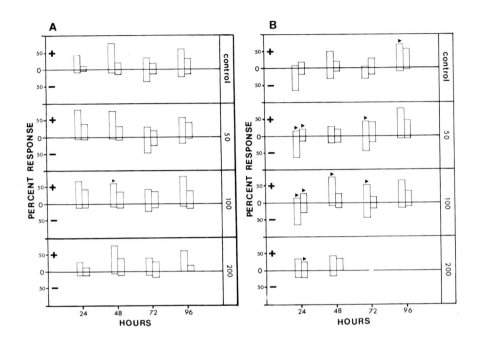

FIGURE 7. Photobehavior of barnacle nauplii exposed to indicated cadmium concentrations (ppb) at 30 °/oo (A) and 15 °/oo (B). See Figure 5 for explanation of figure.

phototaxis or loss of response occurred after 24 hrs (Fig. 5a). Nauplii in the 48 hr assay appeared more sensitive to Cu with increased negative phototaxis evident as low as 10 ppb at 24 hrs (Fig. 6). The combination of Cu exposure and starvation appeared to lessen a reversal to negative phototaxis relative to fed nauplii (Fig. 6). Cu exposure had no consistent effects of photokinesis for fed nauplii (Fig. 6). Starvation suppressed photokinesis in control nauplii; however, photokinesis was observed for exposed groups at 20 and 40 ppb at 24 hrs (Fig. 6).

At 15 $^{\circ}/_{\circ\circ}$, Cu-control nauplii and exposed nauplii exhibited strong positive phototaxis at 24 hrs (Fig. 5b). From 48 to 96 hrs control nauplii exhibited about 50% negative phototaxis. Cu exposed nauplii exhibited less negative phototaxis in most cases, and, in effect, had enhanced positive phototaxis relative to controls (Fig. 5b). Irrespective of the salinity dependent initial control response (+ or -), Cu exposure tended to reverse phototaxis in these nauplii.

Control nauplii used for the Cd assay (Fig. 7, Table 7, 8) were generally less responsive to light. There was virtually no photoresponse of control nauplii to the high intensity stimulus in the 96 hr Cd bioassay at both salinities, with the exception of nauplii at 15 $^{\circ}/_{\circ\circ}$ and 96 hr (Fig. 7b). At low light intensity, however, the phototactic response of nauplii at 15 $^{\circ}/_{\circ\circ}$ was markedly negative at 24 hrs (Fig. 7b). With two exceptions, photokinesis at either light intensity was generally absent at 30 $^{\circ}/_{\circ\circ}$ (Table 7). At 30 $^{\circ}/_{\circ\circ}$ positive phototaxis at the low light intensity was enhanced at exposure levels of 50 and 100 ppb Cd at 24 hrs and 100 ppb Cd at 96 hrs relative to control ($\alpha = 0.05$) (Fig. 7a). At 15$^{\circ}/_{\circ\circ}$ positive phototaxis was enhanced ($\alpha = 0.05$) at an exposure level of 200 ppb Cd at 24 hr, 100 ppb Cd at 48 hr and 50 and 100 ppb at 72 hrs. Photokinesis (Table 8) was also evident at these exposure levels but absent in controls. At 96 hrs both control and exposed nauplii exhibited strong positive phototaxis (Fig. 7b).

In contrast to effects seen with copper exposed nauplii, cadmium exposure did not depress or reverse phototaxis. In some cases, the response to light of cadmium-exposed nauplii, both in terms of positive phototaxis and photokinesis, was enhanced relative to control response.

DISCUSSION

The primary goal of this study was to develop and
evaluate possible short-term pollutant stress indices. We
chose to test copper and cadmium since both metals are
relatively well-studied (Eisler, 1979a) and are significant
toxicants in the marine environment (Davies, 1978).
Copper is an essential trace element in living organisms,
but at increased concentrations it is particularly toxic in
aquatic environments (EIFAC, 1976). The median concentration
of Cu in nearshore ocean water is about 1.5 ppb; values above
3.0 ppb probably reflect anthropogenic additions (Schmidt,
1978). Estuaries with heavy pollution may exhibit particularly high Cu concentrations with up to 27 ppb Cu reported
from Boston Harbor and 65 ppb Cu for Lower New York Bay
(Waldhauer et al., 1978).
Cadmium is a highly toxic metal with no known biolgoical
function (Varma and Katz, 1978). Natural levels of Cd in
nearshore waters range from <0.01 to 0.41 ppb Cd (Preston,
1973; Windom et al., 1976). Concentrations of Cd in highly
polluted water are reported up to 50 ppb Cd in Hong Kong
(Chan et al., 1974) and 130 ppb in Corpus Christi Harbor
(Holmes et al., 1975).
In many cases, Cu appears to be more toxic to aquatic
organisms than Cd (Calabrese et al., 1973; Stebbing, 1976;
Eisler, 1977; see also Tables 9, 10). The toxic effects of
Cd are often slower to develop (Stebbing, 1976; Price and
Uglow, 1979) and lethal effects in fish may not be evident
until after several months (EIFAC, 1977).
For this study, two variables known to effect toxicity
were examined: salinity and diet. The toxicity of both Cu
(Olson and Harrel, 1973; Jones, 1975) and Cd (Westernhagen
et al., 1974; Rosenberg and Costlow, 1976; Theede et al.,
1979) have been reported to increase inversely to salinity.
Sunda et al. (1978) have shown that the protective effect of
increased salinity for Cd toxicity can be explained in terms
of a reduced concentration of free Cd ion as Cd complexation
to chloride ions increases with higher salinity. Cupric ion
activity rather than the total copper concentration also
appears to be a major factor determining Cu toxicity (Sunda
and Guillard, 1976). However, Cu toxicity is highly influenced by complexation with natural organic (or synthetic)
ligands rather than by a direct effect of salinity (chloride
ion) (Engel et al., this volume). Salinity effects seen with

TABLE 9.   Acute toxicity of copper to marine crustacean zooplankton.

Test Organism	LC50 (ug/l Cu)	Time (hr)	Temp C	Salinity o/oo	Reference
Copepods					
Acartia clausi	34–48	48	18	–	Moraitou-Apostolopoulou, 1978
Acartia tonsa –	104–311	24	21	36	Reeve et al., 1976
Acartia tonsa	9–78	72	20	30	Sosnowski et al., 1979
Acartia tonsa	31–55	96	20	10	Sosnowski & Gentile, 1978
Acartia tonsa	17	96	20	30	" "
Calamus plumchrus	2778	24	13	27	Reeve et al., 1976
Labidocera scotti	132	24	21	36	" " "
Cirriped nauplii					
Balanus crenatus	410	6	–	–	Pyefinch & Mott, 1948
Balanus amphitrite (II)	≈ 250	24	20	30	Lang et al., 1979b
Balanus improvisus	≈ 165	24	26	30	" " "
" "	> 200	24	20	30	unpublished study
" "	88	24	20	15	" "
" "	> 160	96	15	30	present study
" "	> 80	96	15	15	" "
Euphausids					
Euphausia pacifica	14–30	24	13	27	Reeve et al., 1976
Decapod larvae					
Homarus americanus	48	96	20	30	Johnson & Gentile, 1979
Homarus gammarus	100–300	48	15	–	Connor, 1972
Crangon crangon	330	48	15	–	" "
Carcinus maenas	600	48	15	–	" "
Paragrapsus quadridentatus	170	96	17	–	Ahsanullah & Arnott, 1978

TABLE 10. Acute toxicity of cadmium to marine crustacean zooplankton.

Test Organism	LC50 (ug/l Cd)	Time (hr)	Temp C	Salinity o/oo	Reference
Copepods					
Acartia clausi	1200-1500	48	14	–	Moraitou-
	600-700	48	22	–	Apostolopoulou et al., 1979
Acartia tonsa " "	90-337	96	20	30	Sosnowski & Gentile, 1978
	122	96	20	10	" "
Nitocia spinipes	1800	96	20-22	7	Bengtsson, 1978
Tigriopus japonicus	4400	72	20-22	30	D'Agostino & Finney, 1974
Cirriped nauplii					
Balanus improvisus	>200	96	15	30	present study
Balanus improvisus	≃160	96	15	15	" "
Mysid					
Mysidopsis bahia	15	96	25-28	15-25	Nimmo et al., 1977
	11	408	25-28	15-25	" "
Decapod larvae					
Homarus americanus	78	96	20	30	Johnson & Gentile, 1979
Paragrapsus quadridentatus	490	96	17	35	Ahsanullah & Arnott, 1978
Callinectes sapidus	320	96	20-22	1	Frank & Robertson, 1979
(juveniles)	4700	96	20-22	15	" "
	11600	96	20-22	35	" "

Cu may reflect reduced concentrations of organic material,
particularly if lower salinity test water is obtained by
dilution of seawater with deionized water (Jones, 1975; Lang
et al., in press). Present results confirm that reduced
salinity increases sensitivity of barnacle nauplii to lethal
and sublethal concentrations of Cd and, to a lesser extent,
Cu.

Results also indicate different control groups can
respond differently to salinity independent of toxic
responses. Marcy (1980) found seasonal differences in
salinity and temperature responses of *Balanus amphitrite*
nauplii. Some embryonic adaptation (Sastry and Vargo, 1977;
Rosenberg and Costlow, 1979) to existing field conditions
could have occurred prior to laboratory acclimation of adults.
In addition, we have observed seasonal changes in water
"quality" (unpublished), a variable which has been shown to
effect survival of copepods (Lewis *et al.*, 1972). Each group
of control nauplii used in the present study exhibited
different salinity responses, photobehavior, and developmental
rates despite similar experimental conditions. Nauplii used
in the 48 hr Cu bioassay appeared to be inherently more
sensitive to Cu relative to nauplii used in the 96 hr Cu
assay. Numerous factors, both in the biological state of
test animals and seasonal water chemistry, may be involved.
Variable sensitivity is perhaps more marked in natural
populations as opposed to multiple generation laboratory
cultures (Sosnowski and Gentile, 1978; Sosnowski *et al.*,
1979).

Another factor which may affect toxicity is diet.
Nutritional stress from poor diet or starvation can increase
susceptibility of test animals to a toxicant (Shealy and
Sandifer, 1975; Winner *et al.*, 1977). Food interaction with
the toxicant can also alter toxicity and the degree of
feeding activity exhibited by test animals can influence
uptake kinetics (Jennings and Rainbow, 1979).

We expect Cu might prove less toxic to fed nauplii due to
possible increased complexing of Cu by algae and lack of
stress from starvation. On the other hand, starved nauplii do
not molt, a process which typically increases sensitivity to
toxicants (Lee and Buikema, 1979). During the 48 hr test
period, few differences were evident between starved and fed
groups. As expected from previous studies (Lang *et al.*,
1980), starvation depressed MLV of control animals and prevented molting. To some extent, Cu exposure appeared to
further depress MLV. At the concentrations tested, no
differences in acute mortality occurred. Continual
starvation, of course, would intensify stress and we expect
marked differences would be evident by 96 hrs. However,

48 hr starvation induced no highly significant effects on parameters tested relative to fed responses.

   Although mortality was recorded in this study, no attempt was made to determine a 96 hr LC50. Concentrations of Cu and Cd which were chosen were expected to be sublethal, based on preliminary studies and existing information for *B. improvisus* (Tables 9, 10). With the exception of 200 ppb Cd at 15 °/$_{oo}$, our treatments proved to be sublethal. Crustaceans are generally considered to be sensitive to both Cd (Eisler, 1971) and Cu (EIFAC, 1976) and larval stages are generally the most sensitive (Connor, 1972). Existing LC50 data (Tables 9, 10) indicate *B. improvisus* nauplii exhibit neither exceptional resistance or sensitivity to acute exposures of these metals relative to other crustacean zooplankton tested.

   Growth or development time to a specific embryonic or larval stage is one of the simplest measures of sublethal stress, although the results are not always easy to interpret as low pollutant concentrations may enhance rather than suppress growth (Waldichuk, 1979). Davies (1978) reviews studies of heavy metal effects on growth of marine plankton. In general, metals retard growth, although transient stimulatory effects may occur at low levels (Stebbing, 1976). For barnacle nauplii the molt from stage II to stage III provides convenient short-term responses. Both a delay in molt and success of molt were considered in this study. The molting process of crustaceans may also reduce the resistance of test animals (Lee and Buikema, 1979; Saroglia and Scarano, 1979) and, thus, it is important to consider both pre- and post-molt sensitivity. Also, a toxicant may induce abnormal molting. Shealy and Sandifer (1975) noted several deformities in molting of grass shrimp larvae following 48 hr exposure to Hg. Crab larvae fed *Artemia* nauplii with high DDT residues exhibited abnormal molting in megalopa and early crab stages (Bookhout and Costlow, 1970).

   Results with *B. improvisus* nauplii indicate both Cd and Cu will delay molting at low to intermediate sublethal concentrations and possibly inhibit molting at higher doses. Although molting is delayed, the Cd-exposed stage III nauplii exhibit normal morphology. In marked contrast, Cu induced greater than 95% abnormal molting at the lowest concentration tested (10 ppb). Interestingly, mortality did not markedly increase during or after molting in Cu-exposed nauplii. Unpublished studies indicate abnormal nauplii can attain at least stage V development under optimum rearing conditions.

   The use of behavioral parameters in toxicological assessment of marine pollutants has been rare (Eisler, 1979b), yet changes in behavior can be an initial response of an animal to environmental perturbation (Olla *et al.*, in press) and thus lends itself well to short term testing. Behavioral changes are easily observable relative to the complex bio-chemical and physiological responses underlying them and may be more readily open to ecological interpretation. Neverthe-less, behavior is difficult to assess quantitatively and often exhibits a high degree of inherent variation (Olla, 1974; Scherer, 1977; Eisler, 1979b). Applications of locomotor response for marine pollution effects measurements have been reviewed by Miller (in press). Simple hand tracking of swimming behavior of crab larvae was incorporated by Vernberg *et al.* (1974) and Mirkes *et al.* (1978) in multiple parameter studies of heavy metal effects. Sharp *et al.* (1979) also noted behavioral changes in shrimp exposed to the pesticide Kelthane, in addition to molting, feeding, respiration, and mortality. Forward and Costlow (1976, 1978), however, have initiated the most rigorous behavioral studies, investigating the swimming behavior recorded on video-tape. Their pollution studies were strongly supported by previous baseline studies on control behavior (see Forward and Costlow, 1978). Simonet *et al.* (1978) has developed a biomonitoring assay based on inhibition of negative phototaxis in mosquito larvae.

These studies and others (Miller, in press) confirm that behavioral changes are often early and sensitive indicators of stress. The technology to rapidly quantify multiple movement parameters of test animals was developed at this laboratory through the use of a video-to-digital processing system called the "Bugwatcher" (Wilson and Greaves, 1979; Greaves and Wilson, 1980). The system is particularly well suited to measure swimming behavior of developing organisms such as fish and invertebrate larvae. However, as this technique is new, baseline information and comparative data on swimming behavior is sparse.

Present results and additional studies in progress indicate altered swimming activity and changes in simple directed responses such as photobehavior may prove to be effective stress indices for zooplankton. Although a directed response in most cases may be a more sensitive indicator relative to random swimming activity, the opposite was seen in barnacle nauplii. Nauplii are essentially continuous swimmers, and with adequate sample numbers of about 30 or more, exhibit a stable MLV at a given temperature, salinity condition. Seasonal and geographical variations exist, (Lang *et al.*, 1979b), however and a control baseline must be established for each experiment. MLV of nauplii is affected almost immediately by the metals tested. At the lower metal concentrations initial hyperactivity was evident here and in earlier studies (Lang *et al.*, 1979b). We suspect that at the

lowest concentrations which induce an immediate response of
hyperactivity, the effect is a transient response followed by
apparent recovery. However, at slightly higher concentrations, the immediate hyperactivity is followed by a
decline in MLV. At still higher concentrations, MLV is
depressed from the onset, a response which appears to indicate
substantial stress.

Photobehavior of barnacle nauplii, as tested in this
study, proved to be less sensitive and more variable.
Swimming of nauplii appears to be impaired before phototaxis
is lost. Effects of Cd and Cu appear dissimilar in that Cu
depresses or reverses naupliar phototactic response while Cd
enhances it. However, differences between the control
responses of these test groups must also be taken into account
when evaluating these contrasting responses. The Cd control
group exhibited less response to light relative to Cu controls. A better understanding of factors influencing phototaxis independent of pollutants is needed. Similarly, photokinesis proved unexpectedly variable and perhaps should be
reassessed in terms of individual naupliar MLV response to a
light stimulus, as opposed to mean velocities calculated for
each treatment group.

The dramatic effects of Cu exposure on molting of nauplii
are believed to be reflected in our swimming speed results.
Nauplii which molt with deformed appendages and reduced
setae swim at slower speeds. The sharp decline in MLV of all
Cu-exposed groups at 72 hrs probably results from increasing
numbers of deformed nauplii. Results with starved nauplii,
however, also indicate that Cu will depress MLV independent
of molting. The ratio of stage II to stage III nauplii is
also a significant factor affecting MLV. Present results
give an interesting chronology of average swimming response
as affected by those several factors. A more precise
picture of metal effects on swimming perhaps would be obtained
if stage II and stage III MLV were considered separately.

In conclusion, it can be stated that barnacle naupliar
molting and swimming behavior hold definite promise as
sensitive, short-term indicators of chemical toxicity. Each
observed effect also has the potential for reducing
ecological fitness of crustacean larvae and holozooplankton.
It is evident that the relative sensitivity of these indices
will vary with the toxicant and the conditions of the test
(Tables 9, 10). These results also demonstrate the value of
using multiple stress indices in toxicant screening and
emphasize the need to conduct the tests under several
contrasting environmental conditions. The assay design should

also permit evaluation of larval batch variability, a problem
often seen in pollutant testing.

SUMMARY

     The effects of exposure of copper and cadmium on newly
released stage II nauplii of the estuarine barnacle *Balanus
improvisus* have been studied. Nauplii are readily obtained
and cultured in the laboratory, exhibit a continuous swimming
pattern, and molt within 96 hrs at 15-20°C; they hold a good
potential as a routine assay organism. Multiple stress
indices were studied and include percent survival, rate of
development (to stage III), normality of molting, spontaneous
swimming speeds, and photobehavior (phototaxis and photokinesis). Motile behavior indices were quantified using a
video-computer behavioral analysis system. Results from
three static assays (15°C) are presented: 96 hr Cu and Cd
exposures using fed nauplii at 15 and 30 °/oo salinity and 48
hr Cu exposure with fed and starved nauplii at 30 °/oo
salinity.
     Exposure levels of Cu range from 10-160 ppb, which are
below the 96 hr LC50. Molting to Stage III is progressively
delayed with increasing Cu concentration. Loss of seatae
and deformed appendages occur in nearly all stage III nauplii
including the lowest concentration tested (10 ppb). Spontaneous swimming speed initially increased from 20 to 80 ppb Cu
but by 72 hrs is significantly depressed at all exposure
levels. Photokinesis and phototaxis are also altered, but
only at concentrations above threshold for changes in
spontaneous swimming speeds. Phototactic behavior is altered
by salinity and a further interaction with Cu occurs. At
30 °/oo control nauplii are photopositive; increasing Cu
concentrations initially decreases percent positive phototaxis
but at the highest levels induces a photoreversal to negative
response. In contrast, control nauplii at 15 °/oo exhibit
a mixed response with about 50% exhibiting negative phototaxis. Cu exposure at this salinity increases positive
response.
     Exposure levels of Cd range from 50-200 ppb. At 30 °/oo
mortality for 96 hrs remains below 50% at all concentrations;
at 15 °/oo a 96 hr LC50 occurs between 100 to 200 ppb Cd.
At 100 ppb Cd development is delayed; no molting abnormalities
were noted. At 24 hrs spontaneous swimming speeds were
increased at 50 to 100 ppb Cd at 30 °/oo, but depressed at
100 ppb Cd at 15 °/oo. After 24 hrs swimming speeds were
consistently depressed at 100 ppb Cd, while at 50 ppb Cd a

mixed response was observed. Unlike results with Cu, Cd tended to enhance positive phototaxis.

Starved nauplii exhibit slower swimming speeds relative to fed nauplii and fail to molt, but for the 48 hr exposure period no obvious differences in Cu toxicity were observed. Toxic effects of Cd are more pronounced at the lower salinity. Decreased salinity has less effect on Cu toxicity. Variability in control responses and Cu sensitivity between different groups of nauplii is seen. Seasonal changes in the biological state of test animals and/or water conditions may be involved.

Embryo-larval life stages of marine organisms typically prove to be most sensitive to pollutants. For short-term screening tests, larval assays may provide better estimates of toxic levels, relative to similar tests using adults. If, in addition to mortality, indices of sublethal pollutant stress are also employed, larval assays may provide a particularly sensitive and rapid test for initial estimates of no effect levels.

ACKNOWLEDGEMENTS

Behavioral tests used in this study were developed in collaboration with Richard B. Forward (Duke University). We would like to acknowledge and thank him for his assistance. John Greaves (Computer Sciences Corporation) and Robert Wilson (Yale University), in addition to developing the basic computer system used, provided valuable technical support for completion of this study. Additional help in computer operations and programming was provided by David Sleczkowski. Heavy metal analyses were performed by Douglas Cullen at the EPA laboratory. Finally we would like to thank Barbara Gardiner for her typing and editing of this manuscript. This is contribution No. 162 of the EPA Environmental Research Laboratory, Narragansett, RI.

LITERATURE CITED

Ahsanullah, M. and G. H. Arnott. 1978. Acute toxicity of copper, cadmium, and zinc to larvae of the crab *Paragrapsus quadridentatus* (H. Milne Edwards), and implications for water quality criteria. Aust. J. Mar. Freshwat. Res. 29: 1-8.

Bengtsson, B. E. 1978. Use of a harpacticoid copepod in toxicity tests. Mar. Poll. 9: 238-241.

Bookhout, C. G. and J. D. Costlow, Jr. 1970. Nutritional effects of *Artemia* from different locations on larval development of crabs. Helgoländer Wiss. Meeresunters. 20: 435-442.

Calabrese, A., R. S. Collier, D. A. Nelson, and J. R. MacInnes. 1973. The toxicity of heavy metals to embryos of the American oyster *Crassostrea virginica*. Mar. Biol. 18: 162-166.

Chan, J. P., M. T. Cheung, and F. P. Li. 1974. Trace metals in Hong Kong waters. Mar. Pollut. Bull. 5: 171-174.

Connor, P. M. 1972. Acute toxicity of heavy metals to some marine larvae. Mar. Pollut. Bull. 3: 190-192.

D'Agostino, A. and C. Finney. 1974. The effect of copper and cadmium on the development of *Tigriopus japonicus*. In: Pollution and Physiology of Marine Organisms. pp. 445-463. Ed. by F. J. Vernberg and W. B. Vernberg. New York: Academic Press.

Davies, A. G. 1978. Pollution studies with marine plankton. Part II. Heavy Metals. Adv. Mar. Biol. 15: 381-508.

Davis, J. C. 1976. Progress in sublethal effect studies with Kraft pulpmill effluent and salmonids. J. Fish. Res. Bd. Can. 33: 2031-2035.

Draggan, S. and J. M. Giddings. 1978. Testing toxic substances for protection of the environment. Sci. Total Environ. 9: 63-74.

Duthie, J. R. 1977. The importance of sequential assessment in test programs for estimating hazard to aquatic life. In: Aquatic Toxicology and Hazard Evaluation. pp. 17-35. Ed. by F. L. Mayer and J. L. Hamelink. Philadelphia American Society for Testing and Materials, Publication No. 634.

EIFAC. 1976. Water quality criteria for European freshwater fish. Report on copper and freshwater fish. European Inland Fisheries Advisory Commission. Tech. Paper No. 27: 34 p.

EIFAC. 1977. Water quality criteria for European fresh-
water fish. Report on cadmium and freshwater fish.
European Inland Fisheries Advisory Commission. Tech.
Paper No. 30: 21 p.

Eisler, R. 1971. Cadmium poisoning in *Fundulus heteroclitus*
(Pisces: Cyprinodontidae) and other marine organisms.
J. Fish. Res. Bd. Can. 28: 1225-1234.

Eisler, R. 1977. Acute toxicities of selected heavy metals
to the softshell clam *Mya arenaria*. Bull. Envir.
Contam. Toxicol. 17: 137-145.

Eisler, R. 1979a. Toxic cations and marine biota: analysis
of research effort during the three-year period 1974-
1976. In: Marine Pollution: Functional Responses. pp
111-149. Ed. by W. B. Vernberg, F. P. Thurberg, A.
Calabrese, and F. J. Vernberg. New York: Academic
Press.

Eisler, R. 1979b. Behavioral responses of marine poikilotherms to pollutants. Phil. Trans. R. Soc. (Ser. B).
286: 507-521.

Engel, D. W., W. G. Sunda and B. A. Fowler. (In press).
Factors affecting trace metal uptake and toxicity of
marine organisms. I. Environmental parameters. In:
Biological Monitoring of Marine Organisms. Ed. by F. J.
Vernberg, A. Calabrese, F. P. Thurberg, and W. B.
Vernberg. New York: Academic Press.

Forward, R. B., Jr. and J. D. Costlow, Jr. 1976. Crustacean
larval behavior as an indicator of sublethal effects of
an insect juvenile hormone mimic. Estuar. Process.
1: 279-289.

Forward, R. B., Jr. and J. D. Costlow, Jr. 1978. Sublethal
effects of insect growth regulators upon crab larval
behavior. Water Air Soil Pollut. 9: 227-238.

Frank, P. M. and P. B. Robertson. 1979. The influence of
salinity on toxicity of cadmium and chromium to the
blue crab, *Callinectes sapidus*. Bull. Environ. Contam.
Toxicol. 21: 74-78.

Greaves, J. O. B. 1975. The bugsystem: the software
structure for reduction of quantized video data of
moving organisms. Proc. Inst. Electrical Electronics
Enqr. 63: 1415-1425.

Greaves, J. O. B. and R. S. Wilson. 1980. Development of an interactive system to study sub-lethal effects of pollutants on the behavior of organisms. 37 p. Narragansett, Rhode Island: Environmental Research Laboratory, U.S. Environmental Protection Agency. (EPA-600/3-80-010).

Guillard, R. R. L. and J. H. Ryther. 1962. Studies on marine planktonic diatoms. I. *Cyclotella nana* Hustedt and *Detonula confervacea* (Cleve) Gran. Can. J. Microbiol. 8: 229-239.

Heritage, J. 1978. Major American toxics disasters. EPA Journal. 8: 8-10.

Holmes, C. W., E. A. Slade, and C. J. McLerran. 1974. Migration and redistribution of zinc and cadmium in marine estuarine system. Environ. Sci. Tech. 8: 255-259.

Jennings, J. R. and P. S. Rainbow. 1979. Accumulation of cadmium by *Artemia salina*. Mar. Biol. 51: 47-53.

Johnson, M. and J. Gentile. 1979. Acute toxicity of cadmium, copper, and mercury to larval American lobster (*Homarus americanus*). Bull. Environ. Contam. Toxicol. 22: 258-264.

Jones, M. B. 1975. Effects of copper on survival and osmoregulation in marine and brackish water isopods (Crustacea). In: Proc. 9th European Marine Biology Symposium. pp. 419-431. Ed. by H. Barnes. Aberdeen: Aberdeen Univ. Press.

Lang, W. H. 1979. Larval development of shallow water barnacles of the Carolinas (Cirripedia: Thoracia) with keys to naupliar stages. 39 pp. Washington, D.C.: Technical Information Division, National Oceanic and Atmospheric Administration (NOAA Tech. Rep. NMFS Circ. 421).

Lang, W. H., R. B. Forward, Jr., and D. C. Miller. 1979a. Behavioral responses of *Balanus improvisus* nauplii to light intensity and spectrum. Biol. Bull. 157: 166-181.

Lang, W. H., R. B. Forward, Jr., D. C. Miller, and M. Marcy. In press. Acute toxicity and sublethal behavioral effects of copper on barnacle nauplii *(Balanus improvisus)*. Mar. Biol.

Lang, W. H., S. Lawrence, and D. C. Miller. 1979b. The effects of temperature, light, and exposure to sublethal levels of copper on the swimming behavior of barnacle nauplii. In: Advances in Marine Environmental Research, Proceedings of a Symposium. pp. 273-289. Ed. by F. S. Jacoff. Narragansett, Rhode Island: Office of Research and Development, U.S. Environmental Protection Agency (EPA-600/9-79-035).

Lang, W. H., M. Marcy, P. J. Clem, D. C. Miller, and M. R. Rodelli. 1980. The comparative photobehavior of laboratory-hatched and plankton-caught *Balnaus improvisus* (Darwin) nauplii and the effects of 24-hour starvation. J. exp. mar. Biol. Ecol. 42: 201-212.

Lee, D. R., and A. L. Buikema, Jr. 1979. Molt-related sensitivity of *Daphnia pulex* in toxicity testing. J. Fish. Res. Bd. Can. 36: 1129-1133.

Lewis, A. G., P. H. Whitefield, and A. Ramnarine. 1972. Some particulate and soluble agents affecting the relationship between metal toxicity and organism survival in the calanoid copepod *Euchaeta japonica*. Mar. Biol. 17: 215-221.

Macek, K. J. and B. H. Sleight, III. 1977. Utility of toxicity tests with embryos and fry of fish in evaluating hazards associated with the chronic toxicity of chemicals to fishes. In: Aquatic toxicology and hazard evaluation. pp. 137-146. Ed. by F. L. Mayer and J. L. Hamelink. Philadelphia: American Society for Testing and Materials. Publication No. 634.

Marcy, M. 1980. Seasonal effects of temperature and salinity on the larval development of the barnacle *Balanus amphitrite amphitrite* (Darwin) (Cirripedia: Thoracica). Unpublished Ms. Thesis. Marine Science Program, Univ. of South Carolina, Columbia, 47 p.

Miller, D. C. In press. Some applications of locomotor response in pollution effects measurement. In: Biological effects of marine pollution and the problems of monitoring. Ed. by A. D. McIntyre and J. B. Pearce.

Rapp. P.-v. Reun. Cons. int. Explor. Mer. 179.

Mirkes, D. Z., W. B. Vernberg and P. J. DeCoursey. 1978.
    Effects of cadmium and mercury on the behavioral
    responses and development of *Eurypanopeus depressus*
    larvae. Mar. Biol. 47: 143-147.

Moraitou-Apostolopoulou, M. 1978. Acute toxicity of copper
    to a copepod. Mar. Pollut. Bull. 9: 278-280.

Moraitou-Apostolopoulou, M., G. Verriopoulos, and P. Lentzou.
    1979. Effects of sublethal concentrations of cadmium as
    possible indicators of cadmium pollution for two
    populations of *Acartia clausi* (Copepoda) living at two
    differently polluted areas. Bull. Environ. Contam.
    Toxicol. 23: 642-649.

Nimmo, D. W. R., D. V. Lightner, and L. H. Bahner. 1977.
    Effects of cadmium on the shrimps, *Penaeus duorarum,
    Palaemonetes pugio,* and *Palaemonetes vulgaris*. In:
    Physiological responses of marine biota to pollutants.
    pp. 131-183. Ed. by F. J. Vernberg, A. Calabrese, F. P.
    Thurberg, and W. B. Vernberg. New York: Academic Press.

Olla, B. L (ed.). 1974. Behavioral bioassays. In: Marine
    bioassays workshop proceedings. pp. 1-31. Chairman,
    G. V. Cox. Washington, D.C.: Marine Technology Society.

Olla, B. L., W. H. Pearson, and A. L. Studholme. In press.
    Applicability of behavioral measures in environmental
    stress assessment. In: Biological effects of marine
    pollution and the problems of monitoring. Ed. by A. D.
    McIntyre and J. B. Pearce. Rapp. P.-v. Reun. Cons. int.
    Explor. Mer. 179.

Olson, K. R. and R. C. Harrel. 1973. Effect of salinity on
    acute toxicity of mercury, copper, and chromium for
    Rangia cuneata (Pelecypods, Mactridae). Contrib. mar.
    Sci. 17: 9-13.

Preston, A. 1973. Heavy metals in British waters. Nature
    242: 95.

Price, R. K. J. and R. F. Uglow. 1979. Some effects of
    certain metals on development and mortality within the
    moult cycle of *Crangon crangon* (L.) Marine Environ. Res.
    2: 287-299.

Pyefinch, K. A. and J. C. Mott. 1948. The sensitivity of barnacles and their larvae to copper and mercury. J. exp. Biol. 25: 276-298.

Reeve, M. R., G. D. Grice, V. R. Gibson, M. A. Walter, K. Darcy, and T. Ikeda. 1976. A controlled environmental pollution experiment (CEPEX) and its usefulness in the study of larger marine zooplankton under toxic stress. In: Effects of pollutants on aquatic organisms. pp. 145-162. Ed. by A. P. Lockwood. Cambridge: Cambridge University Press.

Rosenberg, R. and J. D. Costlow, Jr. 1976. Synergistic effects of cadmium and salinity combined with constant and cycling temperatures on the larval development of two estuarine crab species. Mar. BIol. 38: 291-303.

Rosenberg, R. and J. D. Costlow, Jr. 1979. Delayed response to irreversible non-genetic adaptation to salinity in early development of the brachyuran crab *Rhithropanopeus harrisii,* and some notes on adaptation to temperature. Ophelia 18: 97-112.

Saroglia, M. G. and G. Scarano. 1979. Influence of molting on the sensitivity to toxics of the crustacean *Penaeus kerthurus* (Forskäl). Ecotoxicology Environ. Safety 3: 310-320.

Sastry, A. N. and S. L. Vargo. 1977. Variations in the physiological responses of crustacean larvae to temperature. In: Physiological Responses of Marine Biota to Pollutants. pp. 401-423. Ed. by F. J. Vernberg, A. Calabrese, F. P. Thurberg and W. B. Vernberg. New York: Academic Press.

Scherer, E. 1977. Behavioral assay-principles, results and problems. Environ. Protection Serv. Tech. Rep. No. EPS-5AR-77-1. pp. 33-40. Halifax, Canada.

Schmidt, R. L. 1978. Copper in the marine environment-part I. CRC Critical Reviews in Environmental Control 8: 101-152.

Sharp, J. W., R. M. Sitts and A. W. Knight. 1979. Effects of kelthane on the estuarine shrimp *Crangon franciscorum.* Mar. Biol. 50: 367-374.

Shealy, M. H., Jr. and P. A. Sandifer. 1975. Effects of mercury on survival and development of the larval grass shrimp *Palaemonetes vulgaris*. Mar. Biol. 33: 7-16.

Simonet, D. E., W. I. Knausenberger, L. H. Townsend, Jr., and E. C. Turner, Jr. 1978. A biomonitoring procedure utilizing negative phototaxis of first instar *Aedes aegypti* larvae. Arch. Environ. Contamn. Toxicol. 7: 339-347.

Sosnowski, S. L. and J. H. Gentile. 1978. Toxicological comparison of natural and cultured populations of *Acartia tonsa* to cadmium, copper, and mercury. J. Fish. Res. Bd. Can. 35: 1366-1369.

Sosnowski, S. L., D. J. Germond, and J. H. Gentile. 1979. The effect of nutrition of the response of field populations of the calanoid copepod *Acartia tonsa* to copper. Wat. Res. 13: 449-452.

Sprague, J. B. 1971. Measurement of pollutant toxicity to fish. III. Sublethal effects and "safe" concentrations. Wat. Res. 5: 245-266.

Stebbing, A. R. D. 1976. The effects of low metal levels on a clonal hydroid. J. mar. biol. Ass. U.K. 56: 977-994.

Sunda, W. G., D. W. Engel, and R. M. Thoutte. 1978. Effect of chemical speciation on toxicity of cadmium to grass shrimp, *Palaemonetes pugio*: importance of free cadmium ion. Environ. Sci. Tech. 12: 409-412.

Sunda, W. and R. R. L. Guillard. 1976. The relationship between cupric ion activity and the toxicity of copper to phytoplankton. J. Mar. Res. 34: 511-529.

Theede, H., N. Scholz, and H. Fischer. 1979. Temperature and salinity effects on the acute toxicity of cadmium to *Laomeda loveni* (Hydrozoa). Mar. Ecol. Prog. Ser. 1: 13-19.

Varma, M. M. and H. M. Katz. 1978. Environmental impact of cadmium. J. Environ. Health 40: 324-329.

Vernberg, W. B., P. J. DeCoursey, and J. O'Hara. 1974. Multiple environmental factor effects on physiology and behavior of the fiddler crab, *Uca pugilator*. In: Pollution and physiology of marine organisms. pp. 381-425. Ed. by F. J. Vernberg and W. B. Vernberg. New York: Academic Press.

Waldhauer, R., A. Matte, and R. E. Tucker. 1978. Lead and copper in the waters of Raritan and Lower New York Bays. Mar. Pollut. Bull. 9: 38-42.

Waldichuk, M. 1979. Review of the problem. Phil. Trans. R. Soc. London, B. 286: 399-424.

Walsh, J. 1978. EPA and toxic substances law: dealing with uncertainty. Science 202: 598-602.

Westernhagen, H. Von, H. Rosenthal, and K.-R Sperling. 1974. Combined effects of cadmium and salinity on development and survival of herring eggs. Helgoländer wiss. Meeresunters. 26: 416-433.

Wilson, R. S. and J. O. B. Greaves. 1979. The evolution of the bugsystem: recent progress in the analysis of bio-behavioral data. In: Advances in Marine Environmental Research, Proceedings of a Symposium. pp. 251-272. Ed. by F. S. Jacoff. Narrangansett, Rhode Island: Office of Research and Development, U.S. Environmental Protection Agency (EPA-600/9-79-035).

Windon, H. L., N. S. Gardner, W. M. Dunstan, and G. A. Paffenhofer. 1976. Cadmium and mercury transfer in a coastal marine ecosystem. In: Marine Pollutant Transfer. pp. 135-157. Ed. by H. L. Windom and R. A. Duce. Lexington, Mass.: D. C. Heath and Co.

Winner, R. W., T. Keeling, R. Yeager, and M. P. Farrell. 1977. Effect of food type on the acute and chronic toxicity of copper to *Daphnia magna*. Freshwater Biol. 7: 343-349.

THE ONTOGENY OF RESISTANCE ADAPTATION AND
METABOLIC COMPENSATION TO SALINITY AND TEMPERATURE BY THE
CARIDEAN SHRIMP, *PALAEMONETES PUGIO,* AND MODIFICATION
BY SUBLETHAL ZINC EXPOSURE

Charles L. McKenney, Jr.[1]
Jerry M. Neff[2]

Department of Biology
Texas A&M University
College Station, Texas

INTRODUCTION

Organisms occupying estuaries are those that through the
process of natural selection have developed the capacity to
resist the characteristically fluctuating salinity and temperature conditions of estuarine waters (Vernberg and
Vernberg, 1972a; Lockwood, 1976). A number of recent
studies, however, indicate that exposure to concentrations of
heavy metals which prove sublethal under optimal salinity-
temperature conditions for survival may reduce both salinity
and temperature resistance capacities of estuarine organisms
(Vernberg and Vernberg, 1972b; Vernberg *et al.*, 1973, 1974;
Jones, 1973, 1975a; Rosenberg and Costlow, 1976; McKenney and
Costlow, 1977; McKenney and Neff, 1979).
Adaptations of organisms to estuarine conditions are
not, however, confined to mere tolerance of variations in
temperature and salinity. While the basic mechanisms which
allow for compensatory physiological adaptations to temperature (Hazel and Prosser, 1974; Somero and Hochachka, 1976)

[1]*present address: United States Environmental Protection Agency, Environmental Research Laboratory, Sabine
Island, Gulf Breeze, Florida.*
[2]*present address: Battelle, William F. Clapp Laboratories, Inc., 397 Washington St., Duxbury, Massachusetts.*

and salinity (Giles, 1975; Lockwood, 1976) may be extremely complex, the results may be accompanied by alterations and, upon completion of acclimation, stabilization of energy utilized for metabolic maintenance, (Kinne, 1964, 1971; Schlieper, 1971; Pandian, 1975; Bayne *et al.*, 1976).

While numerous studies have demonstrated metabolic compensation by marine and estuarine invertebrates to both temperature (for a review see Kinne, 1970) and salinity (for a review see Kinne, 1971) acting individually, the combined effects of both factors acting simultaneously on the metabolism of marine invertebrates are poorly understood. Recognizing that early developmental stages of marine invertebrates are generally more sensitive to environmental variations than are adults, several studies have examined metabolic adjustments to both salinity and temperature in larval and juvenile stages of marine crustaceans (Engel and Angelovic, 1968; Vernberg *et al.*, 1973, 1974; Nelson *et al.*, 1977). However, no information exists on metabolic compensation to both factors throughout total larval development: nor has consideration been given to potential alterations of such compensatory mechanisms during larval development following exposure to heavy metals.

In an earlier paper, we reported an area of optimal salinity-temperature conditions within which viability of *Palaemonetes pugio* larvae through metamorphosis was maximal and outside of which total survival through complete larval development of this estuarine shrimp was reduced. Furthermore, resistance adaptations of developing larvae to salinity and temperature were progressively restricted with exposure to increasing sublethal zinc concentrations (McKenney and Neff, 1979). The objectives of the present study were to examine the ontogeny of possible resistance and metabolic capacity adaptations to salinity and temperature through the complete larval development of the caridean shrimp, *Palaemonetes pugio,* and to examine any disruptions in these responses by sublethal zinc exposure.

MATERIALS AND METHODS

Ovigerous female *Palaemonetes pugio* (Holthuis) were collected from East Galveston Bay, Texas at salinities ranging between 19 and 27 °/∘∘ and were transported to the laboratory in College Station, Texas. In the laboratory adults were maintained at a salinity of 19 °/∘∘ and at the experimental temperatures of (20°, 25°, or 30°C) approximating water temperatures at the collection site. Upon hatching,

larvae to be reared in the intermediate salinity (17 or
19 °/oo) were transferred directly to the various test media.
To avoid osmotic shock, freshly hatched larvae to be reared
in either lower or higher salinities were tranferred at three
hour intervals in graded steps of 6-7 °/oo S from the original
intermediate salinity to the appropriate experimental
salinity.

Larvae were individually reared in isolation in compartmented plastic boxes in an approximate liquid volume of 50 ml.
All larvae were fed daily an abundance of freshly hatched
Artemia salina nauplii in an approximately equal portion (2
nauplii per ml). Every third day larvae were transferred to
an acid-cleaned compartmented box containing freshly prepared
media of the appropriate salinity-temperature-zinc condition.

Desired salinities were obtained by diluting 35 °/oo S
seawater (prepared from Instant Ocean Synthetic Sea Salts
containing an undetectable 20 ppb Zn) with distilled water.
Measurements of the pH of the various experimental media
reflected those of natural estuarine water at the respective
salinities. Experimental temperatures were held within
±0.5°C of the desired temperature in constant-temperature
cabinets that also maintained a 12 h light: 12 h dark photoperiod. Zinc solutions were prepared every third day from a
primary stock solution of 1 g Zn^{++} per liter prepared from
$ZnCl_2$. Analysis of zinc solutions by atomic absorption
spectrophotometry revealed no significant difference between
nominal and actual zinc concentrations nor significant loss of
zinc from the compartmented boxes after 96 h at any of the
experimental temperatures (McKenney and Neff, 1979).

The experimental design was a 3x3x2 factorial with
salinities of 7, 19, and 31 °/oo; constant temperatures of
20°, 25°, and 30°C; and nominal zinc concentrations of 0.00
(unexposed controls) and 0.25 ppm Zn. For experiments on
survival of each larval stage, four replicates of 12 larvae
each were reared at each of the 18 salinity-temperature-zinc
conditions. Additional larvae were individually reared at the
various salinity-temperature-zinc conditions for use in the
metabolic studies. Daily records were kept on mortality of
all larvae and the molting frequency of each individual.

Larval oxygen consumption was determined by individually
sealing five larvae from each salinity-temperature-zinc
condition in 5-ml all-glass syringes of fully oxygenated media
at the acclimation salinity. The volumes of media in the
syringe varied from 1-5 ml dependent on the size of the larva.
To avoid any variation in respiratory activity due to molt
cycle, oxygen consumption of larvae was measured on days
representative of their intermolt period. After a 3-4 hour
period, a 75-100 µl sample of the fluid in the syringe was

injected into a Radiometer BMS 3 Mk2 micro blood gas analyzer
with attached PHM71 Mk2 Acid-Base Analyzer and a direct
measurement recorded for partial pressure of oxygen. The
larva was then removed from the respirometer, briefly rinsed
in distilled water, and placed in an oven to dry at 60°C for
48 hours. The dry weight of each larva was subsequently
determined on a Mettler UM7 ultra-microbalance to the nearest
0.01 µg. Differences in partial pressures of oxygen between
the values of control respirometers (syringes with no animal)
and respirometers containing a larva were attributed to
larval uptake of oxygen. These values were corrected for
weight of larva and time in respirometer, and weight-specific
oxygen consumption rates ($\mu g \ O_2 \ mg^{-1}$ dry weight h^{-1}) were
calculated.

For both oxygen consumption rates and arcsine-transformed
percentage survival data of each developmental stage,
multiple linear regression analysis was performed on three
separate models with an increasing order of complexity. The
choice of the most appropriate model was made by a step-down
analysis which indicated whether decreasing complexity significantly increased the residual mean square. Since for both
sets of data the most complex model proved to be the "best
model" (significantly reduced the residual mean square), the
results of this model will be presented here. A more detailed
description of the results of each regression model and the
step-down analysis is available in McKenney (1979).

To further aid in the interpretation of these data,
separate regression equations were produced at each level of
the two discrete independent variables (zinc exposure level
and developmental stage) and employed to generate response
surface curves at a range of levels of the two continuous
independent variables (salinity and temperature) (Box, 1954;
Box and Youle, 1955; Myers, 1971; Alderdice, 1972). Hence,
for each developmental stage both with and without sublethal
zinc exposure, the survival and respiration rates were
expressed as the quadratic function of salinity and temperature by the two factor second-order polynomial equation:

$$\hat{Y} = b_0 x_0 + b_1 x_1 + b_{11} x_1^2 + b_2 x_2 + b_{22} x_2^2 + b_{12} x_1 x_2 \qquad (1)$$

where $\hat{Y}$ = the estimated weight-specific oxygen consumption
rate or arcsine-transformed survival percentage; x_0 = the
intercept; x_1 and x_2 = linear effect of salinity (Sal) and
temperature (Temp), respectively; x_1^2 and x_2^2 = the quadratic
effect of salinity (Sal^2) and temperature ($Temp^2$), respectively; $x_1 x_2$ = the linear x linear interaction between
salinity and temperatures (Sal x Temp); and b_0, b_1, etc. =
the regression coefficients.

RESULTS

Survival of Individual Larval Stages

Percent survival of the various larval stages of
Palaemonetes pugio varied from 31% to 100% over a range of
salinity-temperature conditions with and without exposure to
0.25 ppm Zn (Table 1). Multiple linear regression analysis
indicated the dominant effects of salinity and temperature on
larval survival. Furthermore, the larval stage and zinc-
stage interaction influenced the survival of developing
larvae (Table 2). Pooled data of zinc-exposed and control
larval survival analyzed by stage in a Duncan's Multiple
Range Test indicated a significant reduction of survival
percentages of later larval stages from those of the early
larval stages and significantly lower survival in the last
larval stage. Moreover, a number of interactions significantly modified the survival of developing larvae, including
in order of importance: salinity, stage, and zinc; salinity,
temperature, stage, and zinc; salinity and stage; temperature
and stage; salinity, temperature, and stage; temperature and
zinc; and temperature, stage, and zinc (Table 2).

Since the number of larval stages varied with salinity-
temperature-zinc conditions, response surfaces were developed
from quadratic salinity-temperature equations for categories
of early and late larval stages (larval stages I-IV and
second from the last, next to the last, and final larval
stages, respectively). Regression coefficients for each
separate equation are provided in the appendix of McKenney
(1979). The statistical analyses for these curves are
summarized in Table 3.

In the absence of additional zinc the survival of the
first four larval stages was little affected by salinity and
temperature (Figs. 1 and 2; Table 3). Exposure to 0.25 ppm
Zn, however, significantly reduced the survival of first stage
larvae at low salinity-temperature conditions. Viability of
third stage larvae was affected by a salinity-temperature
interaction in the presence of zinc, but in this case
survival was most reduced at low temperature and high
salinity conditions.

In general, late larval stages were more sensitive to
temperature than early larval stages (Figs. 3 and 4; Table 3).
Salinity and the salinity-temperature interaction initially
reduced survival of developing larvae at the next to the last
stage and in the last larval stage reduced survival under
conditions of high temperature and high salinity. Zinc
addition had the greatest effect on the survival of the last

TABLE 1. Percent survival of the various larval stages of *Palaemonetes pugio* over a range of salinity and temperature conditions when reared with and without the presence of 0.25 ppm Zn throughout the total developmental period.

Larval Stage	Zn (ppm)	20 / 7	20 / 19	20 / 31	25 / 7	25 / 19	25 / 31	30 / 7	30 / 19	30 / 31
I	0.00	100	100	98	98	100	100	99	99	100
	0.25	31	100	100	100	100	100	96	100	100
II	0.00	98	100	98	100	98	94	100	100	100
	0.25	94	100	96	100	100	96	98	99	99
III	0.00	98	100	98	100	100	98	96	100	97
	0.25	100	100	91	100	100	98	97	100	99
IV	0.00	100	98	100	98	100	93	96	99	94
	0.25	100	100	94	100	100	100	99	97	99
V	0.00	100	100	98	100	98	97	92	94	91
	0.25	100	100	97	96	100	98	92	96	97
VI	0.00	100	100	100	98	100	100	94	92	68
	0.25	100	100	100	100	98	98	94	95	96
VII	0.00	98	100	100	98	98	100	98	100	64
	0.25	100	100	97	98	100	98	88	90	74
VIII	0.00	100	100	98	98	100	100			
	0.25	100	100	95	100	100	100			
IX	0.00	100	100	100	100	100	100			
	0.25	100	100	97	100	100	98			
X	0.00	100	93							
	0.25	100	97							
XI	0.00		97							
	0.25	100	96							
XII	0.00		94							
	0.25		92							
XIII	0.00									
	0.25		83							

TABLE 2. Multiple linear regression analysis of the effect of developmental stage, sublethal zinc exposure, salinity, and temperature on survival of developing *Palaemonetes pugio* larvae. Model terms follow those described in Equation (1). Source: source of variation; DF: degrees of freedom; MS: mean square; F value: ratio of treatment mean square to error mean square; PRF: probability of a larger F value. *Significant, **Highly significant.

Source	DF	MS	F value	PR >F
Correctional total	503	99.280		
Model	83	331.072	6.19	.0001**
Stg	6	214.124	4.00	.0007**
Zn	1	333.632	6.24	.0129**
Stg x Zn	6	192.775	3.61	.0017**
Sal	1	819.666	15.33	.0001**
Sal x Sal	1	550.384	10.29	.0014**
Temp	1	1789.912	33.47	.0001**
Temp x Temp	1	1810.905	33.87	.0001**
Sal x Temp	1	5.199	0.10	.7553
Sal x Stg	6	255.703	4.78	.0001**
Sal x Zn	1	41.424	0.77	.3793
Temp x Zn	1	319.210	5.97	.0150*
Temp x Stg	6	192.429	3.60	.0017**
Sal x Stg x Zn	6	525.803	9.83	.0001**
Temp x Stg x Zn	6	113.885	2.13	.0490*
Sal x Sal x Stg	6	355.664	6.65	.0001**
Sal x Sal x Zn	1	0.694	0.01	.9094
Temp x Temp x Stg	6	249.330	4.46	.0001*
Temp x Temp x Zn	1	226.954	4.24	.0400*
Sal x Temp x Stg	6	202.436	3.86	.0009**
Sal x Temp x Zn	1	162.544	3.04	.0820
Sal x Sal x Stg X Zn	6	95.557	1.79	.1002
Temp x Temp x Stg x Zn	6	75.486	1.41	.2086
Sal x Temp x Stg x Zn	6	444.710	8.32	.0001**
Error	420	53.473		

Table 3. Summary of the statistical analysis on the survival of each developmental stage of Palaemonetes pugio as influenced by the quadratic function of salinity (S) and temperature (T) and modifications by continuous zinc exposure. LI, LII, LIII, and LIV-first through the fourth larval form; SL, NL, and LL-second from the last, next to the last, and final larval form.

Developmental Stage	Zn (ppm)	Marked Effect P<.10	Significant Effect P<.05	Highly Significant Effect P<.005	Multiple Correlation
LI	0.00				.297
	0.25		S^2,T^2	S, T, SxT	.858**
LII	0.00	T,T^2			.487
	0.25				.335
LIII	0.00	S^2			.435
	0.25	S^2	SxT		.577*
LIV	0.00				.505
	0.25	T	T^2		.507
SL	0.00	T	T^2		.627*
	0.25	T	T^2,SxT		.717**
NL	0.00		S,T,T^2,SxT		.789**
	0.25	SxT	T,T^2		.564*
LL	0.00	T,T^2,SxT	S,S^2		.752**
	0.25	SxT	S^2	T, T^2	.819**

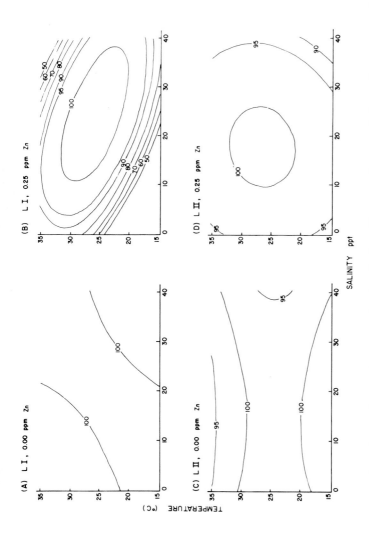

FIGURE 1. Estimation of percent survival of the first and second larval stages of Palaemonetes pugio based on fitted response to observed survival at 9 combinations of salinity and temperature both with and without exposure to 0.25 ppm Zn. LI–first larval stage. LII–second larval stage.

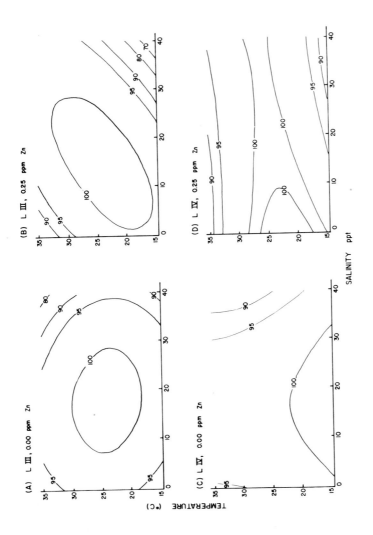

FIGURE 2. Estimation of percent survival of the third and fourth larval stages of Palaemonetes pugio based on fitted response to observed survival at 9 combinations of salinity and temperature both with and without exposure to 0.25 ppm Zn. LIII-third larval stage. LIV-fourth larval stage.

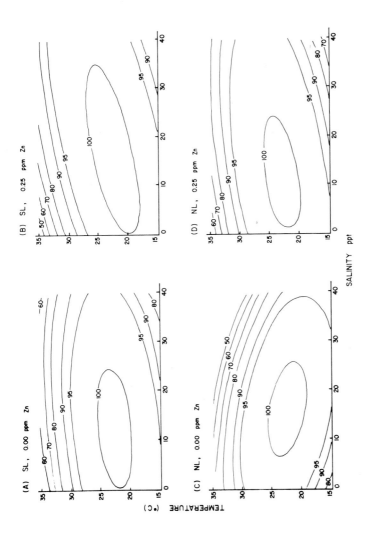

FIGURE 3. Estimation of percent survival of the second to the last larval stage and next to to the last larval stage of Palaemonetes pugio based on fitted response to observed survival at 9 combinations of salinity and temperature both with and without exposure to 0.25 ppm Zn. SL- second from the last larval stage. NL-next to the last larval stage.

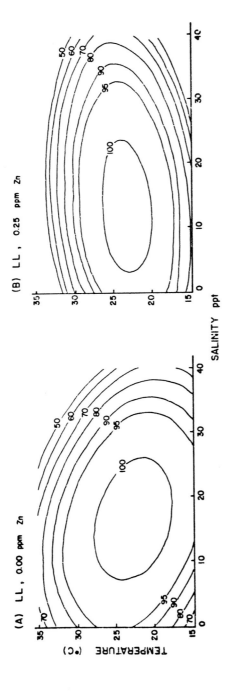

FIGURE 4. Estimation of percent survival of the last larval stage of Palaemonetes pugio based on fitted response to observed survival at 9 combinations of salinity and temperature both with and without exposure to 0.25 ppm Zn. LL–last larval stage.

larval stage by reducing the larval tolerance to temperature.

Oxygen Consumption

Oxygen consumption rates of the various larval stages of
Palaemonetes pugio varied from 2.85 to 10.16 µg O_2 mg^{-1} dry
weight h^{-1} when larvae were reared at a range of salinity-
temperature conditions both with and without sublethal zinc
exposure (Table 4). Multiple linear regression analysis
demonstrated that oxygen consumption of developing larvae was
not only significantly influenced by salinity and temperature,
but that the quadratic function of salinity and temperature
significantly modified larval uptake of oxygen dependent on
stage of development and a zinc-stage interaction (Table 5).
A Duncan's Multiple Range Test by stage of the pooled data for
zinc-exposed and unexposed larvae suggested a significantly
(alpha level = .05) lower oxygen consumption rate among the
last three larval stages and the postlarvae. In addition, a
number of interactions significantly influenced the rate of
oxygen uptake by developing larvae, including in order of
importance: temperature and stage; salinity, temperature and
stage; salinity and stage; salinity, temperature, stage, and
zinc; and salinity, stage and zinc (Table 5).
    By using separate quadratic salinity-temperature
equations for each stage and zinc condition, response-surface
curves were generated for each stage-zinc condition over a
range of salinity-temperature conditions. Regression coefficients for each separate equation are provided in the
appendix of McKenney (1979). A summary of the analysis of
variance of these equations is given in Table 6. All
multiple correlations were significant at the .05 alpha
level.
    First stage larvae increased their rate of oxygen
consumption with an increase in temperature to approximately
25°C; above this temperature, salinity interacted with
temperature to produce increasing respiration rates at higher
temperatures when coupled with a decrease in salinity (Fig.
5A). Sublethal zinc exposure exaggerated this salinity-
temperature interaction in first stage larvae (Fig. 5B), such
that it became more evident at lower temperatures and higher
salinities to the point that above 25°C a lowering of the
salinity dominated over temperature to increase the uptake of
oxygen. Furthermore, at higher temperature-salinity conditions oxygen consumption rates were depressed. No such
interaction was evident in the metabolic rates of second-stage
larvae (Fig. 5C). However, second-stage larvae subjected to a
zinc concentration of 0.25 ppm Zn exhibited both a significant

Table 4. Weight-specific oxygen consumption (μm mg^{-1} hr^{-1}) for the various larval stages of *Palaemonetes pugio* over a range of salinity and temperature conditions when measured with and without the presence of a sublethal zinc concentration (0.25 ppm Zn) throughout development.

Larval Stage	Zn (ppm)	Temp 7	Temp 20 / Sal 19	Temp 20 / Sal 31	Temp 25 / Sal 7	Temp 25 / Sal 19	Temp 25 / Sal 31	Temp 30 / Sal 7	Temp 30 / Sal 19	Temp 30 / Sal 31
I	0.00	4.23±.08	3.31±.06	3.17±.12	6.22±.30	5.62±.70	6.40±.29	10.16±.72	6.15±.16	6.18±.88
	0.25	4.37±.16	3.62±.12	3.97±.03	7.27±.33	5.86±.31	5.84±.13	8.56±.51	6.41±.83	3.94±.53
II	0.00	3.38±.03	4.05±.07	3.97±.25	5.94±.35	5.67±.25	4.48±.29	8.57±.37	4.91±.35	8.08±.34
	0.25	5.00±.24	3.87±.32	3.89±.14	6.96±.37	5.31±.32	5.14±.23	7.18±.47	5.71±.65	8.84±.14
III	0.00	5.12±.13	3.62±.29	4.00±.07	5.59±.25	5.15±.24	5.36±.22	6.49±.48	6.03±.50	6.52±.23
	0.25	4.51±.21	3.89±.21	4.17±.16	5.64±.37	5.17±.18	5.71±.11	6.58±.25	7.98±.55	7.65±.40
IV	0.00	4.69±.08	2.85±.20	3.59±.23	5.79±.23	5.60±.34	5.37±.27	5.47±.15	5.11±.24	5.37±.12
	0.25	5.27±.37	3.14±.27	3.60±.09	5.18±.19	6.25±.26	5.08±.28	4.90±.25	4.61±.24	4.71±.34
V	0.00	5.55±.23	3.85±.27	3.49±.08	5.87±.62	5.12±.18	5.37±.12	5.54±.20	4.47±.22	4.91±.23
	0.25	5.80±.28	4.01±.13	3.41±.04	5.80±.32	5.34±.34	5.07±.19	5.62±.13	4.55±.21	3.59±.33
VI	0.00	5.57±.19	3.90±.22	3.55±.12	5.40±.28	4.37±.09	6.20±.23	5.26±.32	3.49±.14	4.41±.35
	0.25	5.92±.09	3.81±.19	4.37±.24	5.70±.15	5.38±.31	5.29±.21	4.78±.32	4.78±.32	4.45±.32
VII	0.00	5.23±.17	4.77±.14	4.33±.44	4.96±.22	4.08±.25	4.21±.32	5.32±.30	4.56±.45	5.27±.32
	0.25	5.15±.53	3.74±.19	3.43±.08	4.66±.29	3.83±.20	5.21±.09	5.07±.27	5.47±.16	3.01±.32
VIII	0.00	4.68±.21	4.12±.25	4.16±.22	4.37±.27	3.21±.07	4.34±.34			
	0.25	4.72±.51	3.33±.12	4.03±.25	4.60±.17	3.76±.39	4.43±.41			
IX	0.00	4.18±.04	4.03±.21	3.88±.10	5.03±.43	4.09±.25	3.31±.18			
	0.25	4.02±.26	3.83±.15	3.53±.12	4.93±.16	4.10±.15	4.09±.11			
X	0.00	4.41±.24		3.16±.14			3.90±.10			
	0.25	4.69±.39		3.36±.22			4.42±.15			
XI	0.00			3.55±.05			3.62±.28			
	0.25			3.15±.23			3.43±.17			
XII	0.00			3.05±.09			3.36±.06			
	0.25			3.67±.11						
XIII	0.00			3.52±.22						
	0.25									
Post-Larva	0.00	4.08±.21	3.54±.17	2.86±.29	4.62±.29	4.51±.26	5.14±.34	5.01±.15	5.07±.23	4.53±.79
	0.25	3.7±.29	4.03±.33	3.30±.15	5.23±.26	4.43±.16	5.23±.19	5.05±.42	4.83±.28	3.88±.31

TABLE 5. Multiple linear regression analysis of the effects of developmental stage, sublethal zinc exposure, salinity, and temperature on the oxygen consumption of developing *Palaemonetes pugio* larvae. Model terms follow those described in Equation (1). Source: source of variation; DF: degrees of freedom; MS: mean square; F value: ration of treatment mean square to error mean square; PR>F: Probability of a larger F value. *Significant: **Highly Significant.

Source	DF	MS	F value	PR F
Corrected total	650	2.0358		
Model	95	10.2155	16.07	.0001**
Stg	7	6.4295	10.11	.0001**
Zn	1	1.5212	2.39	.1224
Stg x Zn	7	0.7116	1.12	.3489
Sal	1	12.9415	20.36	.0001**
Sal x Sal	1	3.8436	6.05	.0142*
Temp	1	14.7519	23.21	.0001**
Temp x Temp	1	0.6972	1.10	.2954
Sal x Temp	1	0.0000	0.00	.9950
Sal x Stg	7	2.8473	4.48	.0001**
Sal x Zn	1	1.8857	2.97	.0856
Temp x Zn	1	1.4670	2.31	.1293
Temp x Stg	7	6.2099	9.77	.0001**
Sal x Stg x Zn	7	1.9709	3.10	.0034**
Temp x Stg x Zn	7	0.5885	0.93	.4867
Sal x Sal x Stg	7	0.7653	1.20	.2980
Sal x Sal x Zn	1	1.0886	1.71	.1912
Temp x Temp x Stg	7	4.9310	7.76	.0001**
Temp x Temp x Zn	1	0.0128	0.02	.8873
Sal x Temp x Stg	7	5.0030	7.87	.0001**
Sal x Temp x Zn	1	0.5275	0.83	.3627
Sal x Sal x Stg x Zn	7	0.6182	0.97	.4510
Temp x Temp x Stg x Zn	7	0.5118	0.81	.5847
Sal x Temp x Stg x Zn	7	17.3479	3.90	.0004**
Error	555	0.6357		

TABLE 6. Summary of the statistical analysis on the oxygen consumption of each developmental stage of *Palaemonetes pugio* as influenced by the quadratic function of salinity (S) and temperature (T) and modifications by continual sublethal zinc exposure. LI, LII, LII, LIV—first through the fourth larval form; SL, NL, and LL— second from the last, next to the last, and final larval form; and PL— postlarva. *P<.05, ** P<.005.

Developmental Stage	Zn (ppm)	Marked Effect P<.10	Significant Effect P<.05	Highly Significant Effect P<.005	Multiple Correlation
LI	0.00	S², T, SxT			.800**
	0.25			T,T²,SxT	.873**
LII	0.00		S		.810**
	0.25		S²	S,S², SxT	.847**
LIII	0.00	SxT	S²	S	.840**
	0.25				.900**
LIV	0.00	SxT	S²	S,T,T²	.865**
	0.25			T,T²	.726**
SL	0.00	S	T,T²		.780**
	0.25		SxT		.767**
NL	0.00	T,T²	S	S²	.680**
	0.25	T²	S		.675**
LL	0.00	SxT	S		.711**
	0.25		SxT		.860**
PL	0.00		T,T²	T,T²	.750**
	0.25				.668**

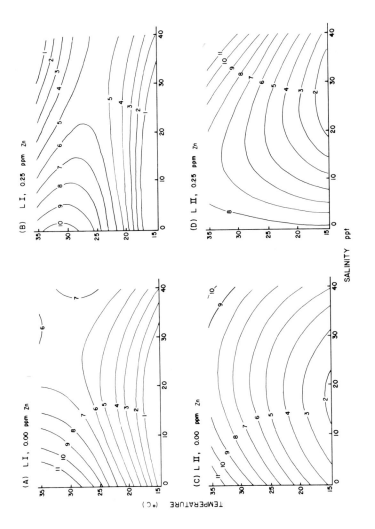

FIGURE 5. Estimation of the weight-specific oxygen consumption rates (μg O_2 mg^{-1} dry weight h^{-1}) of the first and second larval stages of Palaemonetes pugio based on fitted response to observed rates at 9 combinations of salinity and temperature both with and without exposure to 0.25 ppm Zn. LI—first larval stage. LII—second larval stage.

salinity-temperature interaction and a significant influence
by salinity on oxygen consumption rates (Table 6 and Fig. 5D).
At salinities above 20 $^\circ/_{oo}$ increased temperature accounted
for the increase in oxygen consumption, while at lower salinities the influence of salinity dominated over elevated
temperatures.

Salinity acted as the dominant factor affecting oxygen
consumption of third-stage larvae by increasing the respiration rate at lower salinities (Fig. 6A). Sublethal
exposure to zinc masked this relationship (Fig. 6B). Temperature exerted the grestest effect on the oxygen consumption
of the fourth larval stage both with and without zinc
exposure (Table 6 and Fig. 6C and 6D). Temperature increases
to approximately 25°C produced an increase in the uptake of
oxygen, at which point rates were stable (except at salinity
extremes) to a temperature of about 31°C. Above this
temperature rates were depressed by a temperature increase.

Weight-specific oxygen-consumption rates of late larval
stages generally were less influenced by salinity and
temperature (Figs. 7 and 8). Yet, the second to the last
larval stage exhibited increased metabolic rates at temperatures higher and lower than 25°C and at lower salinities
(Fig. 7A). In contrast, this larval stage when exposed to
zinc increased oxygen uptake rates at low salinities and high
temperatures (Fig. 7B). Unlike any other larval stage,
contours for the estimated respiration rates of the next to
the last larval stage demonstrated a salinity-temperature area
at which oxygen uptake was minimal (Fig. 7C). Outside this
area oxygen consumption increased, particularly at lower
salinities. Oxygen uptake was reduced at low temperature and
high salinity conditions for larvae exposed to 0.25 ppm Zn
(Fig. 7D). At temperatures below 30°C, respiration rates for
the last larval stage increased with decreasing salinities
(Fig. 8A). Conversely, oxygen uptake by this stage when
exposed to zinc was depressed and proved most sensitive to
salinity at higher temperatures (Fig. 8B). A dramatic shift
in salinity-temperature metabolic patterns occurred following
completion of metamorphosis (Fig. 8C). Oxygen uptake of
postlarvae was little affected by salinity, but principally
influenced by temperature. Rates of oxygen consumption were
depressed outside the approximate range of 25° to 30°C and
more noticeably so for zinc-exposed larvae at higher
temperatures and higher salinities (Fig. 8D).

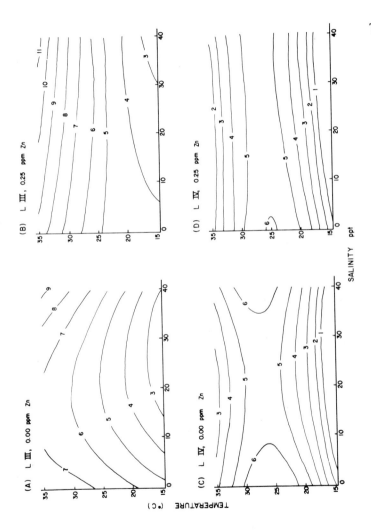

FIGURE 6. Estimation of the weight-specific oxygen consumption rates ($\mu gO_2 \; mg^{-1} dry \; weight \; h^{-1}$) of the third and fourth larval stages of Palaemonetes pugio based on fitted response to observed rates at 9 combinations of salinity and temperature both with and without exposure to 0.25 ppm Zn. LIII-third larval stage. LIV-fourth larval stage.

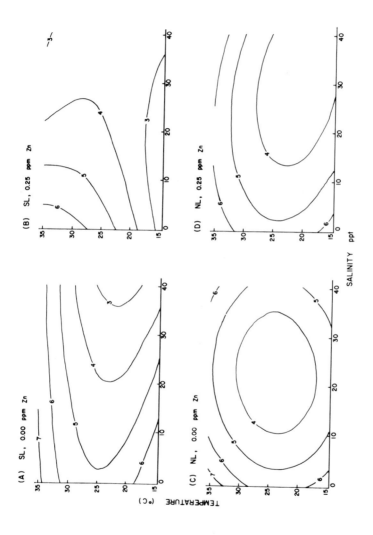

FIGURE 7. Estimation of the weight-specific oxygen consumption rates (μg O_2 mg^{-1} dry weight h^{-1}) of the second from the last and the next to the last larval stages of Palaemonetes pugio based on fitted response to observed rates at 9 combinations of salinity and temperature both with and without exposure to 0.25 ppm Zn. SL-second from the last larval stage. NL-next to the last larval stage.

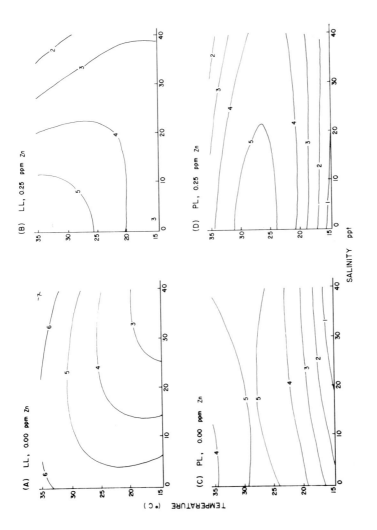

FIGURE 8. Estimation of the weight-specific oxygen consumption rates ($\mu g\ O_2\ mg^{-1}\ dry\ weight\ h^{-1}$) of the last larval stage and postlarva of Palaemonetes pugio based on fitted response to observed rates at 9 combinations of salinity and temperature both with and without exposure to 0.25 ppm Zn. LL–last larval stage. PL–postlarva.

225

DISCUSSION

Survival of Individual Larval Stages

An earlier study demonstrated that not only were
survival patterns of *Palaemonetes pugio* larvae throughout
development modified by salinity and temperature conditions,
but that exposure to concentrations of zinc which were sublethal at optimal salinity-temperature conditions reduced
viability through complete larval development under stressful
levels of salinity and temperature (McKenney and Neff, 1979).
It appears, furthermore, that differential sensitivity to
salinity, temperature, and/or zinc existed among the various
individual developmental stages as indicated by the highly
significant four factor interaciton term, salinity-temper-
ature-zinc-stage. In the absence of additional zinc, early
larval stages were extremely tolerant of a broad range of
salinities and temperatures. Conversely, the late larval
stages beginning with the second to the last larval stage
appeared more sensitive to high temperatures, with this
pattern shifting to reduced viability of the last larval stage
at conditions of high temperature and high salinity. The
greater sensitivity of the last larval stage to environmental
stress may indicate the more delicate nature of this premetamorphic stage, possibly associated with a greater
energetic demand prior to metamorphosis.
These changing salinity-temperature survival patterns
during the ontogeny of *Palaemonetes pugio* were, in addition,
modified by exposure to a sublethal zinc concentration and
these alterations were stage dependent. The first and last
larval stages were most sensitive to zinc exposure over a
broad range of constant salinity and temperature conditions.
In the presence of a sublethal zinc concentration, the first
larval stage was significantly less resistant to low salinities at low temperatures. Following zinc exposure from
hatch throughout complete larval development, the last larval
stage became highly sensitive to temperature extremes prior to
completion of metamorphosis.
With the additional stress of zinc exposure, the reduced
viability of first-stage larvae to low salinity waters could
be indicative of disruption of a less well developed system
for coping with hypo-osmotic stress during this early
developmental stage and, perhaps, heavy metal inhibition of
enzyme systems associated with hypo-osmotic regulation
(Schmidt-Nielsen, 1974; Bryan, 1976a, b). Although osmoregulatory abilities of larval *P. pugio* have not been investigated, postlarval *P. pugio* have been shown to be able to

regulate body fluids both hypo- and hyper-osmotically to environmental salinities outside their isosmotic salinity of 17 °/°° (Roesijadi et al., 1976). Such osmoregulatory abilities of estuarine organisms have been altered by sublethal heavy metal exposure (Thurberg et al., 1973; Jones, 1975a, b; Inman and Lockwood, 1977).

These variations in resistance patterns to salinity and temperature during the larval development of P. pugio reflect the ontogeny of capacity for physiological adaptation to variable environmental conditions. As larval development progresses, the last two larval stages demonstrated reduced resistance to high salinities. This pattern suggests, therefore, an increased capacity for P. pugio to tolerate low salinities accompanied by a reduced capacity to tolerate high salinities. Such a developmental pattern of salinity tolerance in P. pugio agrees well with observations that freshly hatched larvae are released by ovigerous females found more abundantly in higher salinity waters (Wood, 1967), inferring limited hyper-regulatory abilities of early larval stages.

Respiration of Individual Larval Stages

As is common with poikilotherms, the oxygen consumption of early larval stages of P. pugio had a marked temperature dependence. Assuming oxygen consumption rates to be indirect measurements of metabolic rates, the metabolic rates of these early developmental stages were directly related to temperature. As temperature increased oxygen consumption rates were increased. Temperature dependent metabolism has been similarily reported for other decapod larvae (Vernberg and Costlow, 1966; Belman and Childress, 1973; Sastry and McCarthy, 1973; Nelson et al., 1977; Schatzlein and Costlow, 1978). However, as larval development of P pugio continued, compensatory adjustments in metabolism allowed for a reduced temperature dependence in the oxygen consumption rates of the later larval stages. Upon completion of metamorphosis, oxygen consumption rates of postlarval P. pugio were highest at optimal temperatures for resistance and depressed at stressful high and low temperature extremes. Similar to these results, Sastry and Vargo (1977) reported that the earlier larval stages of P. pugio from Rhode Island were more metabolically sensitive to temperature fluctuations than later developmental stages.

The changing capacities of Palaemonetes pugio larvae to adjust metabolic rates to changes in temperature levels during ontogeny may reflect certain biochemical adjustments. Such biochemical mechanisms responsible for metabolic compensations to temperature include alterations in enzyme, substrate, and/ or cofactor concentrations; variations in enzyme activities;

modulation of enzyme activities through the use of multiple
enzyme forms (isozymes); and/or formation of enzyme variants
with unique catalytic efficiencies (Hazel and Prosser, 1974;
Somero and Hochachka, 1976). Although compensatory metabolic
adjustments to temperature appear to be species-specific
among marine crustacean larvae and reflective of individual
tolerance ranges (Sastry and McCarthy, 1973; Sastry and Vargo,
1977), the results of recent biochemical studies on developing
early life stages of several estuarine crustaceans suggest
the application of similar adaptive mechanisms in the
metabolic regulation of these larvae to changes in temperature, including observations of increased enzyme activities
and increased isozyme concentrations later in larval development (Frank *et al.*, 1975; Morgan *et al.*, 1978; Sastry, 1978;
Sastry and Ellington, 1978).

In addition to temperature-sensitive metabolic patterns,
metabolic rates of developing *P. pugio* larvae were differen-
tially altered by the salinity in which they were reared.
These variations in respiratory rates were further modified by
a salinity-temperature interaction for certain larval stages.
Oxygen consumption of the first larval stage was progressively
increased by salinities less than 20 $^{\circ}/_{\circ\circ}$, but only at
temperatures above 25°C. Both lower and higher salinities
than that which is isosmotic for postlarval and juvenile
P. pugio (17 $^{\circ}/_{\circ\circ}$S; Roesijadi *et al.*, 1976) resulted in
elevated respiratory rates of the second and third larval
stages, particularly at higher temperatures. The dominant
influence of temperature on metabolic patterns of the fourth
larval stage masked the effects of salinity except at the
optimal temperature of 25°C where respiratory rates were
elevated at low and high salinity extremes. Unlike salinity-
metabolic patterns of the early larval stages, respiratory
rates of the last three larval stages were dominantly
influenced by changes in salinity. At temperatures below
approximately 30°C, oxygen consumption of the second to the
last and the last larval stages was increased as salinity
decreased. The metabolic pattern of the next to the last
larval stage was unique; a temperature-salinity area existed
within which oxygen consumption was minimum and outside this
area metabolic rates were increased principally by salinities
below 11 $^{\circ}/_{\circ\circ}$. Following completion of metamorphosis postlarval respiration was essentially unaffected by a change in
salinity level.

A number of physiological mechanisms have been described
which aid in the adaption of estuarine organisms to changes in
salinity (i.e., volume regulation, isosmotic intracellular
regulation and anisosmotic extracellular regulation) (Potts
and Parry, 1964; Florkin and Schoffeniels, 1965; Prosser,

1973; Gilles, 1975; Lockwood, 1976; Smith, 1976). Several
studies have described the functioning of some of these
mechanisms in *P. pugio* (Roesijadi *et al.*, 1976; Lucu *et al.*,
1977). Measurements of the hemolymph osmotic concentrations
of juvenile and adult *P. pugio* in salinities of 2 to 32 °/oo
indicated a highly developed hypo- and hyperosmotic regulation
of body fluids as early as the postlarval stage (Roesijadi *et
al.*, 1976). Both hemolymph sodium (Lucu *et al.*, 1977) and
chloride (Roesijadi *et al.*, 1976) concentrations were maintained at relatively stable levels over this wide salinity
range. Water permeability of *P. pugio* was greatest at the
isosmotic salinity (17 °/oo S) and was reduced at salinities
associated with active osmoregulation (Roesijadi *et al.*,
1976). Unfortunately, similar studies describing osmoregulatory abilities of developing larval *P. pugio* have not
been reported.

        In general, the metabolic rate (oxygen consumption rate)
of a euryhaline invertebrate will either (1) increase in sub-
normal salinities and/or decrease in supranormal salinities or
(2) increase in sub- and supranormal salinities. The sal-
inity-metabolic response of stenohaline invertebrates generally involves a decrease in oxygen consumption in sub- and
supranormal salinities, while the respiratory rate of holeuryhaline invertebrates remains basically unaltered by
salinity fluctuations (for a review of this subject see Kinne,
1971). For example, oxygen consumption of the euryhaline
crab, *Panopeus herbstii,* was elevated in low salinities
(Dimock and Groves, 1975). Respiration rates of the stenohaline osmoconformers, *Murex pomum* and *Strombus pugilis,*
declined in stressful low salinities (Sander and Moore, 1978).
Chronic exposures to salinities ranging from 3.5 to 32 °/oo
produced no appreciable variation in the routine metabolic
rate of adult *Palaemonetes vulgaris* (McFarland and Pickens,
1965).

        The respiratory responses of developing *Palaemonetes
pugio* larvae to various salinity conditions basically followed
the patterns described for euryhaline organisms. In general,
oxygen consumption rates of early larval stages beyond the
first stage were increased by both sub- and supranormal
salinities, while those of the later larval stages were increased only by a decrease in salinity. The metabolic-salinity pattern of the postlarval stage was essentially
unaffected by salinity and, therefore, followed the response
pattern of a holeuryhaline organism, similar to that described
for adult *P. vulgaris* (McFarland and Pickens, 1965). This
ability to compensate metabolically to low salinity stress
might be an important factor in the distribution of this
species in upper reaches of estuaries (Bowler and Seidenberg,

1971; Thorp and Hoss, 1975).

Several explanations have been proposed regarding the physiological mechanisms that might be responsible for the observed alterations in respiratory rates with a change in salinity. While most authors have suggested that increased metabolic rates at salinities differing from the isomotic point are indicative of the increased energy cost due to osmoregulation, Potts (1954) produced a theoretical model indicating the metabolic cost of osmoregulation to be less than 2% of the basal metabolic rate. However, Shaw (1959) pointed out that the assumption that the outer surfaces of the organisms are semipermeable ("permeable to water but imper- meable to salts"; Potts, 1954) was invalid. Recent evidence (as reviewed by Smith, 1976; Lockwood, 1976) suggests that the ability to vary surface permeability to both salts and water in response to changing salinities is an important adaptive osmoregulatory mechanism for a number of euryhaline inverte- brates. Spaargaren (1974, 1975a, b), furthermore, has dev- eloped mathematical models, based on empirical evidence, to describe the effect of osmotic stress on the extracellular electrolyte concentration and energy relations in several species of crustaceans. These models suggest that the energy required for ionic regulation by *Carcinus meanas* is only a small part of the basal metabolism within the normal range of salinities experienced by the crab, but that at extreme salinities a substantial part of the total energy consumption is occupied in ionic regulation (Spaargaren, 1975b). More recently, Cameron (1978) calculated that the active Na^+ and Cl^- uptake system in the blue crab *Callinectes sapidus* would account for 20% of the total oxygen demands of the organisms in freshwater. Similarly, variations in respiratory rates of the shore crab *Carcinus meanas* to both salinity and temper- ature were related to changing patterns of osmoregulation (Taylor *et al.*, 1977).

The metabolic response of larvae to zinc was complicated by a significant zinc-salinity-temperature-stage interaction, indicating that zinc disrupted the metabolic patterns of certain developmental stages dependent on the salinity and temperature condition. The most pronounced influence of zinc exposure on the O_2 consumption of developing *Palaemonetes pugio* larvae was a depression of the rates in the first stage, second from the last and the last larval stage under con- ditions of moderate to high temperatures in high salinities and in low salinity-temperature conditions. This disruption of metabolic patterns of late larval stages continued following completion of metamorphosis with the respiratory rates of postlarvae depressed by zinc at high salinities and moderate to high temperatures. These alterations in metabolic capacity are in agreement with reduced resistance adaptations previously described.

Zinc concentrations as high as 480 ppb have been measured in Corpus Christi Bay, Texas (Holmes et al., 1974) and as high as 530 ppb in the Galveston Bay Estuarine System (Texas Water Quality Board, 1975). Since zinc concentrations reported from these estuaries are similar to those to which larvae were exposed in this study, speculation is appropriate. Disruption by heavy metal exposure of compensatory osmoregulatory adaptations have been described for a number of estuarine species (Schmidt-Nielsen, 1974; Jones, 1975a, b; Inman and Lockwood, 1977). Altered serum osmolalities in Carcinas meanas following exposure to cadmium was accompanied by a depressed oxygen consumption rate of gill-tissue (Thurberg et al., 1973). Moreover, heavy metal toxicity is thought to be associated with inhibition of enzyme systems (Vallee and Ulmer, 1972; Bryan, 1976b). The depressed metabolic rates of certain larval stages of Palaemonetes pugio, along with reduced salinity resistance following zinc exposure, might reflect disruptions in osmoregulatory mechanisms in developing larvae resulting from enzyme inhibition by zinc.

It is interesting to note that as larval development of Palaemonetes pugio progresses through completion of metamorphosis, both reduced resistance capacities and reduced metabolic capacities following zinc exposure are associated to a greater extent with high salinities. The hyporegulation of body fluids, as reported for postlarval and juvenile P. pugio (Roesijadi et al., 1976), is a relatively rare process among estuarine or marine invertebrates (Kinne, 1971; Gilles, 1975; Lockwood, 1976). The relatively greater metabolic sensitivity to zinc in higher salinities later in larval development may indicate less well developed physiological mechanisms for hyporegulation than for hyperregulation at this stage of development. Since adaptive capacities and the mechanisms involved are evolutionary products, then the ability to hyperregulate in freshwater during the evolutionary history of a genus might accompany a loss in ability to compensate to high salinities (Kinne, 1963; Hubschman, 1975). The changing ontogenetic adaptive capacities of P. pugio could be indicative of the phylogenetic migration of the genus Palaemonetes from life in marine waters to life in freshwater.

SUMMARY

     Changing patterns in survival and metabolic rates were
measured during the complete larval development of the
estuarine shrimp, *Palaemonetes pugio,* reared from hatch
through metamorphosis in 9 combinations of salinity (7-
31 °/$_\circ$$_\circ$S) and temperature (20-30°C), both with and without
exposure to a sublethal zinc concentration (0.25 ppm Zn).
While early larval stages are extremely tolerant of a broad
range of salinities and temperatures, late larval stages,
beginning with the second to the last larval stage, are more
sensitive to temperature than early larval stages. Furthermore, as larval development progresses, the last two larval
stages demonstrate reduced resistance to high salinities.
These changing salinity-temperature survival patterns during
the ontogeny of *P. pugio* are modified by continuous sublethal
zinc exposure and these alterations are stage dependent with
the first and last larval stages being most sensitive to zinc.
Upon exposure to zinc, the first larval stage is significantly
less resistant to low salinities, especially at low temperatures, and the last larval stage, prior to completion of
metamorphosis, becomes highly sensitive to high temperature
extremes.
     As with resistance, metabolic rates of developing *P.
pugio* larvae were influenced by a salinity-temperature-stage-
zinc interaction. Metabolic rates of early larval stages of
P. pugio have a marked temperature dependence. As larval
development continues, compensatory adjustments in metabolism
allow for a reduced dependence of the respiration of the late
larval stages to temperature. Upon completion of metamorphosis, metabolic rates are maximum in optimal temperatures
for resistance and slightly depressed in stressful high and
low temperature extremes. The metabolic responses of
developing larvae to various salinity conditions basically
follow the pattern described for euryhaline organisms. In
general, metabolic rates of early larval stages beyond the
first stage are increased by both sub- and supranormal
salinities, while metabolic rates of late larval stages are
increased by a decrease in salinity. The metabolic-salinity
pattern of the postlarval stage is essentially unaffected by
salinity and, therefore, follows the response pattern of a
holeuryhaline organism.
     The most pronounced influence of sublethal zinc exposure
on metabolic rates of developing larvae was a depression of
the respiratory rates of the first stage, second from the
last and the last larval stage under conditions of moderate
to high temperatures in high salinities and in low salinity-

temperature conditions. These alterations in metabolic
capacity are in agreement with reduced resistance adaptations
of *P. pugio* larvae reared under continuous exposure to a
sublethal zinc concentration. Since adaptive capacities and
the mechanisms involved are evolutionary products, it appears
that changing ontogenetic adaptive capacities of *P. pugio*
reflect phylogenetic migration from life in marine waters to
life in freshwater with the ability to resist and compensate
metabolically to less saline waters during the ontogeny of the
species accompanying a loss in the ability to compensate to
higher salinities under additional stressful conditions as
imposed by sublethal zinc exposure.

ACKNOWLEDGMENTS

This study was supported by Grant No. ID075-04890 from
the National Science Foundation I.D.O.E. Biological Effects
Program to Jerry M. Neff. We gratefully acknowledge the
patient assistance of Margaret Weirich during portions of this
study.

LITERATURE CITED

Alderdice, D. F. 1972. Factor combinations: responses of
    marine poikilotherms to environmental factors acting in
    concert. In: Marine ecology. Vol. 1. Environmental
    factors, Part 3. pp. 1659-1772. Ed. by O. Kinne.
    London: Wiley-Interscience.

Bayne, B. L., J. Widdows, and R. H. Thompson. 1976. Physio-
    logical integrations. In: Marine mussels: their
    ecology and physiology. pp. 261-291. Ed. by B. L.
    Bayne. Cambridge: Cambridge University Press.

Belman, B. W., and J. J. Childress. 1973. Oxygen consumption
    of the larvae of the lobster *Cancer productus* (Randall).
    Comp. Biochem. Physiol. 44A, 821-828.

Bowler, M. W., and A. J. Seidenberg. 1971. Salinity tol-
    erances of the prawns, *Palaemonetes vulgaris* and *P. pugio*
    and its relationship to the distribution of these species
    in nature. Va. J. Sci. 22, 94.

Box, G. E. P. 1954. The exploration and exploitation of response surfaces: Some general considerations and examples. Biometrics 10. 16-60.

Box, G. E. P., and P. V. Youle. 1955. The exploration and exploitation of response surfaces: An example of the link between the fitted surfaces and the basic mechanism of the system. Biometrics 11, 287-323.

Bryan, G. W. 1976a. Some aspects of heavy metal tolerance in aquatic organisms. In: Effects of pollutants on aquatic organisms. pp. 7-34. Ed. by A. P. M. Lockwood. Cambridge: Cambridge University Press.

Bryan, G. W. 1976b. Heavy metal contamination in the sea. In: Marine pollution. pp. 185-302. Ed. by R. Johnston. London: Academic Press.

Cameron, J. N. 1978. NaCl balance in blue crabs, *Callinectes sapidus,* in fresh water. J. Comp. Physiol. 123, 127-135.

Dimock, R. V., Jr., and K. H. Groves. 1975. Interaction of temperature and salinity on oxygen consumption of the estuarine crab. *Panopeus herbstii.* Mar. Biol. 33, 301-308.

Engel, D. W., and J. W. Angelovic. 1968. The influence of salinity and temperature upon the respiration of brine shrimp nauplii. Comp. Biochem. Physiol. 26. 749-752.

Florkin, M., and E. Schoffeniels. 1965. Euryhalinity and the concept of physiological radiation. In: Studies in comparative biochemistry. pp. 6-40. Ed. by K. A. Munday Oxford: Pergamon Press.

Frank, J. R., S. D. Sulkin, and R. P. Morgan, II. 1975. Biochemical changes during larval development of the xanthid crab, *Rhithropanopeus harrisii.* I. Protein, total lipid, alkaline phosphatase, and glutamic oxaloacetic transaminase. Mar. Biol. 32, 105-111.

Gilles, R. 1975. Mechanisms of ion and osmoregulation. In: Marine ecology. Vol. II. Physiological mechanisms, Part 1. pp. 259-347. Ed. by O. Kinne. London: Wiley-Interscience.

Hazel, J., and C. L. Prosser. 1974. Molecular mechanisms of temperature compensation in poikilotherms. Physiol. Rev. 54, 620-677.

Holmes, C. W., E. A. Slade, and C. J. McLerran. 1974. Migration and redistribution of zinc and cadmium in marine estuarine system. Envir. Sci. Technol. 8, 255-259.

Hubschman, J. H. 1975. Larval development of the freshwater shrimp *Palaemonetes kadiakensis* Rathburn under osmotic stress. Physiol. Zool. 48, 96-107.

Inman, C. B. E., and A. P. M. Lockwood. 1977. Some effects of methyl-mercury and lindane on sodium regeneration in the amphipod *Gammarus duebeni* during changes in the salinity of its medium. Comp. Biochem. Physiol. 58C. 67-75.

Jones, M. B. 1973. Influence of salinity and temperature on the toxicity of mercury to marine and brackish water isopods (Crustacea). Estuar. Cstl. Mar. Sci. 1, 425-431.

Jones, M. B. 1975a. Synergistic effects of salinity, temperature and heavy metals on mortality and osmoregulation in marine and estuarine isopods (Crustacea). Mar. Biol. 30, 13-20.

Jones, M. B. 1975b. Effects of copper on survival and osmoregulation in marine brackish water isopods (Crustacea). In: Proceedings of the ninth European marine biology symposium. pp. 419-431. Ed. by H. Barnes Aberdeen: Aberdeen University Press.

Kinne, O. 1963. Adaptation, a primary mechanism of evolution In: Phylogeny and evolution of crustacea. pp. 27-50. Ed. by H. B. Whittington and W. D. I. Rolfe. Harvard: Spec. Publ. mus. Comp. Zool.

Kinne, O. 1964. The effects of temperature and salinity on marine and brackish water animals. II. Salinity and temperature-salinity combinations. Oceanog. Mar. Biol. Ann. Rev. 2, 281-339.

Kinne, O. 1970. Temperature: animals-invertebrates. In: Marine ecology. Vol. 1 Environmental factors, Part 1. pp. 407-514. Ed. by O. Kinne. London: Wiley-Inter-science.

Kinne, O. 1971. Salinity: animals-invertebrates. In: Marine ecology. Vol. 1. Environmental factors, Part 2. pp. 821-995. Ed. by O. Kinne. London: Wildy-Interscience.

Lockwood, A. P. M. 1976. Physiological adaptations to life in estuaries. In: Adaptation to environment. pp. 315-392. Ed. by R. C. Newell. London: Butterworths.

Lucu, C., G. Roesijadi, and J. W. Anderson. 1977. Sodium kinetics in the shrimp *Palaemonetes pugio*. I. Steady-state and nonsteady-state experiments. J. Comp. Physiol. 115, 195-206.

McFarland, W. N., and P. E. Pickens. 1965. The effects of season, temperature, and salinity on standard and active oxygen consumption of the grass shrimp, *Palaemonetes vulgaris* (Say). Can. J. Zool. 43. 571-585.

McKenney, C. L., Jr. 1979. Ecophysiological studies on the ontogeny of euryplasticity in the caridean shrimp, *Palaemonetes pugio* (Holthuis) and modifications by zinc, 217 p. Ph.D. dissertation, Texas A&M University.

McKenney, C. L., Jr., and J. D. Costlow, Jr. 1977. Interaction of temperature, salinity, and mercury on larval development of the xanthid crab, *Rhithropanopeus harrisii* (Gould). Amer. Zool. 17, 922.

McKenney, C. L., Jr. and J. M. Neff. 1979. Individual effects and interactions of salinity, temperature, and zinc on larval development of the grass shrimp *Palaemonetes pugio*. I. Survival and developmental duration through metamorphosis. Mar. Biol. 52, 177-188,

Morgan, R. P., E. Kramarsky, and S. D. Sulkin. 1978. Biochemical changes during larval development of the xanthid crab *Rhithropanopeus harrisii*. III. Isozyme changes during otogeny. Mar. Biol. 48, 223-226.

Myers, R. H. 1971. Response surface methodology, 246 pp. Boston: Allyn & Bacon, Inc.

Nelson, S. G., D. A. Armstrong, A. E. Knight, and H. W. Li. 1977. The effects of temperature and salinity on the metabolic rate of juvenile *Macrobrachium rosenbergii* (Crustacea: Palaemonidae). Comp. Biochem. Physiol. 56A, 533-537.

Pandian, T. J. 1975. Mechanisms of heterotrophy. In: Marine ecology. Vol. II. Physiological mechanisms, Part 1. pp. 61-250. Ed. by O. Kinne. London: Wiley-Interscience.

Potts, W. T. W. 1954. The energetics of osmotic regulation in brackish- and freshwater animals. J. Exp. Biol. 31, 618-630.

Potts, W. T. W., and G. Parry. 1964. Osmotic and ionic regulation in animals, 423 pp. Oxford: Pergamon Press.

Prosser, C. L. 1973. Comparative animal physiology, 996 pp. Philadelphia: Saunders.

Roesijadi, G., J. W. Anderson, S. R. Petrocelli, and C. S. Giam. 1976. Osmoregulation of the grass shrimp *Palaemonetes pugio* exposed to polychlorinated biphenyls (PCBs). I. Effect on chloride and osmotic concentrations and chloride and water-exchange kinetics. Mar. Biol. 38, 343-355.

Rosenberg, R., and J. D. Costlow, Jr. 1976. Synergistic effects of cadmium and salinity combined with constant and cycling temperature on the larval development of two estuarine crab species. Mar. Biol. 38, 291-303.

Sander, F., and E. A. Moore. 1978. Comparative respiration in the gastropods *Murex pomun* and *Strombus pugilis* at different temperatures and salinities. Comp. Biochem. Physiol. 60A, 99-105.

Sastry, A. N. 1978. Physiological adaptation of *Cancer irroratus* larvae to cyclic temperatures. In: Physiology and behaviour of marine organisms. pp. 57-65. Ed. by D. S. McLusky and A. J. Berry. Oxford: Pergamon Press.

Sastry, A. N., and W. R. Ellington. 1978. Lactate dehydrogenase activity during the larval development of *Cancer irroratus:* effect of constant and cyclic thermal regimes. Experientia 34, 308-309.

Sastry, A. N., and J. F. McCarthy. 1973. Diversity of metabolic adaptations in pelagic larval stages of two sympatric species of brachyuran crabs. Neth. J. Sea Res. 7, 434-446.

Sastry, A. N., and S. L. Vargo. 1977. Variations in the physiological responses of crustacean larvae to temperature. In: Physiological responses of marine biota to pollutants. pp. 401-425. Ed. by F. J. Vernberg, A. Calabrese, F. P. Thurberg, and W. B. Vernberg. New York: Academic Press.

Schatzlein, F. C., and J. D. Costlow, Jr. 1978. Oxygen consumption of the larvae of the decapod crustaceans, *Emerita talpoida* (Say) and *Libinia emarginata* Leach. Comp. Biochem. Physiol. 61A, 441-450.

Schlieper, D. 1971. Physiology of brackish water. In: Biology of brackish water. pp. 211-350. Ed. by A. Remane and C. Schlieper. New York: Wiley-Interscience.

Schmidt-Nielsen, B. 1974. Osmoregulation: effect of salinity and heavy metals. Fedn. Proc. Fedn Am. Socs. Exp. Biol. 33, 2137-2146.

Shaw, J. 1959. Salt and water balance in the east African freshwater crab, *Potamon niloticus* (M. Edw.). J. Exp. Biol. 36, 157-176.

Smith, R. I. 1976. Apparent water-permeability variations and water exchange in crustaceans and annelids. In: Perspectives in experimental biology. Vol. 1. Zoology. pp. 17-24. Ed. by P. S. Davies. Oxford: Pergamon Press.

Somero, G. N., and P. W. Hochachka. 1976. Biochemical adaptations to temperature. In: Adaptation to environment. pp. 125-190. Ed. by R. C. Newell. London: Butterworths.

Spaargaren, D. H. 1974. A study on the adaptation of marine organisms to changing salinities with special reference to the shore crab, *Carinus maenas* (L). Comp. Biochem. Physiol. 47A, 499-512.

Spaargaren, D. H. 1975a. Energy relations in the ion regulation in three crustacean species. Comp. Biochem. Physiol. 51A, 543-548.

Spaargaren, D. H. 1975b. Heat production of the shore-crab *Carcinus maenas* (L). and its relation to osmotic stress. In: Proceedings of the ninth European marine biology symposium. pp. 475-482. Ed. by H. Barnes. Aberdeen:

Aberdeen University Press.

Taylor, E. W., P. J. Butler, and A. Al-Wassia. 1977. The effect of a decrease in salinity on respiration, osmo-regulation and activity in the shore crab, *Carcinus maenas* (L.) at different acclimation temperatures. J. Comp. Physiol. 119, 155-170.

Texas Water Quality Board. 1975. Special report on Houston Ship Channel monitoring program. 9 pp. Report No. SR-3, District 7, mimeogr.

Thorp, J. H., and D. E. Hoss. 1975. Effects of salinity and cyclic temperatures on survival of two sympatric species of grass shrimp (*Palaemonetes*), and their relationship to natural distributions. J. Exp. Mar. Biol. Ecol. 18, 19-28.

Thurberg, R. P., M. A. Dawson, and R. S. Collier. 1973. Effects of copper and cadmium on osmoregulation and oxygen consumption in two species of estuarine crabs. Mar. Biol. 23, 171-175.

Vallee, B. L., and D. D. Ulmer. 1972. Biochemical effects of mercury, cadmium, and lead. An. Rev. Biochem. 41, 91-128.

Vernberg, F. J., and J. D. Costlow, Jr. 1966. Studies on physiological variation between tropical and temperate zone fiddler crabs of the genus *Uca*. IV. Oxygen con-sumption of larvae and young crabs reared in the lab-oratory. Physiol. Zool. 39, 36-52.

Vernberg, W. B., P. J. DeCoursey, and J. O'Hara. 1974. Multiple environmental factor effects on physiology and behaviour of the fiddler crab, *Uca pugilator*. In: Pollution and physiology of marine organisms. pp. 381-426. Ed. by F. J. Vernberg and W. B. Vernberg. New York: Academic Press.

Vernberg, W. B., P. J. DeCoursey, and W. J. Padgett. 1973. Synergistic effects of environmental variables on larvae of *Uca pugilator*. Mar. Biol. 22, 307-312.

Vernberg, W. B., and F. J. Vernberg. 1972a. Environmental physiology of marine animals. 346 pp. New York: Springer-Verlag.

Vernberg, W. B., and F. J. Vernberg. 1972b. The synergistic
    effect of temperature, salinity, and mercury on survival
    and metabolism of the adult fiddler crab, *Uca pugilator*.
    Fish. Bull. U.S. 70, 415-420.

Wood, C. E. 1967. Physioecology of the grass shrimp,
    Palaemonetes pugio, in the Galveston Bay estuarine
    system. Contrib. Mar. Sci. 12, 54-79.

THE EFFECTS OF SALINITY AND MERCURY ON
DEVELOPING MEGALOPAE AND EARLY CRAB STAGES OF THE
BLUE CRAB, *CALLINECTES SAPIDUS* RATHBUN

Charles L. McKenney, Jr.[1]
John D. Costlow, Jr.

Duke University Marine Laboratory
Beaufort, North Carolina

INTRODUCTION

Pelagic larvae of estuarine and marine invertebrates
exert an important influence on the distributional patterns
of benthic invertebrate communities (Mileikovsky, 1971). The
dominant role of salinity on survival, functional capacities, and distribution of estuarine organigms has long been
recognized (for recent reviews see Kinne, 1971; Lockwood,
1976). More recently, the influence of salinity on larval
development of a number of estuarine decapod crustaceans has
been investigated (Costlow *et al.*, 1960, 1962, 1966; Reed,
1969; Ong and Costlow, 1970; Lucas, 1972; Sandifer, 1973a).
Furthermore, recent findings suggest that benthic estuarine
invertebrate communities may be modified by altered
survival, distribution, and settlement of their planktonic
larvae developing in polluted estuarine waters (for a review
see Mileikovsky, 1970). Few studies exist, however, which
have examined the combined effects of salinity and pollutant
on the early pelagic life stages of benthic estuarine
crustaceans (Vernberg *et al.*, 1973; Rosenberg and Costlow,
1976; McKenney and Neff, 1979).

[1]*present address. United States Environmental Protection
Agency, Environmental Research Laboratory, Sabine Island, Gulf
Breeze, Florida.*

Callinectes sapidus Rathbun, the commercially important blue crab, is a member of the family Portunidae of the Brachyura. The native range of this species is reported from Nova Scotia to Uruguay, while more recently it has been introduced to various other parts of the world. The complicated life history of the blue crab includes mating in reduced-salinity waters of the estuary and migration of the females to the mouth of the estuary where the freshly hatched zoeae are released into the more saline waters. Here the zoeae develop in the plankton to the megalopa and successive crab stages which subsequently migrate up the estuary toward the mating grounds (Williams, 1965, 1971; Tagatz, 1968).

Costlow and Bookhout (1959) have described the seven zoeal stages and megalopal stage of *Callinestes sapidus* reared in the laboratory through complete larval development to first crab stage. The distribution of these stages in the field include observations of zoeae and megalopae within almost the entire length of several estuarine systems of the southeastern United States, with peak abundance between 20 and 30 $^\circ/_{oo}$ seawater, and reports of megalopae collected 60 miles offshore (Hopkins, 1943; Nichols and Keney, 1963; Pinschmidt, 1963; Tagatz, 1968; Williams, 1971; Sandifer, 1973b). Survival of megalopae of *Callinectes sapidus* in controlled laboratory conditions was similar at temperatures from 20° to 30°C and at salinities above 10 $^\circ/_{oo}$ after rearing zoeae at 25°C and 30 $^\circ/_{oo}$ (Costlow, 1967). Mortality of megalopae was high at 15°C and increased at salinities from 5 to 10 $^\circ/_{oo}$ with a decrease in temperature. Time for metamorphosis of megalopae was delayed at 15°C and, coupled with increasing salinities from 20 to 40 $^\circ/_{oo}$, this delay was even more pronounced.

Recent studies indicate that the larval stages of *Callinectes sapidus* may serve as useful indicators of the effects of insecticides on estuarine biota (Bookhout and Costlow, 1975; Bookhout *et al.*, 1976; Bookhout and Monore, 1977; Costlow, 1979). Furthermore, in studies of *Callinectes sapidus* megalopae reared to third crab at a range of constant salinities in the laboratory, synergistic interactions of cadmium and salinity affected both survival and developmental rates of these early stages (Rosenberg and Costlow, 1976).

The purpose of this study was to examine the effects of a broad range of salinities and several mercury concentrations on the megalopa and early crab stages of the blue crab, *Callinectes sapidus*. Consideration was given to both lethal and sublethal effects, including any alterations in survival and/or sublethal abnormalities in developmental time periods within each developmental and juvenile stage.

MATERIALS AND METHODS

The method for developing and hatching the eggs of *Callinectes sapidus* was identical to that described by Costlow and Bookhout (1960). Ovigerous females were collected in the Newport River area adjacent to the Duke Marine Laboratory. Pleopods bearing black eggs with developed eyes and a visible heartbeat were removed from gravid females and placed in large glass culture bowls (19.4 cm in diameter) containing filtered seawater at 30 °/₀₀ and 25°C. Setae bearing eggs were removed from pleopods with fine scissors and washed 3-5 times in filtered seawater. A final seawater wash followed further dissociation of the eggs using glass needles. The eggs were then placed in compartmented polypropylene boxes, 32.5 x 22.7 cm, or in 2-liter Erlenmeyer flasks containing filtered 30 °/oo seawater at 25°C. Both the boxes and the flasks containing dissociated eggs were placed on an Eberbach variable speed shaker regulated at 60 oscillations per minute.

At the time of hatching of the first zoea, the larvae were removed to mass culture bowls (19.4 cm in diameter) containing filtered seawater at 30 °/₀₀ salinity and 25°C and larvae were reared through total zoeal development under these optimum conditions in the manner described by Costlow and Bookhout (1959). On reaching the megalopal stage, larvae to be reared at salinities of 20, 30, and 40 °/oo were individually transferred directly into compartmented polypropylene boxes with the appropriate mercury concentration. Megalopae to be reared at a salinity of 10 °/oo were transferred from the rearing salinity of 30 °/oo to 20 °/oo for four hours and then individually moved to 10 °/oo seawater containing an appropriate mercury concentration. The water was changed daily and freshly hatched *Artemia* were added until successful development to the third-crab stage. For megalopa and the first two crab stages, daily records were kept on mortality and successful molts to succeeding stages.

Preparation of the various salinity dilutions was as follows. Seawater collected from the open ocean was filtered and diluted with glass-distilled water to obtain the more moderate salinities. The extreme salinities (10 °/₀₀ and 40 °/₀₀) were prepared by adjusting the salinities of hypo- and hypersaline solutions separated from partially frozen seawater. Experimental temperature regimes for all larvae were held within ± 0.5 C in constant-temperature cabinets maintaining a 12:12 hour photoperiod.

The experimental design was a four by four factorial, involving salinities of 10, 20, 30, and 40 °/oo and theoretical mercury concentrations of 0 (unexposed controls), 5, 10, and 20 ppb. Four replicates of 30 megalopae each were maintained at each of the 16 mercury-salinity conditions. Thus, the study represents responses recorded from a total of 1,920 megalopae.

The data were analyzed by multiple linear regression using the general model

$$\hat{Y} = b_0 x_0 + b_1 x_1 + b_{11} x_1^2 + b_2 x_2 + b_{22} x_2^2 + b_{12} x_1 x_2 \qquad (1)$$

where $\hat{Y}$ = the estimated arcsine $\sqrt{\text{percent survival}}$ or development duration in days for the appropriate larval or juvenile form and x_0 = intercept; x_1 and x_2 = the linear effects of mercury (Hg) and salinity (S), respectively; x_1^2 and x_2^2 = the quadratic effects of mercury (Hg) and salinity (S), respectively; $x_1 x_2$ = the linear x linear interaction between mercury and salinity (Hg x S); and b_0, b_1, etc. = the regression coefficients. Separate equations were produced for each larval or juvenile form and used to generate response surface curves at various levels of salinity and mercury (Box, 1954; Box and Youle, 1955; Myers, 1971; Alderdice, 1972). Where appropriate, *post hoc* examination of significant differences between treatment means was by Duncan's multiple range test (Steel and Torrie, 1960).

A primary stock of 1 ppt Hg was prepared by adding 1.36 grams reagent grade $HgCl_2$ to 1 liter of glass-distilled water. A secondary stock of 1 ppm Hg was prepared from the primary stock every third day. Final mercury solutions were prepared daily in filtered seawater of the appropriate salinity. Analysis of mercury solutions by atomic absorption spectrophotometry revealed no significant loss of mercury from either the primary or secondary stock solutions or exposure media between salinities throughout the study. However, actual mercury concentrations to which the larvae were exposed did demonstrate a loss of mercury from initial concentrations dependent on both time and the concentration of mercury (Table 1). For the purpose of this study, reference to mercury concentrations to which the larvae were exposed are represented as initial levels only.

RESULTS

A summary of the influence of four salinities (10, 20, 30, and 40 °/oo) and four mercury concentrations (0, 5, 10,

TABLE I. Concentration of Mercury (ppb) Experienced by Megalopae and Early-Crab Stages of Callinectes sapidus over a
24-Hour Period.   Each value is the Mean of Three Replicates
and the Standard Error

Nom. con.	Initial concentration	Actual con. after 1 hr	concentration after 4 hr	concentration after 12 hr	concentration after 24 hr
5	4.92±0.11	4.42±0.12	3.18±0.03	2.12±0.02	1.67±0.15
10	9.77±0.18	9.10±0.32	8.76±0.06	6.94±0.25	3.98±0.37
20	20.07±0.23	18.73±0.35	17.97±0.23	14.73±0.44	8.53±0.37

and 20 ppb Hg) on the mean percent survival and developmental
duration of four replicates of Callinectes sapidus from
megalopa to first crab stage, from first crab to second crab
stage, and from second crab to third crab stage is given in
Table 2.   Since significant variations (P = .0001) existed
between replicates within treatments and each replicate
represented the response of larvae from different females, to
best estimate treatment effects all statistical analyses
were performed on treatment means (Sokal and Rohlf, 1969).

Survival

Differential mortality occurred among megalopae of
Callinectes sapidus reared to first crab in various salinities
and mercury levels (Table 2).   Multiple linear regression
analysis performed on arcsine-transformed percentage-survival
data produced a second order polynomial expression explaining
77% of the variation:

$$\hat{Y}_1 = 36.1199 + 3.1531S - 0.0562S^2 + 0.178^2Hg - 0.0468Hg^2-$$

0.0014SxHg

where $\hat{Y}_1$= the estimated arcsine $\sqrt{\text{percent survival}}$ of megalopae
and the other terms follow those of the general model
(Equation 1).

Applying the mean square error term produced from this
analysis to a Duncan's multiple range test assisted in the
interpretation of the data.   In the absence of mercury the
highest survival of megalopae occurred at 30 °/oo seawater.
Though not significantly different from that at 20 and 40°/oo

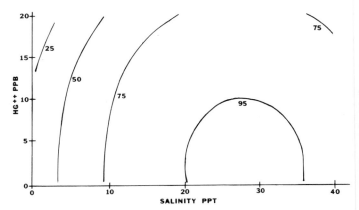

Fig.1. Estimation of the percentage survival of the megalopae of Callinectes sapidus based on the fitted response surface to observed survival at 16 combinations of salinity and mercury.

seawater, survival at 30°/oo was significantly (∝ = .05) higher than megalopal survival at 10 °/oo. Exposure to 5 and 10 ppb Hg allowed for similar survival among megalopae developing in salinities between 20 and 40 °/oo, while exposure to 10 ppb Hg significantly increased the mortality of megalopae developing in 10 °/oo seawater. Megalopae developing in 20 ppb Hg experienced significant reduction in survival at all salinities; this reduction was similar at 20 °/oo and 30°/oo seawater, more pronounced at 40 °/oo seawater, and most prominent for megalopae reared at 10 °/oo seawater.

Predicted percentage values produced by the quadratic equation, following conversion of the transformed values, resulted in response surfaces illustrating the altered survival capacity of megalopae reared at various salinity and mercury conditions (Fig. 1). The highly significant linear and quadratic effects of salinity (P = .0001) on viability of megalopae are demonstrated by the contours. Within the salinity range of 20 to 35 °/oo, survival of megalopae was estimated to be at least 95%, while below 10°/oo survival decreased to 75% and to less than 50% below 4 °/oo. At 10 ppb Hg survival remained at 95%, but only within the much restricted salinity range of 26 to 31 °/oo. Both below and above this optimal salinity range, the significant quadratic effect of mercury (P = 0.01) reduced the survival capacity of the megalopae, as seen by the curvature of the 95% survival envelope. As the salinity was lowered less mercury was required to produce equal mortality among the megalopae. For example, at a salinity of 18 °/oo exposure to 18 ppb Hg resulted in 25% mortality of developing

Table 2. Effect of Salinity and Mercury on Mean Percent and Duration in Days of the Megalopa, First-Crab Stage and Second-Crab Stage of Callinectes sapidus. Each Value is the Mean and Standard Erro of Four Replicates of 30 Initial Megalopae.

Salinity (ppt)	Hg (ppb)	Percent Survival of			Mean Duration of		
		Megalopa	First Crab	Second Crab	Megalopa	First Crab	Second Crab
10	0	82±4	92±8	86±12	9.93±0.45	5.30±0.80	7.04±0.80
	5	77±4	89±6	87±13	10.70±0.96	5.26±0.51	6.38±0.64
	10	68±9	91±7	86±14	11.01±0.62	5.14±0.28	6.70±0.80
	20	54±11	88±8	76±16	12.64±0.97	5.46±0.32	5.68±0.50
20	0	90±7	91±8	86±13	8.42±0.30	4.48±0.25	7.00±1.08
	5	91±3	91±8	84±12	8.61±0.26	4.97±0.36	6.30±0.91
	10	92±4	90±8	85±12	9.70±0.51	4.99±0.35	6.97±0.52
	20	78±3	84±13	89±10	10.85±0.68	4.93±0.33	6.67±0.59
30	0	96±1	91±8	86±11	7.89±0.21	4.77±0.23	6.43±0.71
	5	98±1	92±8	84±12	8.63±0.36	4.76±0.62	7.22±1.10
	10	94±2	89±9	87±12	9.23±0.38	5.35±0.61	6.80±0.96
	20	78±9	85±9	84±13	10.96±0.84	5.26±0.49	6.10±0.53
40	0	88±5	91±8	85±13	9.69±0.38	4.48±0.37	6.76±0.94
	5	84±14	88±11	84±14	9.93±0.67	4.91±0.35	6.38±0.56
	10	82±10	90±9	87±12	10.83±1.20	4.82±0.26	6.24±0.73
	20	68±18	88±5	84±12	11.42±1.18	5.01±0.13	5.91±0.51

megalopae, but exposure to only 8 ppb Hg produced the same
mortality level at 11 $\,^\circ/_{\circ\circ}$.

Inspection of mean survival values of both first-stage
crabs and second-stage crabs revealed similar survival rates
regardless of salinity or mercury conditions (Table 2).
Regression analysis revealed no statistically significant
effect of salinity, mercury or salinity-mercury interactions
on the survival of either the first-crab or second-crab
stages of *Callinectes sapidus*. For this reason, response-
surface curves at various salinities and mercury levels were
not generated for the estimated survival of either stage.

Developmental Duration

The influence of various salinity levels and mercury
concentrations on the developmental duration of megalopae
of *Callinectes sapidus* is shown in Table 2. Multiple
regression analysis of the developmental duration data
produced a second order polynomial expression explaining
85% of the variation:

$$\hat{Y}_2 = 12.9541 - 0.3825S + 0.0074S + 0.1364Hg + 0.007Hg^2 - 0.0010SxHg$$

where $\hat{Y}_2$ = the estimated duration in days for the megalopae
and the other terms follow the general model (Equation 1)
described earlier.

The mean square error term produced from this analysis,
when applied to a Duncan's multiple range test, suggests that
while a minimum mean developmental period of 7.89 days existed
for megalopae reared in 30 $\,^\circ/_{\circ\circ}$ seawater in the absence of
mercury, this time period increased significantly to 9.68 days
in 40 $\,^\circ/_{\circ\circ}$, and to 9.96 in 10 $\,^\circ/_{\circ\circ}$. Megalopae of *Callinectes
sapidus* developing at mercury concentrations of 10 and 20 ppb
required a significantly (P = .004) longer period to complete
metamorphosis. Furthermore, developmental rates of megalopae
were more retarded by mercury exposure at 30 $\,^\circ/_{\circ\circ}$ than at
either lower or higher salinities.

Response-surface curves produced from this equation best
illustrate these relationships (Fig. 2). A minimum developmental duration of 8 days is estimated between salinities of
22 and 29 $\,^\circ/_{\circ\circ}$ seawater. The highly significant linear and
quadratic salinity effect (P = .0001) on the developmental
time required for the megalopae to attain the first-crab
stage is seen as a prolongation of this period at salinities
both above and below this optimal salinity range in the
absence of mercury. At salinities above 37 $\,^\circ/_{\circ\circ}$ and below
15 $\,^\circ/_{\circ\circ}$ the developmental duration of megalopae is extended by

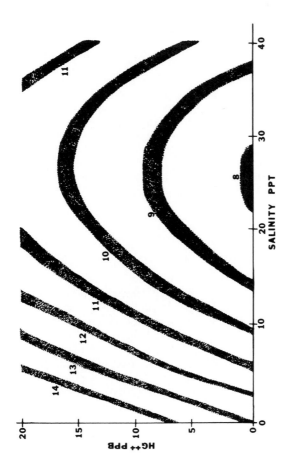

FIGURE 2. Estimation of developmental duration in days for the megalopae of Callinectes sapidus based on the fitted response surface to observed duration at 16 combinations of salinity and mercury.

at least one day. Further reduction in the salinity to 10 and
6 °/oo prolonged the duration of the megalopae by 2 and 3
days, respectively, from the minimum time of 8 days. At the
range of salinities allowing for the maximum developmental
rate (22-29 °/oo) the mean developmental duration is extended
by one day at 8 ppb Hg and by 2 days at approximately 16 ppb.
At both higher and lower salinities, mercury exposure
significantly (P = .003) extended the megalopal stage for an
even longer period. For example, at 10 °/oo and 8 ppb Hg
the megalopae required 11 days to reach the first-crab stage.
At the same salinity, exposure to 14 ppb Hg extended the
minimum developmental period from 8 days to 12 days.

     Although most variation in the developmental duration
data for the first- and second-crab stages of *Callinectes
sapidus* could be attributed to a linear salinity effect
(P = 0.15), regression analysis of the data revealed no
statistically significant effect of either salinity, mercury
or salinity-mercury interaction on developmental rates within
the salinity and mercury levels used in this study. For this
reason no response-surfaces were generated for developmental
duration of the first- and second-crab stages.

DISCUSSION

     Temporal and spatial variations in salinity, character-
istic of estuaries, have resulted in various adaptive adjustments in estuarine organisms (for recent review see
Kinne, 1971; Lockwood, 1976). These compensations to osmotic
stress for estuarine species include adjustments of regulatory
mechanisms throughout all the life-history stages. A
number of studies have considered the adaptations of larval
stages of benthic decapods to these dynamic estuarine
conditions (Costlow *et al.*, 1960, 1962, 1966; Reed, 1969;
Ong and Costlow, 1970; Lucas, 1972; Sandifer, 1973a; Sastry
and Vargo, 1977).

     Viability of *Callinectes sapidus* megalopae through
metamorphosis to first crab stage in the present study was
significantly altered by salinity conditions, as previously
described by Costlow (1967). In the present study it was
estimated that a maximum survival rate of 95% occurred between
20 and 35 °/oo seawater and decreased at both higher and lower
salinities. Survival of first- and second-crab stages was
unaffected by variation of salinity between 10 and 40 °/oo.
Similarly, Rosenberg and Costlow (1976) demonstrated no
significant effect of salinity, acting individually, on
viability of the first two crab stages of *C. sapidus*.

The importance of the free amino acid pool in intracellular osmotic adjustments in response to salinity
variations is well established for estuarine organisms (for
recent reviews see Gilles, 1975, 1979). In general, it is
the nonessential amino acids such as glycine, glutamic acid,
alanine, proline, and aspartic acid which comprise the major
part of the pool and provide the principal functional
responses. In the crab *Carcinus meanas* (Duchateau-Bosson *et
al.*, 1959) and in the shrimp *Crangon crangon* (Weber and Van
Marrewijk, 1972) alanine, proline and glycine are the amino
acids most responsible for intracellular osmotic adjustment.
In the abdominal muscle of the estuarine grass shrimp,
Palaemonetes pugio, glycine accounted for over 50% of the
salinity-sensitive free amino acid pool (Roesijadi *et al.*,
1976).

Alanine and proline were undetectable in the megalopae
of *Callinectes sapidus* reared from hatch in 30 $^\circ/_{oo}$ seawater
at 25°C, but reappeared at high levels in the first-crab
stage (Costlow and Sastry, 1966). In addition, glycine was
found at low levels in megalopae and at high levels in the
first-crab stage. These variations in relative abundance of
potential organic osmotic effectors might indicate a
functional inadequacy on the part of megalopae to adjust to
osmotic stress relative to early crab stages. These
biochemical/physiological considerations could, in part,
account for the reduced salinity plasticity of megalopae
compared with the more euryhaline early crab stages.

Upon exposure to mercury concentrations of 10-20 ppb Hg,
survival of megalopae of *Callinectes sapidus* was significantly
reduced dependent on salinity conditions. As the salinity
was reduced below 20 $^\circ/_{oo}$ less mercury was required to produce
equivalent toxicity among megalopae. Similar to these
results, Rosenberg and Costlow (1976) demonstrated an
increased toxicity of *C. sapidus* megalopa to cadmium at lower
salinities.

The high affinity that several heavy metals, including
mercury, display for amino acids and proteins may directly
alter certain metabolic pathways relying on various enzymes
and coenzymes. Indirectly, those metals having a high
affinity for purines, pyrimidines, and nucleic acids may
alter protein synthesis and modify these same pathways
(Vallee and Ulmer, 1972). The enzyme inhibition by heavy
metals of certain pathways involved in osmoregulation is of
particular interest to estuarine organisms (Schmidt-Nielsen,
1974). Megalopae of *Callinectes sapidus* hyperregulate in
brackish waters in a manner similar to the adult (Costlow,
1968; Kalber and Costlow, 1968). The increased toxicity of
mercury to *C. sapidus* megalopae developing at low salinities

could result from inhibition of these osmoregulatory abilities. Several studies have demonstrated such osmo-regulatory impairments in estuarine crabs following heavy metal exposure. Two species of estuarine crabs, *Carcinus maenas* and *Cancer irroratus,* lost the ability to hyperregulate their hemolymph following exposure to copper (Thurberg *et al.,* 1973). Similarly, cadmium, zinc, mercury and copper significantly reduced the osmoregulatory abilities of a number of estuarine isopods, particularly in dilute seawater (Jones, 1975a, b).

In that the gills represent the initial major site of localization of cadmium upon exposure of adult *Callinectes sapidus* (Hutcheson, 1974), it is plausible that gill pathology accounted for the mechanism of mercury toxicity for the developing megalopae of *C. sapidus* in this study. Low-level cadmium exposure of both the killifish, *Fundulus heteroclitus* (Gardner and Yevich, 1970) and the shrimp, *Paratya tasmaniensis* (Lake and Thorp, 1974) resulted in gross and ultrastructural changes in the gill lamellae. Gills represent the major site of ion exchange in aquatic crustaceans. Gill-tissue damage may have resulted in alterations of normal processes of ion regulation in *C. sapidus* megalopae, including regulation of tissue mercury levels by excretion through the gills.

Hutcheson (1974) speculated that the increased uptake rates of cadmium by adult *Callinectes sapidus* at low salinities reflected a reduced expulsion rate due to a limited amount of energy above that required for maintaining an osmotic gradient in this hyperregulating species. A similar explanation may be responsible, at least in part, for the increased toxicity to mercury for *C. sapidus* megalopae developing at low salinities.

Neither the mercury nor salinity levels used in this study significantly altered the viability of first- or second-crab stages of *Callinectes sapidus*. Similarly, Rosenberg and Costlow (1976) indicated a decreased cadmium toxicity among early crab stages of *C. sapidus* compared to that of the megalopae. It is postulated that the relatively greater sensitivity of the megalopae of *C. sapidus* to both heavy metal toxicity and osmotic stress is associated with the dramatic reorganizations accompanying successful completion of metamorphosis. Unfortunately, limited information exists on the morphological, physiological, and biochemical changes that together comprise metamorphosis. Oxygen consumption and ammonia excretion rates in the American lobster, *Homarus americanus,* increased in each larval stage and decreased in the first postlarval stage (Capuzzo and Lancaster, 1979). Protein utilization, as reflected by O:N ratios, was highest

in the final two premetamorphic larval stages of the caridean
shrimp *Palaemonetes pugio* and was reduced following completion
of metamorphosis to the postlarval stage (McKenney, 1979).
Only future biochemical/physiological studies on developing
decapod larvae subjected to environmental stress will confirm
or deny this hypothesis.

The results of this study demonstrated the significant
effect salinity, acting individually, may have on the developmental rates through metamorphosis of *Callinectes sapidus*
megalopae. It was estimated using response-surface methodology that a minimum developmental period of 8 days existed
between salinities of 22 and 29 $^\circ/_{\circ\circ}$. Similar observations
were made by Costlow (1967) for megalopae of *C. sapidus*
reared, as in this study, through the zoeal stages in optimal
salinity-temperature conditions (25°C, 30 $^\circ/_{\circ\circ}$). In both
studies considerable variation existed between larvae from
different females. Nevertheless, excluding this variation
not associated with treatment effects (Sokal and Rohlf, 1969),
both studies showed that higher and lower salinities prolonged
the duration of megalopae. A reduction in salinity had the
most pronounced influence on megalopal developmental rates, in
which, at 25°C and 10 $^\circ/_{\circ\circ}$, the period for successful completion of metamorphosis was extended to approximately 10
days.

In addition to the differential developmental rates
induced by salinity variations for megalopae of *Callinectes
sapidus* developing at salinities from 10 to 40 $^\circ/_{\circ\circ}$, the time
required for complete metamorphosis was further extended for
megalopae exposed to 10 to 20 ppb Hg. While a combination of
reduced salinity and mercury exposure prolonged the minimum
developmental period for megalopae by as much as 5 days, this
sublethal mercury effect retarded developmental rates to a
greater degree at optimum salinities for survival and development. In contrast to the megalopae, first- and second-crab
stages exposed to these same mercury levels developed to
subsequent stages at rates similar to unexposed controls.
Similarly, Bookhout *et al.* (1976) reported reduced developmental rates of *C. sapidus* megalopae reared at sublethal
levels of methoxychlor. Rosenberg and Costlow (1976)
observed delayed developmental periods in the megalopae of
C. sapidus exposed to another heavy metal, cadmium; the delay,
as in the present study, was most pronounced for megalopae
developing at the optimum salinity of 30 $^\circ/_{\circ\circ}$. Contrary to
our results, first- and second-crab stages exposed to cadmium
also displayed reduced developmental rates, though not to the
extent of megalopae. These retarded developmental rates as a
sublethal response to microcontaminants may represent a
generalized response of brachyuran larvae to stress, as

suggested by Epifanio (1971), or specific interferences with
enzymes or hormones involved in the processes of molting
(Passano, 1961; Tombes, 1970) or metamorphosis (Costlow,
1963, 1968; Tombes, 1970) in *C. sapidus*. The more sensitive
nature of the megalopae to environmental stress than the
early crab stages as indicated by retarded developmental
rates, may reflect variations between certain physiological/
biochemical mechanisms before and after metamorphosis, as
previously discussed.

Recent field observations suggest the role that megalopae
of *Callinectes sapidus* may have as a stage in which immigration and recruitment of the species into estuaries occurs.
From field collections of larval stages of *C. sapidus,* Tagatz
(1968) concluded that some larval stages return to inshore
waters as megalopae. As a result of a ten-year study of the
meroplankton in North Carolina estuaries, Williams (1971)
suggested that substantial numbers of *C. sapidus* migrate
toward the upper portions of estuaries as megalopae. Synthesizing his earlier work (Sandifer, 1973b) and the work of
others, Sandifer (1975) concluded that while early larval
stages of *C. sapidus* may develop in coastal waters outside
the estuary, immigration of megalopae and juveniles from
these waters probably accounts for the major recruitment
mechanism into estuarine adult populations. Additional
support for such a mechanism is offered by the laboratory
findings of Naylor and Isaac (1973), in which barokinetic
behavioral responses indicated an adaptation favoring movement
of *C. sapdius* megalopae up stratified estuaries.

The actual mercury concentrations resulting in reduced
survival rates of *Callinectes sapidus* megalopae in this study
approximate concentrations reported to be released into and to
exist within heavily contaminated estuaries (Klein and Goldberg, 1970; Friberg and Vostal, 1972). Combined with the
already high mortality rate of pelagic zoeal larvae (Thorson,
1950), the additional mortality of megalopae upon exposure to
low ppb mercury would allow for fewer remaining available
for replenishment of the parental population, colonization of
new habitats for the species, or re-establishment of damaged
neighboring adult populations (Mileikovsky, 1971).

The reduced salinity plasticity of mercury-exposed
megalopae would not only increase the mortality of those
found in lower salinity waters of the estuary, but may also
limit the distribution of the parental benthic population
to more saline areas within the estuary. The advantage
offered benthic invertebrates in lower salinity areas of the
estuary, that of reduced interspecific competition from many
parasites and predators (Thorson, 1957), may be diminished by

the reduced salinity plasticity of megalopae exposed to low ppb concentrations of mercury.

Large numbers of *Callinectes* megalopae have been observed to swarm at night in estuarine surface waters (Williams, 1971). The prolongation of the megalopal stage by 2 to 5 days upon exposure to mercury, dependent on the salinity, increases the period during which predation may exert a strong influence on the species. Increased predation on the megalopae would result in fewer being available for recruitment into the parental benthic population.

SUMMARY

1. Megalopae of the blue crab, *Callinectes spaidus,* reared from hatch through zoeal development at 25°C and $30°/_{oo}$ salinity, were maintained through completion of metamorphosis and through the first two crab stages at 16 combinations of salinity (10-40 $°/_{oo}$) and mercury (0-20 ppb) to determine the individual and combined influences of these factors on both survival and rate of development.

2. Individual, but not interactive, effects of salinity and mercury significantly affected viability of developing megalopae. Survival of megalopae was highest at a salinity of 30 $°/_{oo}$ and significantly reduced at 10 $°/oo$. Exposure to 10 ppb Hg significantly increased mortality of megalopae developing in 10 $°/_{oo}$ seawater, but did not significantly affect megalopae at salinities between 20 and 40 $°/_{oo}$. At all salinities significantly fewer megalopae completed metamorphosis when exposed to 20 ppb Hg.

3. Response-surface analysis illustrated the combined effects of both salinity and mercury. Concentrations of mercury, which proved sublethal for megalopae developing at optimal salinity conditions, reduced the salinity range for successful completion of metamorphosis of *Callinectes sapidus.* As the salinity was lowered less mercury was required to produce equal mortality among megalopae.

4. Developmental rates of *Callinectes sapidus* megalopae were significantly influenced by salinity and mercury. Response-surface analysis estimated the minimum developmental duration of 8 days to exist between salinities of 22 and 29 $°/_{oo}$ for megalopae. This period was prolonged by salinities outside this range, particularly by lower salinities, and by exposure to mercury. The developmental duration was extended to 10 days for megalopae developing at 10 $°/_{oo}$ seawater and at this same salinity to nearly 13 days with exposure to 20 ppb Hg.

5. Following completion of metamorphosis, *Callinectes sapidus* was more resistant to altered salinity conditions and to mercury toxicity. Multiple regression analysis revealed no significant effects of either salinity, mercury or interactions of the two on the survival or the developmental duration of the first two crab stages.

6. The reduced viability, restricted euryplasticity, and retarded developmental rates of *Callinectes sapidus* megalopae developing at low ppb concentrations of mercury may reduce recruitment into parent estuarine populations and alter distributional patterns of developing *C. sapidus* larvae.

ACKNOWLEDGEMENTS

This research was supported by the United States Environmental Protection Agency under Grant No. R801305.

LITERATURE CITED

Alderdice, D. F. 1972. Factor combinations: responses of marine poikilotherms to environmental factors acting in concert. In: Marine ecology. Vol. 1. Environmental factors, Part 3. pp. 1659-1772. Ed. by O. Kinne London: Wiley-Interscience.

Bookhout, C. G. and J. D. Costlow, Jr. 1975. Effects of mirex on the larval development of blue crab. Water, Air, Soil Pollut. 4, 113-129.

Bookhout, C. G. and R. J. Monroe. 1977. Effects of malathion on the development of crabs. In. Physiological responses of marine biota to pollutants. pp. 3-19. Ed. by F. J. Vernberg, A. Calabrese, F. P. Thurberg, and W. B. Vernberg. New York: Academic Press.

Bookhout, C. G., J. D. Costlow, Jr. and R. Monroe. 1976. Effects of methoxychlor on larval development of mud-crab and blue crab. Water, Air, Soil Pollut. 5, 349-365.

Box, G. E. P. 1954. The exploration and exploitation of response surfaces: Some general considerations and examples. Biometrics 10. 16-60.

Box, G. E. P., and P. V. Youle. 1955. The exploration and exploitation of response surfaces: An example of the link between the fitted surfaces and the basic mechanism of the system. Biometrics 11, 287-323.

Capuzzo, J. M. and B. A. Lancaster. 1979. Some physiological and biochemical considerations of larval development in the American lobster, *Homarus americanus* Milne Edwards. J. exp. mar. Biol. Ecol. 40, 53-62.

Costlow, J. D., Jr. 1963. The effects of eyestalk extirpation on metamorphosis of the blue crab, *Callinectes sapidus* Rathbun, Gen. Compar. Endocr. 3, 120-130.

Costlow, J. D., Jr. 1967. The effect of salinity and temperature on survival and metamorphosis of megalops of the blue crab *Callinectes sapidus*. Helgolander wiss. Meeresunters 15, 84-97.

Costlow, J. D. Jr. 1968. Metamorphosis in crustaceans. In: Metamorphosis: A problem in developmental biology. pp. 3-41. Ed. by W. Etkin and L. I. Gilbert. New York: Appleton-Century-Crofts.

Costlow, J. D., Jr. 1979. Effect of Dimilin® on development of larvae of the stone crab *Menippe mercenaria,* and the blue crab, *Callinectes sapidus*. In: Marine pollution: functional responses. pp. 355-363. Ed. by W. B. Vernberg, F. P. Thurberg, A. Calabrese, and F. J. Vernberg. New York: Academic Press.

Costlow, J. D., Jr. and C. G. Bookhout. 1959. The larval development of *Callinectes sapidus* Rathbun reared in the laboratory. Biol. Bull. 116, 373-396.

Costlow, J. D., Jr. and C. G. Bookhout. 1960. A method for developing Brachyuran eggs in vitro. Limnol. Oceanogr. 5, 212-225.

Costlow, J. D. Jr. and A. N. Sastry. 1966. Free amino acids in developing stages of two crabs, *Callinectes sapidus* Rathbun and *Rhithropanopeus harrisii* (Gould). Acta Embryol. et Morphol. Exper. 9, 44-55.

Costlow, J. D., Jr., C. G. Bookhout and R. Monroe. 1960. The effects of salinity and temperature on the larval development of *Sesarma cinereum* reared in the laboratory. Biol. Bull. 118, 183-202.

Costlow, J. D., Jr., C. G. Bookhout and R. Monroe. 1962. Salinity-temperature effects on the larval development of the crab, *Panopeus herbstii* Milne-Edwards reared in the laboratory. Physiol. Zool. 35, 79-93.

Costlow, J. D., Jr., C. G. Bookhout and R. Monroe. 1966. Studies on the larval development of the mud crab, *Rhithropanopeus harrisii* (Gould). I. The effect of salinity and temperature on larval development. Physiol. Zool. 39, 81-100.

Duchateau-Bosson, G., M. Florkin and C. Jeuniaux. 1959. Composante amino acide des tissus chez les Crustaces. I. Composante amino acide des muscles de *Carcinus maenas*.L. Lors du passage de l'eau saumatre et au cours de la mue. Arch. int. Physiol. 67, 489-500.

Epifanio, C. E. 1971. Effects of dieldrin in seawater on the development of two species of crab larvae, *Leptodius floridanus* and *Panopeus herbstii*. Mar. Biol. 11, 356-362.

Friberg, L. and J. Vostal. 1972. Mercury in the environment. 215 p. Cleveland: CRC Press.

Gardner, G. R. and P. P. Yevich. 1970. Histological and hematological responses of an estuarine teleost to cadmium. J. Fish. Res. Bd. Can. 27, 1225-1234.

Gilles, R. 1975. Mechanisms of ion and osmoregulation. In: Marine ecology. Vol. II. Physiological mechanisms, Part 1. pp. 259-347. Ed. by O. Kinne. London: Wiley-Interscience.

Gilles, R. 1975. Intracellular organic osmotic effectors. In: Mechanisms of osmoregulation in animals. p. 111-154. Ed. by R. Gilles. Chichester: John Wiley and Sons.

Hopkins, S. H. 1943. The external morphology of the first and second zoeal stages of the blue crab, *Callinectes sapidus* Rathbun. Trans. Am. Microsc. Soc. 62, 85-90.

Hutcheson, M. S. 1974. The effect of temperature and salinity on cadmium uptake by the blue crab, *Callinectes spaidus*. Chesapeake Sci. 15, 273-241.

Jones, M. B. 1975a. Synergistic effects of salinity, temperature and heavy metals on mortality and osmoregulation in

marine and estuarine isopods (Crustacea). Mar. Biol.
30, 13-20.

Jones, M. B. 1975b. Effects of copper on survival and
osmoregulation in marine brackish water isopods
(Crustacea). In: Proceedings of the ninth European
marine biology symposium. pp. 419-431. Ed. by H.
Barnes. Aberdeen: Aberdeen University Press.

Kalber, F. A., Jr. and J. D. Costlow, Jr. 1968. Osmoregulation in larvae of the land crab, *Cardisoma gunahumi*
Latreille. Am. Zool. 8, 411-416.

Kinne, O. 1971. Salinity: animals-invertebrates. In:
Marine ecology. Vol. 1. Environmental factors, Part 2.
pp. 821-995. Ed. by O. Kinne. London: Wiley-Interscience.

Klein, D. H. and E. D. Goldberg. 1970. Mercury in the
marine environment. Envir. Sci. Technol. 4, 765-768.

Lake, P. S. and V. J. Thorp. 1974. The gill lamellae of the
shrimp *Paratya tasmaniensis* Atyidae: Crustacea. Normal
ultrastructure and changes with low levels of cadmium.
8th Internat. Congr. Electron Microsc. 2, 448-449.

Lockwood, A. P. M. 1976. Physiological adaptations to life
in estuaries. In: Adaptation to environment. pp. 315-
392. Ed. by R. C. Newell. London: Butterworths.

Lucas, J. S. 1972. The larval stages of some Australian
species of *Halicarcinus* (Crustacean, Brachyura, Hymenoso-
motidae). II. Physiology. Bull. Mar. Sci. 22, 824-840.

McKenney, C. L., Jr. 1979. Ecophysiological studies on the
ontogeny of euryplasticity in the caridean shrimp,
Palaemonetes pugio (Holthius) and modifications by zinc,
217 p. Ph.D. dissertation, Texas A&M University.

McKenney, C. L., Jr. and J. M. Neff. 1979. Individual
effects and interactions of salinity, temperature, and
zinc on larval development of the grass shrimp
Palaemonetes pugio. I. Survival and developmental
duration through metamorphosis. Mar. Biol. 52, 177-188.

Mileikovsky, S. A. 1970. The influence of pollution on
pelagic larvae of bottom invertebrates in marine nearshore and estuarine waters. Mar. Biol. 6, 350-356.

Mileikovsky, S. A. 1971. Types of larval development in marine bottom invertebrates, their distribution and ecological significance: a re-evaluation. Mar. Biol. 10, 193-213.

Myers, R. H. 1971. Response surface methodology, 246 p. Boston: Allyn & Bacon, Inc.

Naylor, E. and M. J. Issac. 1973. Behavioural significance of pressure responses in megalopa larvae of *Callinectes sapidus* and *Macropipus* sp. Mar. Behav. Physiol. 1, 341-350.

Nichols, P. R. and P. M. Keney. 1963. Crab larvae (*Callinectes*), in plankton collections from cruises of M/V Theodore N. Gill, South Atlantic Coast of the United States, 1953-54. Spec. scient. Rep. U.S. Fish. Wildl. Serv. 448; 1-14.

Ong, K. and J. D. Costlow, Jr. 1970. The effect of salinity and temperature on the larval development of the stone crab, *Menippe mercenaria* (Say), reared in the laboratory. Chesapeake Sci. 11, 16-29.

Passano, L. M. 1961. Molting and its control. In: The physiology of crustacea. pp. 473-536. Ed. by T. H. Waterman, New York: Academic Press.

Pinschmidt, W. C., Jr. 1963. Distribution of crab larvae in relation to some environmental conditions in the Newport River Estuary, North Carolina, 112 p. Ph.D. dissertation, Duke University.

Reed, P. H. 1969. Culture methods and effects of temperature and salinity on survival and growth of dungeness crab (*Cancer magister*) larvae in the laboratory. J. Fish. Res. Bd. Can. 26, 389-397.

Roesijadi, G., J. W. Anderson, and C. S. Giam. 1976. Osmoregulation of the grass shrimp *Palaemonetes pugio* exposed to polychlorinated biphenyls (PCBs). II. Effect on free amino acids of muscle tissue. Mar. Biol. 38, 357-363.

Rosenberg, R., and J. D. Costlow, Jr. 1976. Synergistic effects of cadmium and salinity combined with constant and cycling temperature on the larval development of two estuarine crab species. Mar. BIol. 38, 291-303.

Sandifer, P. A.   1973a.   Effects of temperature and salinity on larval development of grass shrimp, *Palaemonetes vulgaris* (Decapod, Caridea). Fish. Bull. U.S. 71, 115-123.

Sandifer, P. A.   1973b.   Distribution and abundance of decapod crustacean larvae in the York River estuary and adjacent lower Chesapeake Bay, Virginia, 1968-1969.   Chesapeake Sci., 14, 235-257.

Sandifer, P. A.   1975.   The role of pelagic larvae in recruitment to populations of adult decapod crustaceans in the York River estuary and adjacent lower Chesapeake Bay, Virginia.   Estuar. cstl. mar. Sci. 3, 269-279.

Sastry, A. N., and S. L. Vargo.   1977.   Variations in the physiological responses of crustacean larvae to temperature. In: Physiological responses of marine biota to pollutants.   pp. 401-425.   Ed. by F. J. Vernberg, A. Calabrese, F. P. Thurberg, and W. B. Vernberg.   New York: Academic Press.

Schmidt-Nielsen, B.   1974.   Osmoregulation: effect of salinity and heavy metals.   Fedn Proc. Fedn Am. Socs exp. Biol. 33, 2137-2146.

Sokal, R. R. and F. J. Rohlf.   1969.   Biometry.   San Francisco: W. H. Freeman and Co.

Steel, R. G. D. and J. H. Torrie.   1960.   Principles and procedures in statistics.   New York: McGraw-Hill.

Tagatz, M. E.   1968.   Biology of the blue crab, *Callinectes spaidus* Rathbun, in the St. Johns River, Florida.   Fish. Bull. U.S. 67, 17-33.

Thorson, G.   1950.   Reproductive and larval ecology of marine bottom invertebrates. Biol. Rev. 25, 1-45.

Thorson, G.   1957.   Bottom communities (sublittoral or shallow shelf).   In: Treatise on marine ecology and paleoecology. Vol. 1.   pp. 461-534.   Ed. by J. W. Hedgpeth.   New York: Geological Society of America

Thurberg, R. P., M. A. Dawson, and R. S. Collier.   1973. Effects of copper and cadmium on osmoregulation and oxygen consumption in two species of estuarine crabs. Mar. Biol. 23, 171-175.

Tombes, A. S. 1970. An introduction to invertebrate endocrinology, 217 p. New York: Academic Press.

Vallee, B. L., and D. D. Ulmer. 1972. Biochemical effects of mercury, cadmium, and lead. An. Rev. Biochem. 41, 91–128.

Vernberg, W. B., P. J. DeCoursey and J. O'Hara. 1973. Multiple environmental factor effects on physiology and behavior of the fiddler crab, *Uca pugilator*. In: Pollution and physiology of marine organisms. pp. 381–426. Ed. by F. J. Vernberg and W. B. Vernberg. New York: Academic Press.

Weber, R. E. and W. J. A. Van Marrewijk. 1972. Free amino acids and isosmotic intracellular regulation in the shrimp *Crangon crangon*. Life Sci. 11, 589–595.

Williams, A. B. 1965. Marine decapod crustaceans of the Carolinas. Fish. Bull. U.S. 65, 1–298.

Williams, A. B. 1971. A ten-year study of meroplankton in North Carolina estuaries: annual occurrence of some brachyuran developmental stages. Chesapeake Sci. 12, 53–61.

PETROLEUM HYDROCARBONS

APPLICATION OF BIOCHEMICAL AND PHYSIOLOGICAL RESPONSES
TO WATER QUALITY MONITORING

A. N. Sastry

Graduate School of Oceanography
University of Rhode Island
Kingston, Rhode Island

Don C. Miller

Environmental Research Laboratory
U.S. Environmental Protection Agency
Narragansett, Rhode Island

INTRODUCTION

   Water pollution control authorities are increasingly
looking to biomonitoring to detect reductions in water
quality and to provide documentation of the biological consequences of pollution. Pollution has been defined as those
alterations of the environment caused by anthropogenic or
natural changes which render the environment less suitable for
existing life forms (Menzel, 1979). Similarly, Stebbing
(1979) suggested that polluted conditions exist when water no
longer possesses the capacity to sustain naturally occurring
biological processes. The primary question when developing
pollution biomonitors involves deciding which biological
processes are the best to employ.
   Biochemical and physiological processes of aquatic
organisms may be profitably employed as one component of a
monitoring program, as their short response times may aid
detection of pollution in its incipient stages. This is
important if pollution control measures are to be instituted
before appreciable damage occurs. However, interpretation of
changes in biochemical and physiological indices observed

during monitoring is not always straightforward. For example, it may not be apparent whether an altered response is pollutant-induced or whether it is in response to natural environmental change. Extensive baseline information is required which thoroughly describes the normal response of each biological parameter to natural environmental change. Further, the consequences of observed biochemical and physiological changes for whole organism survival or for population success may not always be evident. In this regard, biological processes most applicable to water quality monitoring would be those which are clearly related to survival, growth, reproduction, and successful recruitment, and hence may have an influence on population success.

In this paper we will explore not only the promise of biochemical and physiological measures for pollution monitoring, but also the problems associated with applying these techniques to the field. Aspects of biochemical and physiological responses to natural environmental change are briefly examined, as are some of the reported effects of pollutants on marine organisms, particularly at the biochemical and physiological level. Those showing promise for pollution detection in the field are noted. Finally, a research approach is discussed to address many of the problems involved in field application of biochemical and physiological measures. Examples are cited from recent literature on environmental physiology and the effects of pollution on marine organisms. However, a critical review of these topics is beyond the scope of this paper. Detailed information may be obtained from reviews and symposia on biochemical and physiological effects of pollutants on aquatic organisms (Vernberg and Vernberg, 1974; Vernberg et al., 1977, 1979; Lockwood, 1976; Koeman and Strik, 1975; Corner, 1978; Davies, 1978). Discussions on biochemical and physiological indices applicable to biomonitoring may also be found in the proceedings of several workshops and symposia on marine pollution biomonitoring (Giam, 1977; ICES, 1978; Cole, 1979; McIntyre and Pearce, 1980).

Biological Responses to
Reduced Water Quality

Each level of biological organization has the potential of reflecting a change in water quality, from molecular or cellular, up to the whole organism and through the higher ecological assemblages. Figure 1 depicts a hypothetical sequence of events which may follow an organism's encounter with polluted conditions. Pollutants at very low concentrations might be perceived by an organism's sensory system.

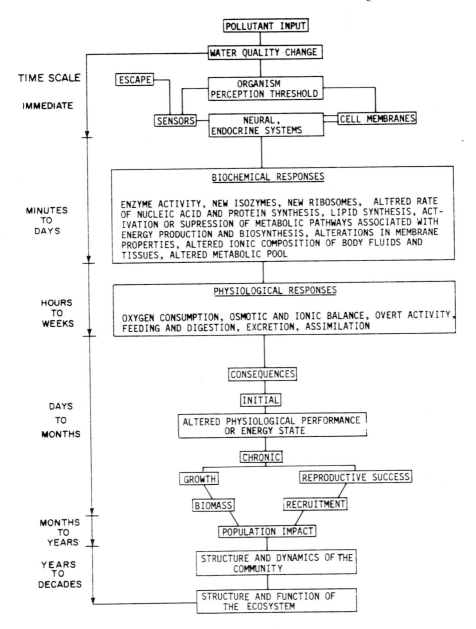

FIGURE 1. A hypothetical time related sequence of possible biological effects of reduced water quality for various levels of biological organization.

If this stimuli is recognized as harmful, avoidance may
follow. In motile organisms, this could lead to escape. In
those forms with little or no mobility, avoidance may take
the form of reducing exposure of external body surfaces
through mucus production, withdrawal into burrows or shells,
or closure of siphons. However some types of pollutants
may not be detected by the sensory system. For example,
pollutants having anesthetic properties, such as the aromatic
components of petroleum hydrocarbons, may block sensory
systems. Also, work on the effects of metals on fish
schooling suggest behavioral accommodation may occur soon
after low levels of contaminants are encountered (Robinson,
personal communication). Little is known concerning sensory
detection of environmental changes *per se*, much less the
neural and hormonal integration of their signals and the
nature of cellular responses to these changes. Our knowledge
of the sites, nature and thresholds for the detection of
pollutants by marine organisms is even more limited.

When exposure to polluted conditions persists, certain
biochemical alterations and responses soon occur. Some of
these are listed in Figure 1. These changes may subsequently
affect physiological processes (e.g., osmotic and ionic
balance, mobilization or utilization of energy stores, etc.)
and lead to perceptible alterations in physiological state.
If reduced water quality continues as a chronic condition,
these alterations in physiological performance may adversely
influence growth or reproductive success of individuals,
changes which have ecological significance due to their
influence on population success. Reduced productivity or
fecundity may also alter community composition and, perhaps,
the function of the total system.

The time required for biological response to environmental
change is usually longer for each level of increasing complexity (Fig. 1). Sensory and cellular detection systems
respond as soon as requisite threshold concentrations are
reached. Biochemical changes may take place within a period
of minutes to days. At the organism level, pollutant-induced
alterations in physiological performance in terms of growth
and reproduction may only be evident after several days, weeks
or months, depending on the season and generation time of the
species in question. Finally, effects at the population level
may require several generations to become established and be
readily evident, depending on the severity of the impact
(Gray, 1979). Accordingly, rapidly responding functional
parameters at the whole organism level and below should be
examined for detection of incipient pollution.

Within a given level of organization, the occurrence and
the intensity of a pollutant-induced response will also vary

with time.   During monitoring the probable time course of each
response must be considered to assure proper interpretation of
observed changes.   Aspects of the time course of biochemical
and physiological response to both natural environmental
change and pollution stress are discussed in the sections
which follow.

Responses to Change in the
Natural Environment

   Successful application of biochemical or physiological
responses to water quality monitoring is dependent on our
ability to distinguish between responses to natural environmental changes and those which are induced by pollutants.   The
response of organisms to natural change alone is complex.
Rate processes measured at a given point in time are a
function of the recent environmental history of the organism
as well as the prevailing environmental conditions.   Organism
responses may also be uniquely influenced by a number of
endogenous factors, including body size, sex, age or
developmental stage, reproductive condition, nutritional
state, and biological rhythms.   Vernberg and Vernberg (1970)
have discussed the manner in which both environmental and
endogenous factors may modify such physiological responses as
respiration, osmotic and ionic regulation, growth and
reproduction.   Many biochemical responses are similarly
influenced by natural environmental changes and endogenous
conditions.
   Another significant problem compounding the interpretation
of biochemical and physiological data concern the temporally
variable nature of biological responses to environmental
change.   Consider, for example, those species which exhibit
physiological compensation.   These organisms live in a
dynamic equilibrium with their environment, making continuous
internal adjustments to assure maintenance of normal biochemical and physiological rate functions in response to
environmental changes.   These compensatory responses involve
a variety of molecular, biochemical, cellular and higher
system changes (Hazel and Prosser, 1974).   At the whole
organism level, the more obvious manifestations of such
environmentally induced biochemical changes are changes in
cellular metabolism, oxygen consumption, neural function,
ciliary or motor activity and physiological performance.
   Three types of temporally related compensatory responses
may be distinguished; the first two are illustrated in Figure
2.   Following an environmental change, there is first a direct
response which occurs within minutes to hours of the change.
This response may involve an initial overshoot of rate

function, followed closely by establishment of a new stabi-
lized rate (Fig. 2). This may be followed by acclimation
after several days or weeks if the environmental change
persists. The third type is evolutionary compensation, or
genetic adaption, to the new environmental conditions, which
may require several generations (Hazel and Prosser, 1974).

Several contrasting patterns of compensatory acclimation,
or long-term physiological adaptation, have been identified
in studies involving temperature change (Precht, 1958;
Prosser, 1958). When an organism is transferred to a new
environmental temperature, one of the following rate-function
responses may occur: complete, partial, overcompensation or
inverse compensation, or no compensation (Fig. 3). Different
classes of enzymes have also shown compensatory acclimation,
no acclimation or increased acclimation (Hazel and Prosser,
1974). The best-studied enzymes which compensate positively
are those involved in oxidative metabolism (e.g., enzymes of
the pentose shunt, TCA cycle, the electron transport
cytochrome system and some synthetic enzymes). Enzymes known
to compensate negatively or not at all include those concerned with nitrogen excretion. Ion transport enzyme response
varies with species and ionic conditions; sometimes it is
compensatory and sometimes it is not.

If environmental alterations are severe and persist for
some time, death may result. If part of a population
survives, it will now be comprised of individuals with
environmentally selected traits, shifting the genetic chara-
ter of the population (MacArthur and Connell, 1966). This
third type of biological response has been termed evolutionary
compensation (Somero, 1969) and can occur over one or more
generations. This response has been reported in populations
inhabiting areas of chronic pollution stress (Bryan, 1976;
Mitton and Koehn, 1976). It would be less likely to occur at
sites of incipient pollution.

The potential for compensatory responses must be considered in water quality monitoring. They may influence both
the magnitude and the time course of responses to natural
environmental changes. Mechanisms also exist enabling
organisms to compensate to low levels of certain pollutants,
as will be discussed below.

Effects of Pollutants on
Marine Biota

In this section we will catalog the principal types of
biochemical and physiological responses reported in marine
organisms following exposure to pollutants. Some examples
are cited to illustrate how pollutants may affect organisms

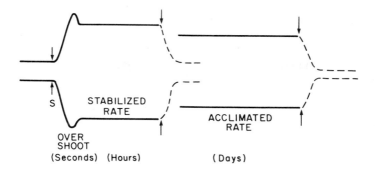

FIGURE 2. *Time-course of a rate function following an organism's exposure (at S) to an environmental change. Upper line is for a stress which results in an increased rate, lower line for a stress resulting in a decreased rate. Dotted lines at second and third arrows indicate rate if returned to original environmental conditions. Recovered rate is dependent on the duration of exposure to the environmental change. (After Prosser, 1959).*

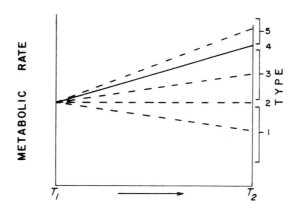

FIGURE 3. *Pecht's classification scheme for five possible types of acclimatization responses. Organism is initially completely acclimated to the environmental temperature, T_1. When exposed to an altered temperature change (T_2), possible response patterns expressed include: Type 1, supraoptimal acclimation; Type 2, perfect acclimation; Type 3, partial acclimation; Type 4, no acclimation and Type 5, inverse acclimation. (After Precht, 1958).*

at a variety of functional levels. It should also be noted
that pollutant effects are not always adverse. At very low
concentrations, stimulatory or enhancing effects can occur.
Organisms may also possess mechanisms to detoxify pollutants,
thus reducing or eliminating adverse effects. Such responses
must also be considered in the selection of biomonitors.

Uptake, Detoxification,
Bioaccumulation and Excretion

   In nature, organisms may take up pollutants from seawater, sediments, suspended particles or from food items. The
rate of uptake will depend not only on the availability of
the pollutant but also on a number of other chemical and
biological factors, many of which are influenced directly or
indirectly by environmental conditions. For example, the
chemical or physical form of a pollutant is of importance
and is subject to modification by temperature, salinity,
dissolved oxygen, pH, light, suspended particle and bottom
sediment characteristics, and the presence of other
pollutants. Most of these factors also directly influence
biological rate processes, including feeding, respiratory
ventilation, and metabolism, which in turn also influence
contaminant uptake. Other biotic factors, such as developmental stage, age, sex, size, and reproductive state, may
also influence uptake of pollutants (Bryan, 1979).
   Considerable evidence is accumulating that certain
marine organisms possess mechanisms to detoxify pollutants
(Brown, 1976). Detoxification of heavy metals can be achieved
through binding with metallothionein type proteins (Ridlington
and Fowler, 1979) or in the case of methylmercury, through
demethylation, with mercury being converted to a less toxic
inorganic form (Bryan, 1979). Mixed function oxygenase
enzyme systems capable of metabolizing organochlorines
(Addison, 1976) and petroleum hydrocarbons (Burns, 1976; Lee
et al., 1977, 1979) have also been reported in marine
organisms. Several weeks exposure to low pollutant levels
may be required to induce the function of these systems. If
detoxification systems exist in species used in biomonitoring,
their effectiveness as water quality monitors may be significantly reduced.
   Detoxified material may be stored as intracellular
granules in membrane-bound vesicles, in skeletal material, or
in the case of many organic compounds, in lipid stores.
Pollutants may also be lost from an organism by several
routes, including diffusion across the general body surface
and gills, by mucus secretion and via the feces and urine.

Lipophilic contaminants may accumulate in yolked eggs and be subsequently eliminated as the eggs are released.

In summary, the concentration of pollutants which are ultimately accumulated by an organism will be influenced by the general bioavailability of pollutants in the environment, the extent of the organism's exposure to it, and the uptake rate in relation to the ability of the organism to detoxify, store, or excrete the pollutant. At the present time, bioaccumulation data alone are of little value in predicting specific biological impacts at any level of organization. They can, however, provide useful information about pollutant routes and reservoirs within a marine system.

Pollutant Stress Effects
Biochemical Responses

Some biochemical consequences of non-lethal pollutant exposure in marine organisms are listed in Table 1. Effects on enzymes, levels of steroid hormones and hematological properties have received particular attention. Effects of stress on cellular metabolic energy level, using the adenylate energy charge method, have also been explored.

Enzyme activity may be altered by exposure to heavy metals, organophosphate pesticides, petroleum hydrocarbons and sodium pentachlorophenol. DDT and PCBs have been shown to inhibit $Na^+K^+ATPase$, an enzyme important in ion transport. Inhibition of enzyme activity by DDT or PCBs may also impair osmoregulatory capacity. *In vitro* inhibition of ATPase by methoxychlor in the crab, *Cancer magister,* was reported, although no impairment of osmotic or ionic regulation was observed (Caldwell, 1974). Fox and Rao (1978) have found that microsomal Ca^{++} ATPase from blue crab, *Callinectes sapidus,* was inhibited by sodium pentachlorophenol under *in vitro* and *in vivo* conditions. Interference with Ca^{++} ATPase activity may affect active transport of calcium. Pollutant stress can also induce free lysosomal enzyme activity. Moore (1977) found that destabilization of intracellular lysosomes can occur, releasing hydrolases into the cytoplasm, causing autolytic cellular damage.

Studies of the effects of pollutants on the blood of aquatic organisms have demonstrated changes in plasma hematocrit, "leucocrit", glucose and chloride levels (Iwama *et al.*, 1976; McLeay, 1973; McLeay and Gordon, 1977; Eisler and Edmunds, 1966; Larsson, 1975). Certain contaminants are also known to have an immunosuppressive action in marine invertebrates (Waldichuk, 1979).

TABLE 1. Some examples of biochemical effects of pollutants on marine organisms.

BIOCHEMICAL PROCESS	PHYSIOLOGICAL FUNCTION	POLLUTANT	REFERENCE
Enzymes:			
Acetyl choline esterase	*Nerve impulse conduction*	*Phosphrous and carbonate insecticides*	*Cornish, 1971; Weiss and Gakstatter, 1964*
Delta amino leval- inic acid dehydro- genase	*Haemoglobin synthesis*	*Mercury*	*Jackim, 1974*
Mixed function oxygenases	*Detoxification*	*Organic compounds, polycyclic aromatic hydrocarbons*	*Kahn, et al., 1972, Lee, et al., 1979; Kurelac, et al., 1977; Stegeman, 1978.*
Tricarboxylic acid cycle enzymes: fumarase, isocitrate, succinic dehydrogenase	*Oxygen consumption*	*Pentachlorophenol*	*Rao, et al., 1979*
Glucose metabolism: pyruvate kinase, lactic dehydrogenase	*Glucose metabolism*	*Pentachlorophenol*	*Rao, et al., 1979*
Glutamate-pyruvate transaminase	*Lipid metabolism*	*Pentachlorophenol*	*Rao, et al., 1979*
Na, K, Mg	*Osmotic and ionic regulation oxidative metabolism, transport phenomenon*	*PCBs, DDT, Pentachloro- phenol*	*Kinter, et al., 1972; Caldwell, 1974; Rao, et al., 1979*
Lysosomal enzymes	*Cellular damage*	*Metal and organic pollutants*	*Allison, 1969; Moore, 1977; Moore, et al., 1978*

Table 1 con't.

BIOCHEMICAL PROCESSES	PHYSIOLOGICAL FUNCTION	POLLUTANT	REFERENCE
Adenylate energy charge	Measure of metabolic energy	Toluene, Cd, hydro-carbons	Skjoldal and Bakke, 1978; Giesy, et al., 1978; Rainer, et al., 1979
Amino acids: Free amino acids Taurine:glycine		Cd, oil	Schafer, 1961 Jeffries, 1972, Roesijadi and Anderson, 1979
Heavy metal binding of proteins	Detoxification	Cd, Hg	Brown, et al., 1976, Brown and Parsons, 1978; Jeffries, et al., 1979
Steroid hormones:	Reproduction	PCBs	Freeman and Idler, 1975
Blood plasma:	Plasma glucose level, hemato-crits and lymphocyte changes	Kraft pulp mill effluent	McLeary and Gordon, 1977
Hematological responses:	Elevated serum levels of K, Na, Ca, cholesterol	Eldrin	Eisler and Edmunds, 1966

Physiological Responses

A variety of physiological processes, including sensory
responses, ciliary and motor activity, metabolism, osmotic
and ionic regulation, endocrine function, growth, reproduction
and development have been reported to be affected in organisms
exposed to pollutants (Table 2). Pollutant impact on
swimming capacity has been demonstrated by stamina tests with
salmonid fishes (Howard, 1973) and the mysid shrimp,
Mysidopsis bahia (Cripe, this volume). Effects on ciliary
function have been observed by measuring feeding rates of
bivalve molluscs (Bayne *et al.*, 1979; Gonzales *et al.*, 1979).
Sensory function, which is usually studied behaviorally,
can be affected by petroleum hydrocarbons, pesticides, and
copper (Olla *et al.*, 1980; Lang *et al.*, this volume).

Oxygen consumption rates of whole organisms or isolated
tissues may be altered by pollutants and have been employed
to measure organism condition (Vernberg and Vernberg, 1972;
Thurberg *et al.*, 1973, 1974). However, because metabolism is
also influenced by a number of environmental and endogenous
factors, it may be difficult to relate observed changes to
pollutant stress with certainty. Interpretation may be
aided by coupling oxygen consumption with other physiological
measures, such as nitrogen excretion.

Nitrogen excretion rates which exceed the normal range
indicate stress and possibly a fundamental nutritional
disturbance. The ratio of oxygen consumption to nitrogen
excretion (O:N) provides an index of the metabolic balance
between protein, carbohydrate and lipid substrates (Corner and
Cowey, 1968). A high O:N value indicates a predominance of
lipid and/or carbohydrate metabolism, with little protein
degradation. Bayne (1975) suggests the O:N ratio as a useful
index of stress for *Mytilus edulis*. Good correlations have
been found with this index and pollution stress in the field
(Widdows *et al.*, 1980).

Regulation of osmolality and blood and cellular ionic
concentration may be affected by DDT and PCBs (Kinter *et al.*,
1972; Nimmo and Blackman, 1972). Copper and cadmium have also
been shown to impair the capacity for osmoregulation in some
estuarine crabs and isopods (Thurberg *et al.*, 1973; Jones,
1975).

Growth rates are often reduced by pollutants, although at
very low pollutant levels, stimulation may occur (Stebbing,
1976; Davies, 1978). Growth does provide a well integrated
measure of whole organism function and ecological fitness.
However, as growth will reflect pollutant effects on a number
of different biochemical and physiological processes, it is
often more informative to evaluate pollutant effects on

TABLE 2. Sublethal physiological responses of marine
organisms to pollutants.

PHYSIOLOGICAL PROCESS	FUNCTION	ECOLOGICAL RELEVANCE
Sensory physiology	Chemoreception	location of habitat, food and mates
	Mechanoreception	migration, feeding and prey avoidance
	Photoreception	vertical migration, water column position
Ciliary and motor activity	Filtration rate, locomotor activity	feeding, predator avoidance
Metabolism	Respiration, excretion, digestion, assimilation	energy state
Osmoregulation	Membrane function, ion fluxes	salinity tolerance
Endocrines	Growth, metabolism, reproduction	general fitness
Growth	Scope for growth	biomass
Reproduction	Gametogenesis, fertilization, embryogenesis, larval development	recruitment

specific processes associated with bioenergetics, i.e. feeding, respiration, excretion and assimilation. If the variance of these processes is not large, they can be integrated as an estimate of the scope for growth and provide an expression of an organism's energy state and its capacity for growth and egg production (Warren and Davis, 1967; Bayne, 1975; Bayne *et al.*, 1976). The water soluble fraction of crude oils has been shown to suppress scope for growth in *Mya arenaria* (Gilfillan *et al.*, 1976) and *Crangon crangon* (Edwards, 1978). Suppression has also been reported for *Mytilus edulis* following transplant along a pollution gradient (Widdows *et al.*, 1980).

Many events associated with the reproduction and early development of organisms are sensitive to pollutants. Gametogensis may be hindered if pollutant stress has reduced the build up of sufficient energy stores. Lipophilic contaminants accumulated by the adult may also accumulate in yolk material during oogenesis and impair embryonic development (Rossi and Anderson, 1978). Pollutants may also directly block fertilization or some step of embryonic or larval development. Reviews of this topic have been published by Rosenthal and Alderdice (1976) for fish eggs and larvae and by David (1972) for marine organisms generally. High sensitivity of invertebrate larvae to pollutants has been demonstrated by several workers (Bookhout *et al.*, 1972; Calabrese *et al.*, 1973; Vernberg *et al.*, 1974). High sensitivity of developmental stages and fry of freshwater fish has led to routine utilization of this portion of the life cycle in toxicity testing (Macek and Sleight, 1977; McKim, 1977).

BIOCHEMICAL AND PHYSIOLOGICAL INDICES
SUGGESTED FOR MONITORING

The potential of specific biochemical and physiological indices for field monitoring has been extensively explored by several workshops and symposia (ICES, 1978; Cole, 1979; McIntyre and Pearce, 1980). Some conclusions of the most recent workshop are summarized here. The ICES biochemical techniques panel (Lee *et al.*, 1980) cited five tests as "possibly suited" for biomonitoring. Two of these, adenylate energy charge and lysosomal membrane integrity, may serve to detect a general decline in water quality. Measurements of fish steroid hormone biosynthesis may detect chronic effects. The two remaining techniques signal specific types of

pollutant problems. Mixed function oxygenase reactions
indicate uptake of certain organic compounds, such as hydrocarbons and some halogenated biphenyl isomers. For certain
metals, determination of the ratio of heavy metal binding
proteins to the metalloenzyme pool is suggested to describe
the relative loading of this detoxification system. There
are other measures which may be useful in future monitoring
applications, but to date have been little studied as
pollutant stress measures. These include RNA/DNA changes,
variations in lipid or carbohydrate metabolism, and membrane,
immunological or neuro-chemical responses.

The panel emphasized that all biochemical techniques
require considerable field investigation before application
to biomonitoring. Also, some techniques have not been well
evaluated by different investigators. The adenylate energy
charge method has been little tested with pollution stress,
although initial studies employing natural environmental
stress suggest this technique may be a useful measure of
organism condition, at least for molluscs. Lysosomal membrane fragility has been studied more extensively, although
primarily with one species, *Mytilus edulis*. Findings to date
make this index attractive for monitoring as its response
time is short, there is a direct correlation between
magnitude of membrane disruption and stress level, and good
correlations exist between this response and other function-
ally related indices, e.g. cell ultrastructure, tissue
histology and whole organism bioenergetic parameters.

The biochemical and physiological techniques panels
both noted the desirability of choosing methods which are
complementary, including techniques from other disciplines
(e.g. chemistry, genetics, cytology, pathology, behavior and
ecology), to obtain a comprehensive assessment of organism
condition. Following this approach, the physiology working
group recommended whole organism growth and reproduction as
basic measures of organism performance (Bayne *et al.*, 1980).
These are integrated responses to prevailing environmental
conditions in both a functional and temporal sense. They
also have ecological relevance, as each may be related to
population success. Measurement of respiration, feeding and
assimilation rates was recommended to compliment the above
parameters, as these functions jointly influence the energy
available for growth, activity, and reproduction. Additional
parameters cited as potentially useful for monitoring include
oxygen consumption and nitrogen balance or excretion, with
these rates combined as an index (i.e. O:N ratio) or expressed
as components of other measures of performance. Other
techniques which may be applicable to monitoring, but require
further laboratory and field evaluation, include measures of

osmotic and ionic regulation, hematology, and reproduction.

The list of techniques which are sufficiently understood to permit immediate use in pollution monitoring is quite short. Most require considerable additional research. This is due, in part, to the fact that much of our present knowledge of organism response to pollutants has been derived from laboratory studies. However, laboratory pollutant exposures may be very imperfect simulations of the field situation. Important disparities between the conclusions of laboratory toxicological studies and field observations may arise from differences in (a) the chemical form of the contaminant, (b) contaminant concentration, (c) the phase in which it occurs (dissolved or associated with particulates or food) and (d) the duration and temporal pattern of the exposure.

In the field, animals probably encounter pollutants most frequently in association with particles or food, while in laboratory studies toxicants are usually added in a dissolved form. Field exposures may be intermittent or relatively persistent; they also usually involve low pollutant concentrations for long time periods. The latter is rarely the case in the laboratory. Some laboratory practices, such as the use of constant environmental conditions rather than more adaptive fluctuating regimes (Costlow and Bookhout, 1971; Sastry, 1979) and use of inadequately tested diets (Mehrle et al., 1977) may also compromise laboratory results. Some contrasting conclusions from field and laboratory studies on pollutant uptake and depuration are described by Bryan (1979). Disparities should also be expected regarding biochemical and physiological responses to pollutants.

DEVELOPMENT OF WATER QUALITY BIOMONITORS

It is apparent that extensive field and laboratory research is required to develop an ecologically meaningful set of biochemical and physiological stress indices for water quality monitoring. An approach involving six steps is outlined here. This involves: (1) delineation of the biomonitoring program purpose, location and scope; (2) preliminary assessment of the locality to characterize the dominant aquatic communities, types of pollutants expected and their probable routes of transport through the natural system; (3) selection of potential biomonitoring species; (4) extensive field and laboratory studies to elucidate organism responses to pollution and predict the biological significance of observed effects; (5) development of a

definitive response profile for each species; and upon
initiation of monitoring, (6) document the consequences of
observed biochemical and physiological effects at the
individual and population level, and evaluate the adequacy of
the overall program to detect reduced water quality.

The starting point in developing a biomonitoring program
is a clear statement of its purpose and scope, including
delineation of the water bodies to be monitored and the level
of water quality protection desired. Specification of the
protection level, which is usually described in biological
terms, defines the sensitivity required of the biomonitors.
Knowledge of the specific water bodies to be monitored is
necessary if preliminary site assessment is to be conducted
prior to species selection.

Selection of the monitoring species is a critical
decision point in biomonitor development. Preliminary
studies of the monitoring sites should identify species having
a high potential for exposure to pollutants. Ecological and
taxonomic diversity should be considered, since the biomonitors are intended to signal the onset of pollution stress
for the entire community. Practical considerations will
encourage using species which are relatively abundant, easily
collected and of sufficient size to permit biochemical or
physiological measurement of condition. Sessile or only
slightly motile species are preferred since it is easier to
ascertain their previous environmental history, including any
pollutant exposure. These organisms would also be more
amenable to holding in live cars should transplant techniques
be employed. Biological information should be available for
the species selected to assist in evaluating the implications
of observed pollutant stress. Finally, each potential study
species should be field evaluated to determine its practical
feasibility as a monitor.

When the list of potential biomonitoring species has
been reduced to a small and manageable number, extensive
field and laboratory studies must thoroughly elucidate their
responses to both natural environmental variations and to
pollutants. A number of research topics are indicated in
Table 3. This work should initially be field centered.
Characteristic responses should be documented for populations
living in, or transplanted to, clean and contaminated areas.
Routine environmental data should be collected, along with
chemical analyses for pollutants in the water, major food
organisms, suspended material and sediments to describe
conditions at each study site. Pollutant body burden data of
monitoring species, will indicate contaminant bioavailability.
A transplant approach is outlined in Table 3. Only organisms
collected from an ecologically comparable location should be

used in transplant experiments, as there is a possibility for
genetic differences if the microhabitats differ appreciably
(Koehn *et al.*, 1973). The natural environmental conditions
at each transplant site should also be quite comparable;
otherwise the test groups may not possess genotypes having
comparable environmental fitness.

    Performance profiles for selected biochemical and physio-
logical indices should be described for each experimental
field group. Information should be collected on the time
course of pollutant effects as clean water organisms are
placed into contaminated waters. The ability of organisms to
detoxify pollutants and to recover from any initial adverse
effects should be investigated in pollutant-exposed organisms.
Reciprocal transplants between the clean and contaminated
sites would provide information for each of these questions.
Data would also be generated on the time course of pollutant
effects as clean water organisms are placed into contaminated
waters. Conversely, for the pollutant-stressed group, data
would be provided on recovery of biochemical and physiological
systems.

    After pollutant effects have been identified in the field,
each response deemed practical for biomonitoring should be
further examined in controlled cause and effect laboratory
studies (Table 3). Pertinent topics include identifying
effective dose thresholds, pollutant response time courses
and elucidating the influence of major environmental and
biotic variables on organism response, using conditions likely
to occur during monitoring. Also, the basic functional
systems impacted by pollutants should be examined, so the
mechanism underlying an observed effect can be elucidated.
Indices with known cause and effect properties will greatly
assist in relating observed effects to ecological fitness. In
organisms capable of detoxifing and excreting contaminants,
laboratory studies should identify the minimum exposure
conditions needed to induce these systems as well as those
excessive concentrations under which they fail.

    The above research should permit a compilation of effects
response profiles for the biomonitoring species which are
well understood from both the biological and ecological
standpoint. These profiles could take the form of com-
puterized diagnostic programs, analgous to those employed in
clinical medicine. The normal response range for a number of
indices would be tabulated for each species, along with
correction factors for the modifying influences of major
environmental and biotic variables.

    The final step of the development project is initiated as
monitoring begins. The adequacy of the effects indices to
predict changes in water quality must be evaluated along with

TABLE 3. A field and laboratory research approach to develop biochemical and physiological indices for water quality monitoring.

I. *Field Studies*

CLEAN WATER POPULATION	*FIELD TRANSPLANT*	*MONITOR OVER TIME*	*RECIPROCAL TRANSPLANT*
Test Group 1	*To Clean Site*	*a) Environmental Con-*	*To Polluted Sites*
Test Group 2..n	*To Polluted Sites*	*ditions*	*To Clean Sites*
		b) Contaminant Body Burdens	
		c) Biochemical and Physio- logical Responses	
		d) Detoxification Systems	

II. *Laboratory Studies*

Establish Sublethal Dose for Selected Pollutants	*Conduct Short and Long Term Expo- sures*	*Determine:*
		-Detoxification Capacity, Bioaccumulation, Excretion Dynamics
		-Biochemical and Physiological Effects:
		a) Effective Dose Threshold
		b) Response Time Course
		c) Influence of Environmental and Biotic Variables
		d) Mechanism of Effect
		e) Consequences for Whole Organism

documentation of the actual short- and long-term consequences
at the population and higher levels of organization of the
biochemical and physiological pollutant effects observed.

SUMMARY

Biochemical and physiological responses should provide
an early warning of reductions in water quality, considering
their typically short response times to natural environmental
changes. Environmental physiology response models are
applicable to development of pollutant indices, for there are
many parallels between physiological responses to natural
change and pollutant stress. However, interpretation of the
significance of alterations in biochemical and physiological
responses to pollutants may not always be straightforward,
since these responses are subject to modification by a
number of environmental and endogenous factors. The capacity
of organisms for biochemical and physiological compensation
to environmental change, for example, may markedly modify a
stress response over time.
    A number of reports of biochemical and physiological
change following pollutant stress are cited here. The
utility of some of these responses for biomonitoring is
largely conjectural, however, as they have been little studied
in field organisms. A research approach is discussed to
apply biochemical and physiological responses to long-term
stress, documenting the influence of biotic and abiotic
variables on these responses, identifying cause and effect
relationships and effective dose thresholds, determining the
potential of biomonitoring species to detoxify, store and
excrete pollutants, and evaluating the ecological consequences
of observed pollutant effects.

LITERATURE CITED

Addison, R. F. 1976. Organochlorine compounds in aquatic
    organisms: their distribution, transport and physio-
    logical significance. In: Effects of Pollutants on
    Aquatic Organisms. pp. 127-143. Ed. by A. P. M. Lock-
    wood. Cambridge: Cambridge Univ. Press.

Allison, A. C. 1969. Lysosomes and cancer. In: Lysosomes in Biology and Pathology. V. 2, pp. 178-204. Ed. by J. T. Dingle and H. B. Fell. New York: North Holland-American Elsevier.

Bayne, B. L., J. Anderson, D. Engel, E. Gilfillan, D. Hoss, R. Lloyd and F. Thurberg. 1980. Physiological techniques for measuring the biological effects of pollution in the sea. Rapp. P.-v. Reun. Cons. int. Explor. Mer. 179: 88-99.

Bayne, B. L. 1975. Aspects of physiological condition in *Mytilus edulis* L., with special reference to the effects of oxygen tension and salinity. In: Proc. Ninth Europ. Mar. Biol. Symp. pp. 213-238. Ed. by H. Barnes. Aberdeen: Aberdeen Univ. Press.

Bayne, B. L., D. R. Livingstone, M. N. Moore, and J. Widdows. 1976. A cytochemical and a biochemical index of stress in *Mytilus edulis* L. Mar. Pollut. Bull. 7: 221-224.

Bayne, B. L., M. N. Moore, J. Widdows, D. R. Livingstone, and P. Salkeld. 1979. Measurement of the responses of individuals to environmental stress and pollution: studies with bivalve molluscs. Phil. Trans. R. Soc. Lond. B. 286: 563-581.

Bayne, B. L., J. Widdows, and R. J. Thompson. 1976. Physiological integrations. In: Marine Mussels: their ecology and physiology. pp. 261-291. Ed. by B. L. Bayne. Cambridge: Cambridge Univ. Press.

Bookhout, C. G., A. J. Wilson, Jr., T. W. Duke, and J. I. Lowe. 1972. Effects of mirex on the larval development of two crabs. Water, Air, Soil Pollut. 1: 165-180.

Brown, D. A. and T. R. Parsons. 1978. Relationship between cytoplasmic distribution of mercury and toxic effects to zooplankton and chum salmon (*Oncorhynchus keta*) exposed to mercury in a controlled ecosystem. J. Fish. Res. Bd. Can. 35: 880-884.

Brown, G. W., Jr. 1976. Biochemical aspects of detoxification in the marine environment. In: Biochemical and Biophysical Perspectives in Marine Biology. V. 3, pp. 320-406. Ed. by D. C. Malins and J. R. Sargent. New York: Academic Press.

Bryan, G. W. 1976. Some aspects of heavy metal tolerance in aquatic organisms. In: Effects of Pollutants on Aquatic Organisms. pp. 7-34. Ed. by A. P. M. Lockwood. Cambridge: Cambridge Univ. Press.

Bryan, G. W. 1979. Bioaccumulation of marine pollutants. Phil. Trans. R. Soc. Lond. B. 286: 483-505.

Burns, K. A. 1976. Microsomal mixed function oxidases in an estuarine fish, *Fundulus heteroclitus,* and their induction as a result of environmental contamination. Comp. Biochem. Physiol. 53B: 443-446.

Calaberse, A., R. S. Collier, D. A. Nelson, and J. R. McInnes. 1973. The toxicity of heavy metals to embryos of the American oyster *Crassostrea virginica*. Mar. Biol. 18: 162-166.

Caldwell, R. S. 1974. Osmotic and ionic regulation in decapod crustacea exposed to methoxychlor. In: Pollution and Physiology of Marine Organisms. pp. 197-223. Ed. by F. J. Vernberg and W. B. Vernberg. New York: Academic Press.

Cole, H. A. (Organizer). 1979. The assessment of sublethal effects of pollutants in the sea. Phil. Trans. R. Soc. Lond. B. 286: 399-633.

Corner, E. D. S. 1978. Pollution studies with marine plankton. Part I. Petroleum hydrocarbons and related compounds. Adv. Mar. Biol. 15: 289-380.

Corner. E. D. S. and C. B. Cowey. 1968. Biochemical studies on the production of zooplankton. Biol. Rev. 43: 393-426.

Cornish, H. H. 1971. Problems posed by observations of serum enzyme changes in toxicology. Chemical Rubber Company. CRC Critical Reviews in Toxicology 1: 1-32.

Costlow, J. D., Jr. and C. G. Bookhout. 1971. The effect of cyclic temperatures on larval development in the mud crab, *Rhithropanopeus harrisii*. In: Fourth European Marine Biology Symposium. pp. 211-220. Ed. by D. J. Crisp. Cambridge Univ. Press.

Cripe, J. M., D. R. Nimmo, and T. L. Hamaker. This volume.
    Effects of two organophosphate pesticides on swimming
    stamina of the mysid, *Mysidopsis bahia*. In: Biological
    Monitoring of Marine Organisms. Ed. by A. Calabrese,
    F. P. Thurberg, F. J. Vernberg and W. B. Vernberg. New
    York: Academic Press.

Davies, A. G. 1978. Pollution studies with marine plankton.
    Part II. Heavy metals. Adv. Mar. Biol. 15: 381-508.

Davies, C. C. 1972. The effects of pollutants on the
    reproduction of marine organisms. In: Marine Pollution
    and Sea Life. pp. 305-311. Ed. by M. Ruivo. London:
    Fishing News (Books) Ltd.

Edwards, R. R. C. 1978. Effect of water soluble oil
    fraction on metabolism, growth and carbon budget of the
    shrimp *Crangon crangon*. Mar. Biol. 46: 259-265.

Eisler, R. and P. M. Edmunds. 1966. Effects of endrin on
    blood and tissue chemistry of a marine fish. Trans.
    Amer. Fish. Soc. 95: 153-159.

Fox, F. R. and K. R. Rao. 1978. Characteristics of a Ca^{2+}
    -activated ATPase from the hepatopancreas of the blue
    crab, *Callinectes sapidus*. Comp. Biochem. Physiol.
    59B: 327-331.

Freeman, H. C. and D. R. Idler. 1975. The effect of poly-
    chlorinated biphenyl on steroidogenesis and reproduction
    in the brook trout (*Salvelinus fontinalis*). Can. J.
    Biochem. 53: 666-670.

Giam, C. S. 1977. Pollutant Effects on Marine Organisms.
    Lexington, Mass.: D. C. Heath Co. 211 p.

Giesy, J. P., C. Duke, R. Bingham, and S. Denzer. 1978.
    Energy charges in several molluscs and crustaceans:
    natural values and responses to cadmium stress. Bull.
    Ecol. Soc. Amer. 59: 66.

Gilfillan, E. S., D. Mayo, S. Hanson, D. Donoban, and L. C.
    Jiang. 1976. Reduction in carbon flux in *Mya arenaria*
    caused by a spill of no. 6 fuel oil. Mar. Biol. 37:
    115-123.

Gonzalez, J. G., D. Everich, J. Hyland, and B. D. Melzian. 1979. Effects of no. 2 heating oil on filtration rate of blue mussels, *Mytilus edulis* Linne. In: Advances in Marine Environmental Research. pp. 112-234. Ed. by F. S. Jacoff. Cincinnati: U.S. Environmental Protection Agency.

Gray, J. S. 1979. Pollution induced changes in populations. Phil. Trans. R. Soc. Lond. B. 286: 545-561.

Hazel, J. R. and C. L. Prosser. 1974. Molecular mechanisms of temperature compensation in poikilotherms. Physiol. Rev. 54: 620-677.

Howard, T. E. 1973. Effects of kraft pulp mill effluent on the swimming stamina, temperature tolerance and respiration on some salmonid fish. Ph.D. Dissertation. Univ. Strathclyde, Glasglow, Scotland. 183 p.

I.C.E.S. 1978. On the feasibility of effects monitoring. International Council for the Exploration of the Sea, Cooperative Research Report No. 75. Charlottenlund, Denmark. 42 p.

Iwama, G. K., G. L. Greer, and P. A. Larkin. 1976. Changes in some hematological characteristics of coho salmon (*Oncorhynchus kisutch*) in response to acute exposure to dehydroabietic acid (DHAA) at different exercise levels. J. Fish. Res. Bd. Can. 33: 285-289.

Jackim, D. 1974. Enzyme responses to metals in fish. In: Pollution and Physiology of Marine Organisms. pp. 59-65. Ed. by F. J. Vernberg and W. B. Vernberg. New York: Academic Press.

Jeffries, H. P. 1972. A stress syndrome in the hard clam, *Mercenaria mercenaria*. J. Invert. Pathol. 20: 242-251.

Jeffries, J. R., P. S. Rainbow, and A. Q. Scott. 1979. Studies on the uptake of cadmium by the crab *Carcinus maenas* in the laboratory. II. Preliminary investigations of cadmium binding proteins. Mar. Biol. 50: 141-149.

Jones, M. B. 1975. Effects of copper on survival and osmo-regulation in marine and brackish water isopods (Crustacea). In: Proc. 9th Europ. Mar. Biol. Bymp. pp. 419-431. Aberdeen: Aberdeen University Press.

Kahn, M. A. W., W. Coello, A. A. Khan, and H. Pinto. 1972.
    Some characteristics of the microsomal mixed-function
    oxidase in the freshwater crayfish, *Cambarus*. Life
    Sci. 11: 405-415.

Kinter, W. B., L. S. Merkens, R. H. Janicki, and A. M.
    Guarino. 1972. Studies on the mechanism of toxicity of
    DDT and polychlorinated biphenyls (PCBs): Disruption of
    osmoregulation in marine fish. Environ. Health Persp.
    1: 169-173.

Koehn, R. K., F. J. Turano and J. B. Mitton. 1973.
    Population genetics of marine pelecypods. II. Genetic
    differences in microhabitats of *Modiolus demissus*.
    Evolution 27: 100-105.

Koeman, J. H. and J. J. T. W. A. Strik. 1975. Sublethal
    Effects of Toxic Chemicals on aquatic animals. Amsterdam:
    Elsevier. 234 p.

Kurelec, B., S. Britvić, M. Rijavec, W. E. G. Muller, and
    R. K. Zahn. 1977. Benzo(a)pyrene monooxygenase
    induction in marine fish-molecular response to oil
    pollution. Mar. Biol. 44: 211-216.

Lang, W. H., D. C. Miller, P. J. Ritacco, and M. Marcy.
    1980. The effects of copper and cadmium on the behavior
    and development of barncale larvae. In: Biological
    Monitoring of Marine Organisms. Ed. by A. Calabrese,
    F. P. Thurberg, F. J. Vernberg and W. B. Vernberg. New
    York: Academic Press.

Larsson, A. 1975. Some biological effects of cadmium on
    fish. In: Sublethal effects of toxic chemicals on
    aquatic animals. pp. 3-13. Ed. by J. H. Koeman and
    J. J. T. W. A. Strik. Amsterdam: Elsevier.

Lee, R., J. Davies, H. C. Freeman, A. Ivanovici, M. N. Moore,
    J. Stegeman and J. F. Uthe. 1980. Biochemical techniques
    for monitoring biolgoical effects of pollution in the
    sea. pp. 48-55. Rapp. P.-v. Reun. Cons. int. Explor.
    Mer. 179.

Lee, R. F., E. Furlong, and S. Singer. 1977. Metabolism of
    hydrocarbons in marine invertebrates: Aryl hydrocarbon
    hydroxylase from the tissues of the blue crab, *Callinectes
    sapidus* and the polychaete worm, *Nereis* sp. In:
    Pollutant Effects on Marine Organisms. pp. 111-124. Ed.

by C. S. Giam. Lexington, Mass: D. C. Heath.

Lee, R. G., S. C. Singer, K. R. Tenore, W. S. Gardner, and
    R. M. Philpot. 1979. Detoxification system in
    polychaete worms: importance in the degradation of
    sediment hydrocarbons. In: Marine Pollution: functional
    responses. pp. 23-37. Ed. by W. B. Vernberg, F. P.
    Thurberg, A. Calabrese, and F. J. Vernberg. New York:
    Academic Press.

Lockwood, A. P. M. 1976. Effects of Pollutants on Aquatic
    Organisms. Cambridge: Cambridge Univ. Press. 193 p.

Macek, K. J. and B. H. Sleight, III. 1977. Utility of
    toxicity tests with embryos and fry of fish in evaluating
    hazards associated with the chronic toxicity of chemicals
    to fishes. In: Aquatic Toxicology and Hazard Evaluation.
    pp. 137-146. Ed. by F. L. Mayer and J. L. Hamelink.
    Philadephia: Am. Soc. for Testing and Materials.

MacArthur, R. H. and J. H. Connell. 1966. The Biology of
    Populations. New York: Wiley. 200 p.

McIntyre, A. D. and J. B. Pearce. 1980. Biological effects
    of marine pollution and the problems of monitoring.
    Rapp. P.-v. Cons. int. Explor. Mer. 179.

McKim, J. M. 1977. Evaluation of tests with early life
    stages of fish for predicting long-term toxicity. J.
    Fish. Res. Bd. Can. 38: 1148-1154.

McLeay, D. J. 1973. Effects of a 12-hr and 25 day exposure
    of kraft pulp mill effluent on the blood and tissues of
    juvenile coho salmon (*Oncorhynchus kisutch*). J. Fish.
    Res. Bd. Can. 30: 395-400.

McLeay, D. J. and M. R. Gordon. 1977. Luecocrit: a simple
    hematological technique for measuring acute stress in
    salmonid fish, including stressful concentrations of
    pulpmill effluent. J. Fish. Res. Bd. Can. 34: 2164-2175.

Mehrle, P. M., F. L. Mayer, and W. W. Johnson. 1977. Diet
    quality in fish toxicology: effects on acute and chronic
    toxicity. In: Aquatic Toxicology and Hazard Evaluation.
    pp. 269-280. Ed. by F. L. Mayer and J. L. Hamelink.
    Philadelphia: Am. Soc. for Testing and Materials.

Menzel, W. 1979. Clams and snails [Mollusca: Pelecypoda (except oysters) and Gastropoda]. In: Pollution Ecology of Estuarine Invertebrates. pp. 371-396. Ed. by C. W. Hart and S. H. Fuller. New York: Academic Press.

Mitton, J. B. and R. K. Koehn. 1976. Morphological adaptation to thermal stress in a marine fish, *Fundulus heteroclitus*. Biol. Bull. 151: 548-559.

Moore, M. N. 1977. Lysosomal responses to environmental chemicals in marine organisms. In: Pollutant Effects on Marine Organisms. pp. 143-154. Ed. by C. S. Giam. Lexington, Mass.: D. C. Heath Co.

Moore, M. N., D. M. Lowe, and P. E. M. Fieth. 1978. Lysosomal responses to experimentally injected anthracene in the digestive cells of *Mytilus edulis* Mar. Biol. 48: 297-302.

Nimmo, D. R. and R. R. Blackman. 1972. Effects of DDT on cations in the hepatopancreas of penaeid shrimp. Trans. Amer. Fish. Soc. 101: 547-549.

Olla, B. L., J. Atema, R. Forward, J. Kittredge, R. J. Livingston, D. W. McLeese, D. C. Miller, W. B. Vernberg, P. G. Wells and K. Wilson. 1980. The role of behavior in marine pollution monitoring. In: Biological Effects of Marine Pollution and the Problems of Monitoring. Ed. by A. D. McIntyre and J. B. Pearce. Rapp. P.-v. Reun. Cons. int. Explor. Mer. 179.

Precht, H. 1958. Concepts of the temperature adaptation of unchanging reaction systems of cold-blooded animals. In: Physiological Adaptation. pp. 50-78. Ed. by C. L. Prosser. Washington, D.C.: Amer. Physiol. Cos.

Prosser, C. L. 1958. General summary: the nature of physiological adaptation. In: Physiological Adaptation. pp. 167-180. Ed. by C. L. Prosser. Washington, D.C.: Amer. Physiol. Soc.

Prosser, C. L. 1959. The origin after a century: Prospects for the future. Amer. Sci. 47: 536-550.

Rainer, S. F., A. M. Ivanovici, and V. A. Wadley. 1979. Effect of reduced salinity on adenylate energy charge in three estuarine molluscs. Mar. Biol. 54: 91-99.

Rao, K. R., F. R. Rox, P. J. Conklin, A. C. Cantelmo, and
    A. C. Brannon. 1979. Physiological and biochemical
    investigations of the toxicity of pentachlorophenol to
    crustaceans. In: Marine Pollution: functional
    responses. pp. 307-339. Ed. by W. B. Vernberg, F. P.
    Thurberg, A. Calabrese and F. J. Vernberg. New York:
    Academic Press.

Ridlington, J. W. and B. A. Fowler. 1979. Isolation and
    partial characterization of a cadmium-binding protein
    from the American oyster, (*Crassostrea virginica*).
    Chem. Biol. Interactions 25: 127-138.

Roesijadi, G. and J. W. Anderson. 1979. Condition index
    and free amino acid content of *Macoma inquinata* exposed
    to oil-contaminated marine sediments. In: Marine
    Pollution: functional responses. pp. 69-83. Ed. by
    W. B. Vernberg, F. Thurberg, A. Calabrese and F. J.
    Vernberg. New York: Academic Press.

Rosenthal, H. and D. F. Alderdice. 1976. Sublethal effects
    of environmental stressors, natural and pollutional,
    on marine fish eggs and larvae. J. Fish. Res. Bd. Can.
    33: 2047-2065.

Rossi, S. S. and J. W. Anderson. 1978. Effects of no. 2
    fuel oil on water-soluble-fractions on growth and
    reproduction in *Neanthes arenaceodentata* (Polychaeta:
    Annelida). Water, Air, Soil Poll. 9: 155-170.

Sastry, A. N. 1979. Metabolic adaptation of *Cancer irroratus*
    developmental stages to cyclic temperatures. Mar. Biol.
    51: 243-250.

Schafer, R. D. 1961. Effects of pollution on the free
    amino acid content of two marine invertebrates. Pac.
    Sci. 15: 49-55.

Skjoldal, H. R. and T. Bakke. 1978. Relationship between
    ATP and energy charge during lethal metabolic stress of
    the marine isopod *Cirolana borealis*. J. Biol. Chem.
    253: 3355-3356.

Somero, G. N. 1969. Enzymatic mechanisms of temperature
    compensation: Immediate and evolutionary effects of
    temperature on enzymes of aquatic poikilothersm. Amer.
    Nat. 103: 517-530.

Stebbing, A. R. D. 1976. The effects of low metal levels on a clonal hydroid. J. mar. biol. Ass. U.K. 56: 977-994.

Stebbing, A. R. D. 1979. An experimental approach to the determinants of biological water quality. Phil. Trans. R. Soc. Lond., B. 286: 465-481.

Stegeman, J. J. 1978. Influence of environmental contamination on cytochrome P-450 mixed-function oxygenases in fish: implications for recovery in the Wild Harbor marsh. J. Fish. Res. Bd. Can. 35: 668-674.

Thurberg, F. P., A. Calabrese, and M. A. Dawson. 1974. Effect of silver on oxygen consumption of bivalves at various salinities. In: Pollution and Physiology of Marine Organisms. pp. 67-78. Ed. by F. J. Vernberg and W. B. Vernberg. New York: Academic Press.

Thurberg, F. P., M. A. Dawson, and R. Collier. 1973. Effects of copper and cadmium on osmoregulation and oxygen consumption in two species of estuarine crabs. Mar. Biol. 23: 171-175.

Vernberg, F. J., A. Calabrese, F. P. Thurberg, and W. B. Vernberg. 1977. Physiological Responses of Marine Biota to Pollutants. New York: Academic Press. 462 p.

Vernberg, F. J. and W. B. Vernberg. 1970. The Animal and the Environment. New York: Holt Rinehart and Winston. 398 p.

Vernberg, W. B. and F. J. Vernberg. 1972. The synergistic effects of temperature, salinity, and mercury on survival and metabolism of the adult fiddler crab, Uca pugilator. Fish. Bull. 70: 415-420.

Vernberg, F. J. and W. B. Vernberg. 1974. Pollution and Physiology of Marine Organisms. New York: Academic Press. 492 p.

Vernberg, W. B., F. P. Thurberg, A. Calabrese and F. J. Vernberg. 1979. Marine pollution functional responses. New York: Academic Press. 454 p.

Vernberg, W. B., P. J. DeCoursey, and J. O'Hara. 1974.
    Multiple environmental factor effects on physiology and
    behavior of the fiddler crab, *Uca pugilator*. In:
    Pollution and Physiology of Marine Organisms. pp. 381-
    425. Ed. by F. J. Vernberg and W. B. Vernberg. New
    York: Academic Press.

Waldichuk, M. 1979. Review of the problems. Phil. Trans.
    T. Soc. Lond., B. 286: 399-424.

Warren, C. E. and G. E. Davis. 1967. Laboratory studies on
    the feeding, bioenergetics, and growth of fish. In:
    The Biological Basis of Freshwater Fish Production.
    pp. 175-214. Ed. by S. D. Gerking. Oxford: Blackwekk
    Sci. Publ.

Weiss, C. M. and R. H. Gakstatter. 1964. Detection of
    pesticides in water by biochemical assay. J. Wat.
    Pollut. Control Fed. 36: 240-253.

Widdows, J., D. K. Phelps, and W. Galloway. 1980.
    Measurement of physiological condition of mussels trans-
    planted along a pollution gradient in Narragansett Bay.
    Mar. Environ. Poll. In Press.

THE EFFECTS OF CHRONIC LOW CONCENTRATIONS OF
NO. 2 FUEL OIL ON THE PHYSIOLOGY OF A TEMPERATE
ESTUARINE ZOOPLANKTON COMMUNITY IN THE MERL MICROCOSMS

Sandra L. Vargo[1]

University of Maryland
Center for Environmental and Estuarine Studies
Chesapeake Biological Laboratory
Solomons, Maryland

INTRODUCTION

Large oil spills such as those of the *Amoco Cadiz* and the
Argo Merchant are the most obvious form of oil pollution in
the sea. Although they can do tremendous biological and
economic damage, these spills occur as a single pulse of oil
into the environment. Chronic low level oil pollution from
sources such as storm sewers, marinas, and ship discharges
into estuaries may be of equal or greater importance. (Van
Vleet and Quinn, 1977; Hyland and Schneider, 1976). The
biological impact of chronic low levels is difficult to
assess since subtle physiological effects can be masked by
natural seasonal and spatial variability.

Large scale microcosms, such as the tanks at the Marine
Ecosystems Research Laboratory (MERL) are appropriate devices
for assessment of the effects of chronic pollutants on
estuarine ecosystems. The microcosms were constructed as
analogues of coastal estuarine systems and can be experimentally manipulated to test pollutant effects on benthic-
pelagic coupled systems. This facility is particularly
valuable for studying pelagic communities since encapsulation

[1]*present address: Department of Marine Science, University of South Florida, St. Petersburg, Florida.*

removes spatial variability due to patchiness in the natural
environment and allows for repeated sampling of the same
species assemblage. Populations reproduce, change seasonally
and maintain themselves in the tanks (Heinle et al., 1978).
Links between functional and structural responses can also be
examined, particularly the role of physiological responses as
predictors of population and community stress (Wilson, 1975).

The effects of oil on natural zooplankton populations
have been neglected despite their role in the food web. Most
investigators have concentrated on setting lethal limits for
single species (Lee and Nicol, 1978; Tatem et al., 1978;
Lee and Nicol, 1977; Benville and Korn, 1977; Byrne and
Calder, 1977; Vanderhost et al., 1976; Barnett and Kontogiannis, 1975; Mironov, 1969). The LC_{50} concentrations for
crude and refined oil generally range from 1-100 ppm. However, acute toxic responses are of limited value in making
ecological predictions (Wilson, 1975). When functional
processes such as reproduction, development time, feeding, and
respiration are measured in the laboratory, exposure times are
usually short (24-48 hrs), similar to the situation near an
oil slick (Lee et al., 1978; Dillon et al., 1978; Edwards,
1978; Berdugo et al., 1977; Donahue et al., 1977; Ustach,
1977; Gyllenberg and Lindqvist, 1976; Spooner and Corkett,
1974). Few studies involve long term exposure over the
entire life cycle of the organism. Such exposures may have
greater impact on a species population dynamics than short
term exposure to high oil levels (Lee, 1978; Laughlin et al.,
1978; Linden, 1976; Venezia and Fossato, 1977).

One of the problems inherent in assessing effects of
long term exposure is developing appropriate criteria for
predicting damage to the population or community. At low
levels of oil, physiological changes occur prior to changes in
population numbers and community structure. Definite linking
of such functional changes to subsequent structural changes
would offer more quickly obtained criteria for probable
ecological effect. Van Overbeek and Blondeau (1954) and
Goldacre (1968) report that oil affects the cell membrane.
Therefore membrane mediated processes such as respiration and
excretion rates should be good indicators of an organism's
response to oil. These rates are also a good indication of an
animal's metabolic state, especially when combined as an O:N
ratio (Mayzaud, 1973). Reeve et al. (1977) reported zooplankton respiration rate was not particularly sensitive to
copper, but the toxic action of this metal may not be membrane
specific.

The effects of oil on respiration rate of crustaceans
has been reported by Ponat (1975), Gyllenberg and Lindqvist
(1976), Lee et al. (1978) and Tatem (1977). They found much

higher oil concentrations (>1 ppm) than those that will be discussed in this paper were required to change respiration rate. There are no reports in the literature on the effects of oil on excretion rates although changes in excretion, in addition to indicating physiological changes in the zooplankton community, are also important in recycling nutrients to primary producers.

METHODS

The MERL microcosms are large insulated fiberglass tanks 2 m in diameter and 6 m high with a 5.1 m water column (13.1 m^3) and (in a bottom tray) 30 cm of a soft bottom benthic community from Narragansett Bay. The tanks are mixed on a 2 hr. on, 4 hr. off cycle. Water from the Bay is automatically introduced into the tanks at a rate which yields a 30 day residence time. Further details of the microcosms are described in Pilson *et al.* (1979).

Three tanks were used as controls and three tanks received oil during the two oil experiments described in this paper. The oiled tanks were dosed twice weekly with an oil-water dispersion (OWD) containing 100-200 ppm No. 2 fuel oil (Gearing *et al.*, 1979). The OWD was added during mixing cycles and no visible surface slick was seen. In the first or "high" oil experiment (Feb. 14, 1977-Aug. 1, 1977) concentrations in the tanks averaged 181 ppb with a range of ∿98 ppb immediately before an addition to ∿265 ppb immediately after addition. In the second or "low" oil experiment (March 6, 1978-July 2, 1978) the concentrations averaged 91 ppb and ranged from ∿52 ppb immediately before to ∿113 ppb immediately after an addition, decreasing to ∿100 ppb 18 hrs. after addition (Gearing *et al.*, 1979).

The microcosms supported a zooplankton community similar in seasonal species succession to Narragansett Bay (Heinle and Vargo, 1978). Chronically and acutely exposed animals were collected from oiled and control tanks using a vertically towed 0.5 m, #20 mesh (80 μm) plankton net. Zooplankton were collected immediately before use. The tow was sieved and the >150 μm fraction, designated mixed zooplankton, used. *Acartia clausi* and *Acartia tonsa* adults were present in very low numbers in the oiled tanks, therefore these species were obtained from tows taken either in a control tank or the bay. Thus, neither *Acartia* sp. had been previously exposed to oil.

Respiration and excretion rates of both mixed zooplankton and *Acartia* sp. were measured by incubating them in 300 ml

BOD bottles for 24 hrs. at tank temperature. Changes in
dissolved oxygen and ammonia concentrations were used as a
measure of respiration and excretion rate. Animal densities
were kept as low as possible, allowing 6.0 to 10.0 ml per
animal depending on the temperature. Control bottles with no
animals were run at the same time to account for changes due
to microbial activity. After incubation, subsamples were
siphoned from control and experimental bottles. Dissolved
oxygen concentrations were measured by the modified Winkler
method (Carritt and Carpenter, 1966) and ammonia concentrations by the Solorzano technique (Solorzano, 1969). The
animals were filtered at low vacuum onto preweighed, oven
dried (60°C) 12 μm Nucleopore filters and dried at 60°C for
dry weight determinations. Oven drying of the Nucleopore
filters at 60°C was required to obtain consistent filter
weights (Vargo, unpublished).

The simplest experimental design, comparing the rates of
control tank animals incubated in control tank water with
oiled tank animals in oiled tank water, was discarded after a
few trials since variability between tanks in species composition and food supply precluded direct comparisons. A
crossover design (Fig. 1) was adopted which allowed comparison

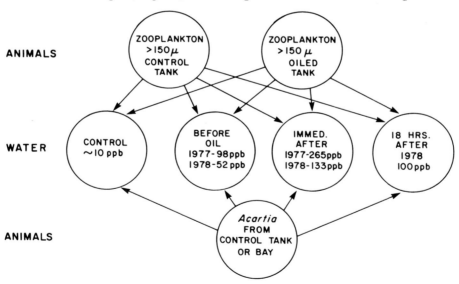

FIGURE 1. The effect of oil on zooplankton respiration
and excretion rate was determined by incubating chronically
and acutely exposed animals in the various water types shown.
The control water is from a control tank and the before,
immediately after, and 18 hrs after oil addition water from
the same oil tank.

of acute and chronic responses of the total zooplankton
community and the acute response of *Acartia*.   Acute responses
were obtained from control tank animals incubated in oiled
water siphoned immediately before and after OWD addition.
Chronic responses were measured by incubating oiled tank
animals in the same sources of water as used to estimate
acute response.   Both oil and control tank animals were
incubated in control tank water to assess differences in rates
caused by intertank differences in phytoplankton biomass and
composition and to give an estimate of recovery from oil
stress in chronically exposed animals.   Additionally, during
the "low" oil experiment, control and oil tank animals were
incubated in water from oiled tanks 18 hrs. after OWD
addition.   *Acartia* sp. incubated in these same combinations
(Fig. 1) allowed us to assess the impact on a single numerically dominant species in the community.

The data was analyzed using the multivariate analysis of
variance (MANOVA) procedure available in the SAS 76 software
package.   Treatment (oil concentration), time, and chronic
versus acute exposure were tested for effect.   This was done
for single dates as well as the complete series for each
experiment.   Effects were judged significant if the calculated F was significant (p < .05).

RESULTS

"High" Oil (98-265 ppm) Experiment

Oil additions began on February 14, 1977 and ceased
August 1, 1977.   From May through June the zooplankton community was dominated by *Acartia clausi* which was gradually
replaced in June by harpacticoid nauplii and subsequently by
Acartia tonsa in late July and early August.   Rate measurements were made on seven dates from May 15 through August 1 to
assess the effects of oil on physiological function.   Weekly
measurements were made from August 1 to August 21 to assess
recovery.   Background oil levels in the oiled tanks were
reached about one week after the last OWD addition (∿10 ppb).

When respiration rate measurements for all ten dates
were analyzed together no significant relationship was found
between treatment or source of animals.   A high within tank
variability (large coefficeint of variance) for respiration
measurements on several of the dates precluded finding a
significant relationship.   There was, however, a significant
increase in community respiration rate with time, as expected,

since temperature increased from 11°C to 23°C from May through
August. When analyzed individually three of the ten dates
showed statistically significant differences in respiration
rate depending on treatment (p < .01) and source of the
animals (p < .01). Respiration rate decreased with increasing
oil concentration in water taken immediately after an OWD
addition and chronically exposed animals from the oiled tanks
had a lower respiration rate than acutely exposed animals in
the same treatment (Fig. 2a). Comparison of the respiration
in control tank water and oiled tank water before an addition
shows that oiled tank animals do not return to normal levels
even at the lowest oil concentrations. Increased respiration
for control tank animals and oiled tank animals incubated in
water collected before oil additions may be a reflection of
the higher food levels in the oiled tanks, (Elmgren et al., In
press).

Excretion rate measurements showed far less variability
than respiration rate measurements. When all dates were

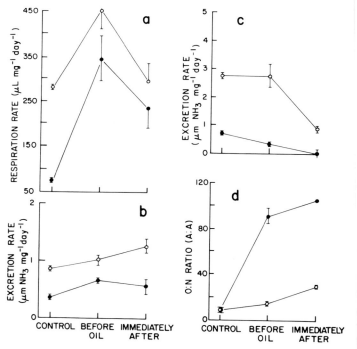

FIGURE 2.   Changes in respiration and excretion rates
and O:N ratio of acutely (O) and chronically (●) exposed
zooplankton in control and oil tank water during the "high"
oil experiment.   A. Respiration rate of zooplankton (Aug.
25, 1977).   B. Excretion rate of zooplankton (May 14, 1977).
C. Excretion rate of zooplankton (Aug. 25, 1977).   D. O:N
ratio of zooplankton (Aug. 25, 1977).

analyzed there was a significant effect of oil over time (p < .001), treatment, (p < .0001), and source of the animals (p < .0001). Excretion rate also increased with time as temperature increased. The changes in excretion rate with oil showed two different patterns: it either remained constant or decreased with increasing oil concentration (Fig. 2b and 2c). In either case, excretion rate in control water was not always significantly different from that measured in water collected before an oil addition, but was significantly different than the rate measured in water collected immediately after an addition. Regardless of the pattern in response which the animals exhibited, the excretion rate of chronically exposed animals was significantly lower than acutely exposed animals incubated in water from the same source.

No differences were found in the excretion rate of animals incubated in control tank water and oil tank water collected one week after the last oil addition, i.e. no difference due to the source of the water; but, animals from the former oil tanks still had a significantly (p < .003) lower excretion rate than the control tank animals, i.e. a difference due to source of animals. Three weeks after the last addition, however, there was no longer any significant difference between the two communities.

The O:N ratios for animals from both oil and control tanks were significantly affected by treatment and source of animals. The O:N ratio of the animals from both sources increased with increasing oil concentration indicating a switch to a greater predominance of carbohydrates and lipids as metabolic substrates. In addition, chronically exposed animals have a much higher ratio than acutely exposed animals (Fig. 2d) when incubated in water collected immediately before and after oil additions.

A. clausi, a winter-spring numerical dominant in both the tanks (Heinle *et al.,* unpublished) was used for respiration and excretion measurements in only two experiments, one on March 6 (3°C) and one on May 14, 1977 (11°C). The presence of oil did not significantly affect its respiration rate at either temperature. Excretion rate, however, showed a marked effect due to oil (p < .02). Zooplankton excretion in control tank water was significantly lower than in oil tank water before an oil addition (Fig. 3a). Immediately after oil addition, excretion rate decreased to a level below that of the control. Although excretion rate varied, there was no significant treatment effect of oil on the O:N ratio due to the variance in respiration rate.

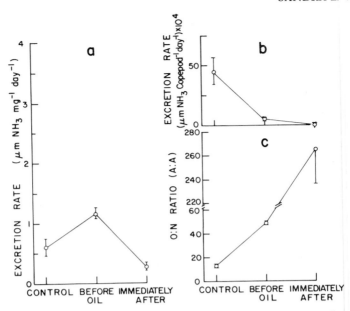

FIGURE 3. *Changes in excretion rate of* Acartia *in control and oil tank water during the "high" oil experiment.*
A. Excretion rate of A. clausi *(May 14, 1977).* *B. Excretion rate of* A. tonsa *(August 21, 1977).* *C. O:N ratio of* A. tonsa *(August 21, 1977).*

A. tonsa, the summer numerical dominant in the tanks and bay (Heinle et al., unpublished) was present for the majority of the respiration and excretion rate determinations, both during the oil additions and recovery period. Respiration rate of A. tonsa was not significantly affected by oil but its excretion rate was significantly lower than controls when incubated in water collected before and after oil addition (Fig. 3b). Depression of excretion was greatest immediately after oil addition. Lower excretion rates of animals incubated in oil tank water were detectable until three weeks after oil additions ceased, after which no difference due to source of water were seen. The O:N ratio for A. tonsa showed a response pattern similar to that for mixed zooplankton, i.e., increasing as oil concentration increased (Fig. 3c). This indicated that this species also switched to predominantly carbohydrate-lipid metabolism.

"Low" Oil (52-133 ppb)

Respiration measurements were made in the spring and early summer during the second oil experiment. The zooplankton community present during this time was similar in composition to that found during the same period in the "high" oil experiment although total number of species was greater (Table 1).

The changes in respiration and excretion rates of the mixed zooplankton differed in a number of aspects from the "high" oil experiment. Within one week after the first oil addition, respiration rates of mixed zooplankton in oil tank water collected immediately after an oil addition were significantly higher than for animals incubated in control tank water but rates in water collected before oil addition and 18 hours after an oil addition were indistinguishable from that in the control water (Fig. 4a). Respiration rates of chronically exposed animals were not significantly different from acutely exposed animals and recovery occurred between oil additions. Six weeks after the first oil addition, when water temperature had increased from 1°C to 7.5°C, respiration measurements showed a somewhat different response to oil additions. Respiration rates of chronically and acutely exposed animals were again indistinguishable but the great differences in respiration rates of animals in control tank water compared to rates in oil tank water collected immediately after an oil addition had disappeared (Fig. 4b). Statistical analysis indicates that the respiration rate in water collected immediately after an addition was significantly lower than the control value ($p < .04$). As in the earlier determination the respiration rate appeared to return to control values between additions.

The excretion rates of mixed zooplankton were affected by oil in a manner similar to the first response pattern shown by respiration rate. Excretion rates increased significantly immediately after an oil addition, but could not be distinguished from the controls at other oil levels, and returned to control tank rates between additions. No difference was found in the response pattern of chronically and acutely exposed animals (Fig. 4c).

There is a significant treatment effect on the O:N ratio (Fig. 4d), which reflects the relative changes in respiration and excretion rate. This ratio is significantly lower than that in the controls immediately after an addition indicating increased utilization of protein as a metabolic substrate. There is no difference in the ratio between chronically and acutely exposed animals and the ratio returns to control levels between additions.

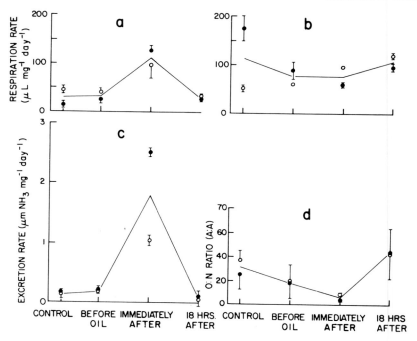

FIGURE 4.   Changes in respiration and excretion rates
and O:N ratio of acutely (o) and chronically (•) exposed
zooplankton in control and oil tank water during the "low" oil
experiment.   A. Respiration rate of zooplankton (March 13,
1978).   B. Respiration rate of zooplankton (April 10, 1978).
C. Excretion rate of zooplankton (March 13, 1978).   D. O:N
ratio of zooplankton (March 13, 1978).

        Respiration and excretion rate measurements were made for
A. clausi only, at different oil levels during the second
experiment.   This species was the numerical dominant through
the spring and early summer in the tanks and bay (Heinle et
al., unpublished).   Although A. clausi was the numerical
dominant, its total abundance was low and insufficient animals
were available for measuring excretion rates of chronically
exposed animals.   Therefore all "low" oil measurements
established acute effects.   The respiration rate of the
species was significantly altered by oil but the pattern of
response varied with water temperature and the length of time
oil had been added to the tanks.   One week after the first oil
addition the respiration rate of A. clausi was significantly
higher in water collected immediately after an addition than
in control water ($p < .01$) and declined below the control
rate, in water collected 18 hours after an addition ($p < .01$).
Six weeks after oil additions started and at warmer water
temperatures (7.5°C), A. clausi responded differently

(Fig. 5a). Respiration at all oil concentrations was higher than in the control water and the rate measured in water collected immediately after an oil addition was lower compared to the other two oil concentrations.

The excretion rate of *A. clausi* was depressed in water collected immediately after an oil addition compared to that in control water and two other oil concentrations. This difference is statistically significant (p   .05), (Fig. 5b). When respiration and excretion rates are combined in an O:N ratio there is also a significant effect due to oil. The ratio is significantly higher than control values before an oil addition and is further elevated immediately after oil additions. O:N ratios of animals in control tank water and water from oil tanks collected 18 hrs. after addition are indistinguishable (Fig. 5c).

DISCUSSION

The two oil experiments in the MERL microcosms allow us to compare the chronic effects of two low levels of No. 2 fuel on the physiological functions of a coastal marine zooplankton community. During both "high" and "low" oil experiments the

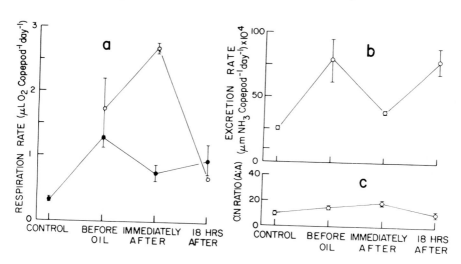

FIGURE 5. Changes in respiration and excretion rate and O:N ratio for A. clausi during the "low" oil experiment. A. Respiration rate on March 6, 1978 (first oil addition) (o) and April 10, 1978 (●). B. Excretion rate (April 10, 1978). C. O:N ratio (April 10, 1978).

respiration rate of mixed zooplankton was significantly
altered at the different concentrations before and after the
twice weekly oil addition. Altered respiration rates varied
in magnitude and pattern of change with oil concentration
between the two experiments. The difference in magnitude
of respiration rate between acutely and chronically exposed
animals in the "high" oil experiment has at least two possible
explanations. Oil tank animals may have acclimated to oil
stress, possibly by changes in species composition or modification of the toxic effects of oil, or irreversible
damage to the chronically exposed zooplankton may have
occured. Differences in species composition do not seem
large enough to account for the large difference in respiration rate betweem the zooplankton communities in the control
and oil tanks (Table 1). Therefore, either modification of
the toxic effect of oil or irreversible damage, perhaps to
membranes, seem more reasonable explanations for the lower
respiration rates of oil tank zooplankton compared to control
tank zooplankton in all oil concentrations. The lower respiration rates of oil tank animals compared to control tank
animals when incubated in control tank water indicate that
at least over 24 hrs., no recovery occurs, supporting the
hypothesis that irreversible damage has occurred. This
difference in magnitude for respiration rates of chronically
and acutely exposed animals was not found in the "low" oil
experiment. In fact, recovery to normal rates occurred
at the lower oil concentrations between additions. Therefore,
the effect of oil appeared to be reversible in the "low" oil
experiment.

Differences in the pattern of change in respiration rate
with oil concentrations between the two experiments consisted
mainly of whether stimulation or depression of respiration
occurred in water collected before and immediately after an
oil addition (Figs. 2a and 4a). During the "high" oil
experiment respiration rate increased in water collected
before an oil addition (98 ppb) then decreased in water
collected immediately after an addition (265 ppb). In
contrast, in the "low" oil experiment respiration rate increased in water collected immediately after an oil addition
(133 ppb). Species composition of the zooplankton assemblages
in the oil and control tank was very similar in both experiments, therefore it appeared valid to compare the change in
respiration rate in water collected before an oil addition
relative to the rate in control tank water in the "high"
oil experiments (98 ppb) with the change in respiration rate
in water collected immediately after an oil addition relative
to the rate in the control tank water in the "low" oil
experiment. When this comparison is made it appears that oil

TABLE 1. Zooplankton species composition in order of abundance during the "high" and "low" oil experiments in control and oil tanks.

| "High" oil Experiment | | | | "Low" Oil Experiment | |
| Spring-Early Summer | | Summer | | Winter-Spring | |
Control	Oil	Control	Oil	Control	Oil
Acartia clausi	Acartia clausi	Acartia tonsa	Acartia tonsa	Acartia clausi	Acartia clausi
Longipedia helgolandica	harpacticoid nauplii	Oithona similis	Oithona colcarva	Pseudocalanus minutus	Pseudocalanus minutus
harpacticoid nauplii	Pseudocalanus minutus	Oithona colcarva	Pseudodiaptomus coronatus	Eurytemora herdmani	Eurytemora herdmani
Oithona similis	Temora longicornis	Pseudodiaptomus coronatus	Acartia clausi	harpacticoid nauplii	harpacticoid nauplii
Oithona colcarva	Sappharella	Longipedia helgolandica	harpacticoid nauplii	Centropages hamatus	Centropages hamatus
Pseudocalanus minutus	barnacle nauplii	Acartia clausi	Sappharella	Pseudodiaptomus coronatus	Oithona similis
Temora longicornis	polychaete larvae	harpacticoid nauplii	Longipedia helgolandica	Temora longicornis	Temora longicornis
Sappharella	veligers	barnacle nauplii	polychaete larvae	Sappharella	Sappharella
barnacle nauplii		polychaete larvae	veligers	Longipedia helgolandica	Tortanus
polychaete larvae		veligers	crab zoea	Tortanus	barnacle nauplii
veligers		crab zoea		Podon	barnacle cyprids
				barnacle nauplii	polychaete larvae
				barnacle cyprids	veligers
				polychaete larvae	
				veligers	

concentrations in the range of 98-133 ppb stimulate respiration rate while higher concentrations cause depression.
These results are compatible with those of Gyllenberg and
Lindqvist (1976) for *A. bifilosa* which showed an initial
increase in respiration with oil stress followed by a decrease
as narcosis occurred. Unfortunately, no oil concentrations
were reported. Lee *et al.* (1978) found an increase in
respiration rate in *Lucifer faxoni,* a sergestid shrimp, in
concentrations of No. 2 fuel up to 6 ppm, then a decrease at
higher concentrations. Tatem (1977) found a single exposure
to 3 ppm No. 2 fuel oil caused a depression of respiration
rate in *Palaemontes pugio.* Edwards (1978) reported that the
respiration rate of *Crangon* exposed to crude oil increased in
estimated oil concentrations from 950-1900 ppb and decreased
in concentrations from 1900-3800 ppb. These authors used
much higher concentrations than those in the MERL microcosms,
a water-soluble fraction of either No. 2 fuel oil or crude
oil, and only measured acute responses. Keeping these
differences in mind it appears the zooplankton community in
the MERL tanks was much more sensitive to oil than some other
species. The respiration rate of zooplankton in the MERL
tanks was affected by concentrations 10-60 times less than
those used by the investigators mentioned above. In addition,
the OWD of No. 2 fuel oil is lower in aromatics than the
water soluble fraction (Gearing *et al.* 1979). Aromatics are
considered to be the more toxic component in oil (Anderson *et
al.,* 1974). Thus, the changes in zooplankton respiration
rate with oil stress in the MERL microcosms are not unprece-
dented but occur at much lower concentrations of oil than
previously reported.

The permanence of changes in respiration and excretion
rates caused by oil can be assessed during the three week
recovery period. One week after oil additions ceased, the
oil concentrations in the oiled microcosms were reduced to
background levels (10 ppb) (Gearing *et al.,* 1979). However,
the respiration rate of mixed zooplankton was still significantly affected by source of water and animals indicating
differences still existed between the control and oil tank
water and acutely and chronically exposed animals. It was
three weeks after the last oil addition before this difference
disappeared indicating the effect of oil in the "high" oil
experiment was indeed irreversible. After three weeks at
summer temperatures it is likely that the majority of the
animals then present in the oil tanks were produced after the
additions of oil ceased and had therefore never been exposed
to oil.

The excretion rate for mixed zooplankton, like respiration rate, was significantly affected by the different oil

concentrations before and after an oil addition. This change in excretion rate varied in magnitude and pattern of change with oil concentration for the two experiments. During the "high" oil experiment variability in replicate measurements of excretion rate was much less than for respiration rate, therefore differences due to oil concentration were significant on most of the dates on which excretion rate was measured. Acclimation to oil stress by the chronically exposed animals was again evident in the "high" oil experiment. Chronically exposed animals had a consistently lower excretion rate than acutely exposed animals at the same oil concentration. Recovery in control tank water did not occur which indicates that the change in excretion rate, as that in respiration rate, is irreversible at least over 24 hrs. in the "high" oil experiment.

     Unlike respiration, two patterns of change in excretion rate with oil concentration are shown in the "high" oil experiment. In May excretion rates show an increase in the before oil addition water and a slight further increase or leveling off in water collected immediately after an addition. By the end of July, approximately five months after oil addition began the pattern is quite different. The excretion rate in before oil addition water was the same as in control water but then declined sharply in water collected immediately after an oil addition. This change in pattern may be due to an alteration in the effect of oil on excretion rate with temperature (which over this period has risen from 11°C to 23°C) or a change in species composition in all the tanks, i.e., a shift from a winter-srping community to a summer community. Seasonal change in food supply is another possibility, although unlikely, since phytoplankton biomass in the oil tanks was considerably greater than in the control tanks during July (Fig. 6a). Thus, the decline in excretion rate in water collected immediately after an addition compared to the excretion rate in control tank water cannot be due to lack of food. Whatever the correct explanation may be, it should be noted that there was only one response pattern to oil concentration for respiration, but two for excretion. However, excretion rate was similar to respiration rate in that, regardless of the pattern of change with oil concentration, no recovery was apparent between additions. During the three week recovery period after oil additions ceased, zooplankton from either control or oil tanks showed no significant differences in excretion rate when incubated in water from oil tanks compared to the excretion rate in water from control tanks. The chronically exposed animals from the oil tanks still, however, had an overall lower excretion rate than acutely exposed animals in the same water.

It took three weeks for the difference in magnitude of
excretion rate between chronically and acutely exposed animals
to disappear. In contrast to the "high" oil experiments,
there was no difference in the magnitude of the excretion
rate between acutely and chronically exposed animals during
the "low" oil experiment, the same difference in effect
observed for respiration rate, and the response pattern for
excretion rate to oil concentration was similar on all dates
that excretion rate was measured (Fig. 4c). The pattern of
change was more distinct but similar to one of the patterns of
change found in the "high" oil experiment, with excretion rate
showing a sharp increase in water collected immediately after

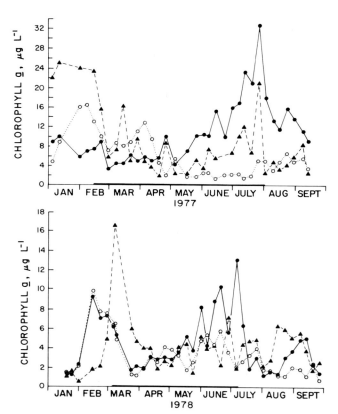

*FIGURE 6. Mean phytoplankton biomass in the three
control (o) and three oil (●) tanks in the "high" oil
experiment (A) and "low" oil experiment (B). The darkened
portion of the x-axis shows the period of oil addition in each
experiment.*

an oil addition (133 ppb) while excretion rate at the two
other oil concentrations was not significantly different
from the control rate. This was similar to the slight
increase in excretion rate in water collected before oil
addition (98 ppb) in the "high" oil experiment compared to the
control. THus, between 98-133 ppb total hydrocarbons it
appears that excretion rate is stimulated as is respiration
rate. Recovery in excretion rate compared to the controls
occurred between additions, another similarity to respiration
rate.

The similarity of pattern of change in excretion rate
with oil concentration over the dates excretion rate was
measured may give a clue as to why the pattern of change
did not alter with time as it did the "high" oil experiment.
The range of temperatures (2°C to 17°C) over corresponding
months during the "high" and "low" oil experiments was
similar. Food levels, measured as chlorophyll, were again
higher in the oil tanks compared to the controls but the
difference between the oil and control tanks was not as
great (Fig. 6b) as in the "high" oil experiment. Stimulation
of excretion rate by oil tank water was, however, greater
(Fig. 2c and 4c) during the "low" oil experiment despite
greater similarity in food availability. The species
composition of the zooplankton community did not change, however, during the "low" experiment. The community, over all
the dates excretion rate was measured, was a typical winter-
spring community dominated by *A. clausi*. This gives added
weight to the role of species composition in determining the
response pattern of excretion rate in the "high" oil
experiment.

Respiration and excretion rates for organisms may be
combined as an O:N ratio. Mayzaud (1976) has used this ratio
for zooplankton to given an indication of the biochemical
nature of the metabolic substrates being utilized by the
organisms and the physiological state of the organisms. In
this study the O:N ratio also shows the relative differences
in the impact of oil on respiration and excretion.

In the "high" oil experiment the O:N ratio is altered by
both oil concentration and source of the animals. The ratio
increases with increasing oil concentration, due to a relatively large increase in respiration rate and a smaller
decrease in excretion rate in water collected before an oil
addition compared to respiration and excretion rates in
control tank water. In water collected immediately after an
oil addition, excretion rate is greatly reduced compared to
the control yielding a further increase in O:N ratio.
Increases in O:N ratio indicate a switch from proteins as
metabolic substrates to a greater dominance of carbohydrates

and lipids.   This may be a reflection of the differences in
biochemical composition of phytoplankton in the oiled tanks or
a disruption of nitrogen metabolism in the zooplankton by oil.
The phytoplankton in the oiled microcosms may have been
nitrogen limited (Elmgren *et al.*, in press).   Nitrogen
stress alters the biochemical composition of the phytoplankton
resulting in an accumulation of carbohydrates (Haug *et al.*,
1973).   This change in biochemical composition is a possible
explanation for the increase in the O:N ratio in water
collected before an oil addition compared to that in control
tank water although there are also differences in phytoplankton species composition and biomass in the oil tanks
compared to the controls.   However, the abrupt increase in
O:N ratio in water collected immediately after an oil
addition seems to indicate that a disruption of nitrogen
metabolism is a more likely explanation, at least at higher
oil concentrations.

The O:N ratio of zooplankton chronically exposed to oil
varied to a greater extent than the O:N ratio of acutely
exposed animals over the same range of oil concentrations.
O:N ratios in control tank water were very similar for both
oil and control tank animals.   Chronically exposed oil tank
animals recovered the balance between their respiration
and excretion rates in control tank water, although both
rates were lower in the oil tank animals than in the control
tank animals.   The much higher ratios for chronically
exposed animals compared to acutely exposed animals at the
same oil concentrations indicate a greater shift to carbohydrates and lipids or an increased sensitivity of nitrogen
metabolism in the chronically exposed animals.

In the "low" oil experiment the O:N ratio did not show
acclimation to oil.   The O:N ratio decreased in water
collected immediately after an oil addition compared to the
O:N ratio in water collected before an oil addition and
therefore was different from the response pattern observed
during the "high" oil experiment.   The lower O:N ratio
indicates a greater proportion of proteins are being used as
metabolic substrates.   As in the "high" oil experiments, the
phytoplankton community in the oil tanks may be nitrogen
limited.   However, the O:N ratio returned to control levels
bewteen oil additions which it did not in the "high" oil
experiment, indicating a direct effect of oil on nitrogen
metabolism is a more likely cause for the changes in O:N
ratio than differences in the biochemical composition of the
phytoplankton.

The zooplankton community is made up of a number of
interacting species.   The responses of these constituent
species to oil may be similar to the total community response

or quite different but interrelated in such a way that they yield the community response. *A. clausi* is the winter-spring numerical dominant in Narragansett Bay and the tanks occurring from February to July (Heinle *et al.*, 1978). It is present over a range of temperatures from 1°C to 20°C, reproducing up to approximately 15°C (Gonzales, 1974; Jeffries, 1962). Respiration and excretion rate measurements were made on this species two times during the "high" oil experiment and four times in the "low" experiment. In the "high" oil experiment, partially due to high variability in the rate measurement, there was no significant effect of oil on respiration rate. Excretion rate was significantly higher in water collected before an oil addition than in control water and lower in water collected immediately after an oil addition. The O:N ratio was sufficiently variable that no significant changes due to oil concentration were detectable. The excretion rate of the whole zooplankton community during the "high" oil experiment was higher in water collected before an oil addition compared to that in control water but remained constant with the higher oil concentrations in water collected immediately after an oil addition. THus, the change in excretion rate of *A. clausi* was similar to the change in community excretion rate at one oil concentration but at higher concentrations compensating differences in response by other species held the community excretion rate relatively constant.

In the "low" oil experiment lower variability in the respiration rate measurements allowed detection of significant changes in respiration rate with oil concentration. The pattern of change in respiration rate with oil concentration varied with time from the initial oil addition but in both patterns of change the respiration rate of *A. clausi* was always higher in oil tank water than in control tank water. This increase in respiration rate may be a reflection of the greater phytoplankton abundance (Fig. 6b) in the oiled microcosms although the difference in abundance were less than during the "high" oil experiment. Mayzaud (1973) and Nival *et al.* (1974) have both shown that both respiration and excretion rate are higher in fed than starved copepods. In the "low" oil experiment excretion rate of *A. clausi* decreased in water collected immediately after an oil addition relative to the excretion rate in water before an oil addition which is the reverse of that exhibited by the whole community. Such decreases in excretion rate may be a result of the changes in feeding behavior reported by Berman and Heinle (1980), ranging from complete to partial suppression of feeding.

Reflecting these differences between the respiration and excretion rates for *A. clausi* in the "low" oil experiment, the O:N ratio for *A. clausi* increases with increasing oil concentration indicating a switch to a higher proportion of lipids and carbohydrates, as metabolic substrates the reverse of the pattern of change shown by the total community which indicated a switch to proteins (Figs. 3d and 5c). Therefore the effects of oil on the respiration and excretion rates of one species from a community, although it was the numerical dominant, did not adequately predict the effect of oil on the respiration and excretion rates of the total community.

Respiration and excretion rates for *A. tonsa* were only measured during the "high" oil experiment. Although the respiration rate of *A. tonsa* was not significantly affected by oil concentration, its excretion rate decreased significantly with increasing oil concentration. The resultant O:N ratio indicates a shift to a higher proportion of lipids and carbohydrates as metabolic substrates. These changes in excretion rate and O:N ratio with oil concentration are similar to those of the total community. In this instance, the effects of oil on respiration and excretion rates of a constituent species are a good indication of the effects on the respiration and excretion rates of the total community (Fig. 2c and 3b). The decrease in excretion rate with increasing oil concentration for *A. tonsa* appears to be a direct effect of oil and not related to food supply. Phytoplankton biomass was higher in the oil than in control tanks, therefore absence of food cannot have caused the decrease in excretion rate. Although Berman and Heinle (1980) found suppression of feeding by *A. tonsa* in water collected immediately after an oil addition, i.e. high oil concentrations, feeding was normal in water collected before an oil addition. Therefore, the lower excretion rate of *A. tonsa* in water before an addition compared to the rate in control water cannot be accounted for by a change in feeding behavior.

Changes in physiological function are certainly of interest to the investigators studying the direct toxic effects of oil on organisms, such as damage to membranes but for ecological assessment the principal question is what impact do these changes in physiological function have on the zooplankton community structure. It is possible that other homeostatic mechanisms, such as compensating physiological changes in individual species within the community, allow it to continue to adapt to oil stress. Total zooplankton abundance and species counts in the MERL tanks indicate that this was not the case. Total zooplankton abundance decreased in the oil tanks compared to the control tanks in both the

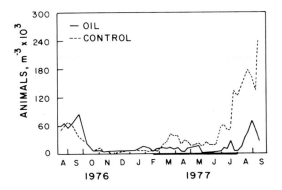

FIGURE 7. Mean total zooplankton abundance in the control and oil tanks during the "high" oil experiment. The darkened portion of the x-axis shows the period of oil addition in each experiment.

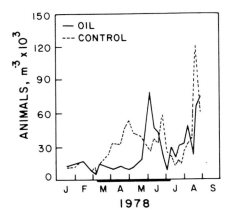

FIGURE 8. Mean total zooplankton abundance in control and oil tanks during the "low" oil experiment. The darkened portion of the x-axis shows the period of oil addition.

"high" and "low" experiments (Fig. 7 and 8). Individual species such as A. clausi and A. tonsa declined in the oil tanks compared to the control tanks in both experiments (Fig. 9, 10 and 11). Therefore changes in physiological responses are translated into structural changes for the whole community. The relative time scale of these changes is of interest. Changes in physiological response were detectable immediately as is shown by comparing the acute and chronic responses in the two experiments although acclimation to oil may alter the pattern of change with oil concentration.

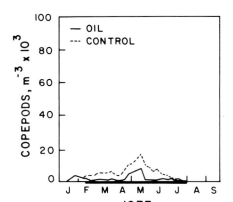

FIGURE 9. Mean abundance of A. clausi in control and oil
tanks during the "high" oil experiment. The darkened
portion of the x-axis shows the period of oil addition.

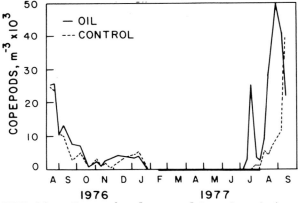

FIGURE 10. Mean abundance of A. clausi in control and
oil tanks during the "low" oil experiment. The darkened
portion of the x-axis shows the period of oil addition.

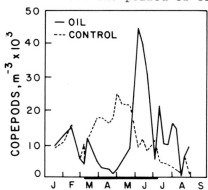

FIGURE 11. Mean abundance of A. tonsa in control and oil
tanks during the "high" oil experiment. The darkened portion
of the x-axis shows the period of oil addition.

However, three to four weeks for an individual species and six weeks for the whole community are required for structural changes to take place as is demonstrated by differences in total zooplankton abundance between oil and control tanks. Therefore, anyone interested in detecting early effects of pollutants is more likely to find such effects by measuring changes in physiological response rather than changes in community structure.

SUMMARY

1.   Chronic low levels of an oil-water dispersion (OWD) of No. 2 fuel oil affected the respiration and excretion rates of zooplankton in the MERL microcosms.
2.   An average concentration of 180 ppb caused irreversible changes in the respiration and excretion rates of chronically exposed animals while an average concentration of 90 ppb did not.
3.   The respiration and excretion rates of the zooplankton community were stimulated by oil concentrations between 98-133 ppb, but decreased at higher concentrations. The O:N ratio for the community decreases at lower oil concentrations and increases at higher concentrations indicating a switch from proteins to carbohydrates and lipids as metabolic substrates.
4.   The physiological responses of numerically dominant species in the community in one case were the same as that of the community (*A. tonsa*) and in the other not (*A. clausi*).
5.   Physiological changes shown by the community and its species lead to changes in total abundance and species composition but the time scale of these structural changes is much longer than that for the functional changes. Therefore, monitoring of physiological changes gives an earlier indication of the effect of oil on the community.

ACKNOWLEDGEMENTS

This research was supported by the Environmental Protection Agency under grant number R803902020 to the Marine Ecosystems Research Laboratory.

LITERATURE CITED

Anderson, J. W., J. M. Neff, B. A. Cox, H. E. Tatem, and
    G. M. Hightower. 1974. Characteristics of dispersions
    and water soluble extracts of crude and refined oils and
    their toxicity to estuarine crustaceans and fish. Mar.
    Biol. 27: 75-88.

Barnett, C. J. and J. E. Kontogiannis. 1975. The effect of
    crude oil fractions on the survival of a tidepool
    copepod, *Tigriopus californicus*. Environ. Pollut.
    8(1): 45.

Benville, P. E. Jr. and S. Korn. 1977. The acute toxicity
    of six monocyclic aromatic crude oil components to
    striped bass (*Morona saxatilis*) and bay shrimp (*Crago
    franciscorum*) Calif. Fish and Game 63(4): 204-209.

Berdugo, V., R. P. Harris, and S. C. O'Hara. 1977. The
    effect of petroleum hydrocarbons on reproduction of an
    estuarine planktonic copepod in laboratory cultures.
    Mar. Pollut. Bull. 8(6): 138-143.

Berman, M. and D. R. Heinle. 1980. Modification of the
    feeding behavior of marine copepods by sublethal con-
    centrations of water-accomodated fuel oil. Mar. Biol.
    (in press).

Byrne, C. J. and J. A. Calder. 1977. Effect of the water-
    soluble fractions of crude, refined, and waste oils on
    the embryonic and larval stages of the Quahog clam
    Mercenaria sp. Marine BIol. 40: 225-231.

Carritt, D. C. and J. H. Carpenter. 1966. Comparison and
    evaluation of currently employed modifications of the
    Winkler method for determining dissolved oxygen in
    seawater; a NASCO report. J. of Marine Res. 24(3):
    286-318.

Dillon, T. M., J. M. Neff and J. F. Warner. 1978. Toxicity
    and sublethal effects of No. 2 fuel oil on the supra-
    littoral isopod *Lygia exotica*. Bull. Environm. Contam.
    Toxicol. 20: 320-327.

Donahue, W. H., R. T. Wang, M. Welch, and J. A. C. Nicol. 1977. Effects of water-soluble components of petroleum oils and aromatic hydrocarbons on barnacle larvae. In: Environ. Pollut. (13). Applied Science Publishers Ltd., England.

Edwards, R. R. C. 1978. Effects of water-soluble oil fractions on metabolism growth, and carbon budget of the shrimp *Crangon crangon*. Mar. Biol. 46: 259-265.

Elmgren, R., G. A. Vargo, J. F. Grassle, J. P. Grassle, D. R. Heinle, G. Lang Lois, and S. L. Vargo. (In press) Trophic interactions in experimental marine ecosystems perturbed by oil. Symposium on Microcosms in Ecological Research, Savannah River Ecology Laboratory, 8-10 Nov. 1978. (In press).

Gearing, J. N., P. J. Gearing, T. Wade, J. G. Quinn, H. B. McCarthy, J. Farrington, and R. F. Lee. 1979. The rates of transport and fates of petroleum hydrocarbons in a controlled marine ecosystem and a note on analytical variability. In: Proceedings of the 1979 Oil Spill Conference (Prevention, Behavior, Control, Cleanup) pp. 555-564, American Petrol. Institute.

Goldacre, R. J. 1968. The effects of detergents and oils on the cell membrane. Field Studies 2 (Suppl.): 131-138.

Gonzalez, J. G. 1974. Critical thermal maxima and upper lethal temperatures for the calanoid copepods *A. tonsa* and *A. clausi*. Marine Biol. 27: 219-223.

Gyllenberg, G. and G. Lindqvist. 1976. Some effects of emulsifiers and oil on two copepod species. Acta zoologica Fennica 148, Societas Pro Fauna et FLora Fennica (ed.).

Haug, A., S. M. Myklestad, and E. Sakshaug, 1973. Studies on the phytoplankton ecology of the Trondheimsfojod. I. The chemical composition of phytoplankton populations. J. Exp. Mar. Biol. Ecol. 11: 15-26.

Heinle, D. R., S. L. Vargo, R. Lambert, and M. Berman. 1978. Effects of petroleum hydrocarbons on the physiology, feeding behavior, and population dynamics of copepods. Final Report to EPA from the Marine Ecosystems Research Laboratory, University of Rhode Island, Kingston, R.I.

Hyland, J. L. and E. D. Schneider. 1976. Petroleum hydro-
    carbons and their effects on marine organisms, popu-
    lations communities and ecosystems. In: Sources,
    Effects, and Sinks of Hydrocarbons in the Aquatic
    Environment. Proc. Symp. AIBS, Washington, D.C. August
    9-11, 1976. p. 463-506.

Jeffries, H. P. 1962. Succession of two *Acartia* species in
    estuaries. Limnol. Oceanog. 7(3): 354-364.

Laughlin, R. B. Jr., L. G. Young, and J. M. Neff. 1978. A
    long-term study of the effects of water-soluble fractions
    of No. 2 fuel oil on the survival, development rate, and
    growth of the mud crab *Rhithropanopeus harrisii*. Mar.
    Biol. 47(1): 87-95.

Lee, W. Y. and J. A. C. Nicol. 1977. The effects of water
    soluble fractions of No. 2 fuel oil on the survival and
    behavior of coastal and oceanic zooplankton. Environ.
    Pollut. 12: 279-291.

Lee, W. Y. 1978. Chronic sublethal effects of the water
    soluble fractions of No. 2 fuel oils on the marine isopod
    Sphaeroma quadridentatum. Mar. Environ. Res. 1(1): 5-18.

Lee, W, Y, and J. A. C. Nicol. 1978. The effect of naph-
    thalene on survival and activity of the amphipod
    Parhyale. Bull. Environ. Contam. Toxicol. 20: 233-240.

Lee, W. Y., K. Winters and J. A. C. Nicol. 1978. The bio-
    logical effects of the water-soluble fractions of a No. 2
    fuel oil on the planktonic shrimp, *Lucifer faxoni*.
    Environ. Pollut. 15: 167-183.

Linden, O. 1976. Effects of oil on the amphipod *Gammarus
    oceanicus*. Environ. Pollut. 10: 239-250.

Mayzaud, P. 1973. Respiration and nitrogen excretion of
    zooplankton. II. Studies of the metabolic character-
    istics of starved animals. Mar. Biol. 21: 19-28.

Mayzaud, P. 1976. Respiration and nitrogen excretion of
    zooplankton IV. The influence of starvation on the
    metabolism and the biochemical composition of some
    species. Marine Biology 37: 47-58.

Mironov, O. G. 1969. Effect of oil pollution upon some
    representatives of Black Sea zooplankton. Zoologichiskii
    Zhurnal 48: 980-984.

Nival, P., G. Malara, M. R. Charra, I. Palazzoli, and S.
    Nival. 1974. Studies on the respiration and excretion
    of several copepods in the upwelling off the coast of
    Morocco. J. exp. mar. Biol. Ecol. 15: 231-260 (in
    French).

Pilson, M. E. Q., C. A. Oviatt, G. A. Vargo, and S. L. Vargo.
    1979. Replicability of MERL microcosms: Initial ob-
    servations. In: Advances in Marine Research, F. S.
    Jacoff (ed). pp. 359-381.

Ponat, A. 1975. Investigations on the influence of crude oil
    on the survival and oxygen consumption of *Idotea baltica*
    and *Gammarus salinus*. Kieler Meeresforch. 31: 26-31.

Reeve, M. R., M. A. Walker, K. Darcy, and T. Ikeda. 1977.
    Evaluation of potential indicators of sublethal toxic
    stress on marine zooplankton (Feeding, fecundity,
    respiration, and excretion): Controlled ecosystem
    pollution experiment. Bull. Mar. Sci. 27: 105-113.

Solorzano, L. 1969. Determination of ammonia in natural
    waters by the phenol-hypochlorite method. Limnol.
    Oceanogr. 14: 799-801.

Spooner, M. F. and C. J. Corkett. 1974. A method for testing
    the toxicity of suspended oil droplets on planktonic
    copepods used at Plymouth. In: Ecological Aspects of
    Toxicity Testing of Oils and Dispersants. L. R. Beynon
    and E. B. Cowell (ed.) 149 p. John Wiley and Sons,
    N.Y. p. 69-74.

Tatem, H. E. 1977. Accumulation of naphthalenes by grass
    shrimp: effects on respiration, hatching, and larval
    growth. In: Fate and Effects of Petroleum Hydrocarbons
    in Marine Organisms and Ecosystems (Wolfe, O. A. ed)
    Pergamon Press, N.Y. pp. 201-209.

Tatem, H. E., B. A. Cox, and J. W. Anderson. 1978. The
    toxicity of oils and petroleum hydrocarbons to estuarine
    crustaceans. Est. Coast. Mar. Sci. 6: 365-374.

Ustach, J. F. 1977. Effects of sublethal oil levels on the reproduction of a copepod, *Nitocra affinis*. Sea Grant Publ. UNC-SG 76-10.

Vanderhorst, J. R., C. I. Gibson, and L. J. Moore. 1976. Toxicity of No. 2 fuel oil to coon stripe shrimp. Mar. Poll. Bull. 7: 106-107.

Van Overbeek, J. and R. Blondeau. 1954. Mode of action of phytotoxic oils. Weeds, 3: 55-65.

Van Vleet, E. S. and J. G. Quinn. 1977. Input and fate of petroleum hydrocarbons entering the Providence River and upper Narragansett Bay from waste-water effluents. Environ. Sci. Technol. 11: 1086-1092.

Venezia, L. D. and V. U. Fossato. 1977. Characteristics of suspensions of Kuwait oil and Corexit 7664 and their short- and long-term effects on *Tisbe bulbisetoa* (Copepods: Harpacticoida). Marine Biology 42: 233-237.

Wilson, K. W. 1975. The laboratory estimation of the biological effects of organic pollutants. Proc. R. Soc. Lond. B. 189: 459-477.

EFFECTS OF CRUDE OIL ON GROWTH AND MIXED FUNCTION OXYGENASE ACTIVITY IN POLYCHAETES, *NEREIS* SP.

Richard F. Lee
James Stolzenbach
Sara Singer
Kenneth R. Tenore

Skidaway Institute of Oceanography
P. O. Box 13687
Savannah, Georgia

INTRODUCTION

The mixed function oxygenase (MFO) system, responsible for the metabolic modification of a variety of foreign organic compounds in vertebrates, has been detected in marine crabs and polychaetes (Lee, 1976; Lee, et al., 1979; Singer and Lee, 1977). The function of the system is to convert lipophilic foreign compounds to more water-soluble metabolites, facilitating their elimination from the animal. MFO activity in mammals and fish can be induced by exposure of the animals to chlorinated hydrocarbons or polycyclic aromatic hydrocarbons (Bickers, et al., 1972; Gelboin, 1972; Gruger, et al., 1976; Malins, 1977; Philpot, et al., 1976). Studies by Payne (1976) showed that fish from petroleum contaminated areas had higher levels of MFO than fish from cleaner nearby areas. No clear evidence has as yet been presented for such induction in marine invertebrates.

The objectives of our studies included the following: (1) changes in MFO activity in the polychaetes, *Nereis succinea* and *Nereis virens*, after exposure to crude oil or certain aromatic hydrocarbons; (2) the effects of oil on growth rates of *Nereis succinea* from coastal Georgia; (3) comparisons between the biomass per individual of a population of *Nereis virens* from oil contaminated and clean sites in coastal Maine.

Polychaete annelids are important members of the benthic
marine community. They and other deposit feeding invertebrates play an important role in the oxidation and recycling
of sediment organic matter (Tenore, *et al.*, 1977; Rhoads,
1967). Certain polychaete species are the dominant animals in
areas of oil spills, refinery effluents, oil field brine
effluents and natural oil seeps (Armstrong, *et al.*, 1979;
Baker, 1976; Reish, 1971; Sanders, *et al.*, 1972; Spies, *et
al.*, 1979). The cosmopolitan species, *Capitella capitata*,
dominated the benthos after the West Falmouth oil spill
(Sanders, *et al.*, 1972) and in parts of Long Beach harbor
receiving refinery wastes (Reish, 1971). Such species as
Arenicola marina and *Nereis diversicola*, though reduced in
numbers after oil spills, re-establish in the sediments in
spite of the continued presence of oil (Prouse and Gordon,
1976; Levell, 1976; Baker, 1976). The recent AMOCO CADIZ oil
spill had little effect on polychaetes, particularly *Arenicola
marina* (Chasse, 1978).

MATERIALS AND METHODS

Growth Experiments

Juvenile *Nereis succinea* were collected from intertidal
mudflats of an estuary (Georgia, USA) in March, 1979, and
held in the laboratory in filtered (no particles larger than
1 μm), temperature-controlled (20°C) sea water and supplied
with a diet of mixed cereal. Experiments were conducted in
trays ($0.1m^2$) that received 200 μl/min of filtered, temperature controlled seawater. Each tray was layered to a depth
of 3 cm with fine sand (<0.3mm). All of the trays initially
received 5 gms of mixed cereal.

At the start of the experiment ten individuals were
randomly chosen from the holding tray, dried (90°C for 24
hours), and ashed (500°C for 12 hours) for initial biomass
determinations. Ten worms, randomly selected from the
holding trays, were weighed and placed in each tray. Three
days were allowed for the worms to adjust to the trays and
make burrows in the sediment. Trays received 4 different
levels of food with or without absorbed Kuwait crude oil
(American Petroleum Institute standard). Crude oil (1.5ml)
was added with the food which was then carefully spread on
the sediment. Worms in both oiled and control trays were fed
twice a week. Four feeding regimes (2, 3, 6, and 8 gms/m^2-
day) of cereals were used with control trays with 3 trays for
each food concentration while for oil treatments 2 trays were

used for each food concentration. The food was 42.5% carbon
and 2.6% nitrogen as determined with a Perkin-Elmer Model
240 elemental analyzer. During feeding, the air and water
flow was stopped for 30 to 60 minutes to allow particles to
settle to the sediment. After 44 days the worms were sieved
from the sediment, allowed to void their guts for 2 hours and
weighed. Tissue samples from each food concentration were
dried and ashed to determine dry weight and ash free dry
weight conversion factors. Average growth of individual worms
was determined for each tray. Population production for each
tray was calculated by multiplying the number of worms times
the average growth of an individual.

Three randomly selected sediment samples from each oil
tray and from one control tray were taken at 14 day intervals
to determine concentrations of crude oil accumulated in the
sediment. Sediment was mixed with an equal volume of water
and the slurry was saponified with 4 N NaOH by heating at
95°C for two hours. The saponified sample was mixed with 5 ml
hexane. The hexane extracts were taken just to dryness by a
flow of dry nitrogen and the residue dissolved in methanol,
which was passed down a Sephadex LH-20 column to obtain a
polycyclic aromatic hydrocarbon fraction (Giger and Blumer,
1974). This fraction was taken to dryness and a portion (5μl)
of the concentrate was analyzed with a high pressure liquid
chromatograph (Model 7000B-Micrometrics) with a fluorescence
detector (Amino Fluoro Monitor-American Instrument Co.). The
chromatograph was fitted with a 4mm x 25 cm Spherisorb ODS
column. The column was eluted with 65% methanol in water with
a flow rate of 2 ml/min at 50°C. Calibration curves based on
peak areas were prepared for the two compounds of interest to
quantitate results. Compounds analyzed were phenanthrene and
chrysene which were 35 μg/g and 9μg/g, respectively in the
whole oil.

Mixed Function Oxygenase Assays

Tissues of *Nereis virens*, collected from mudflats in
Maine were dissected, weighed and homogenized in 0.15 M KCl
buffered with 0.05 M Tris buffer (pH 7.5) using a Potter-
Elvehjem homogenizer.

The intestines were used for determination of MFO activity
and the amount of cytochrome P-450. Cell debris and nuclei
were removed by centrifugation at 700 x g for 10 minutes at
4°C. Mitochondria were collected by centrifugation of the
supernatant at 8,000 x g for 10 minutes. The supernatant from
the mitochondrial fraction was centrifuged at 140,000 x g for
60 minutes in an ultracentrifuge (Beckman Model L-40) to
sediment microsomes. The protein content of each fraction,

after resuspension in the buffered KCl solution, was determined by the method of Lowry et al., (1959), using bovine serum albumin as the reference standard.

The assay for mixed function oxygenase activity was similar to that of Wattenberg, et al., (1962) with modifications described by Nebert and Gelboin (1968). Each assay mixture, in a total volume of 1 ml, contained 50 μmoles of Tris buffer, pH 7.5, 0.6 μmoles of NADPH, 3 μmoles of $MgCl_2$, 0.2 ml of cell homogenate (1-2 mg protein) and 0.01 μmoles of benzo(a)pyrene (added in 50μl of methanol). The mixture was shaken at 30°C for 30 minutes. The reaction was stopped by addition of 1 ml of cold acetone followed by 3 ml hexane. The organic phase was extracted with 3.0 ml of 1 N NaOH. The concentration of hydroxylated benzo(a)pyrene in the alkali phase was determined with a fluorometer with activation at 396 nm and emission at 522 nm. Enzyme activities were determined in triplicate and compared with a boiled enzyme control. One unit of hydroxylase activity was defined as that amount catalyzing the formation in a 60 minute incubation at 30°C of hydroxylated product causing fluorescence equivalent to that of 1×10^{-12} moles of 3-hydroxybenzo(a)pyrene.

For a determination of cytochrome P-450, cuvettes contained between 1 and 2 mg microsomal protein in 0.05 M Tris buffer at pH 7.5 containing 0.15 M KCl. The sample cuvette was bubbled for 20 seconds with carbon monoxide and then reduced by addition of a few grains of sodium dithionite. The reference cuvette was reduced by sodium dithionite addition. Spectra were run at room temperature (26°C) on an Aminco DW-2A spectrophotometer at a full scale readout of 0.01 absorbance unit. The amount of P-450 was determined by the method of Omura and Sato (1964).

Clams were maintained for 6 days in water containing benz(a)anthracene (10 μg/liter). These clams, containing benz(a)anthracene, were fed to *Nereis virens* for contaminated food studies, and after 48 hours the animals were dissected. The intestines were used for determination of MFO activity and the amount of cytochrome P-450. Controls were worms fed uncontaminated clams.

Whole *Nereis succinea* fed oil contaminated food were assayed for mixed function oxygenase activity by the methods described above. In a separate experiment 5 mg of benz(a)-anthracene were mixed with 1 ml of corn oil and added to the food twice a week. The food concentration was 6 g/m^2-day.

RESULTS

Growth in the control trays increased as food concentrations increased up to a concentration of 100 mg nitrogen/m^2-day after which it leveled off. A polynominal regression was performed to describe the growth pattern. At higher food concentrations (above 50 mg nitrogen/m^2-day) the growth rate of worms in the oiled trays was significantly lower than those in control trays (Figure 1). A paired t test performed on the control and oil treated trays showed a significant difference between the means of the two groups ($0.05 < P < 0.1$). The worms in oiled trays appeared to show a linear increase in growth with higher food concentrations ($r^2 = 0.86$). Observations of the food-crude oil mixture in the trays showed no obvious clumping of food particles and food was layered in a similar manner in both control and oiled trays. Oil was clearly observed on the food particles. The worms were not disturbed by the presence of oil and normal feeding behavior was observed in oiled trays. An oil slick temporarily formed during feeding but was removed when air and water flow were resumed. Analysis of the sediment for phenanthrene and chrysene and extrapolation back to the whole oil indicated that approximately 0.2 g (0.1 mg oil/g sediment) of crude oil was present in the sediment of the oiled trays after 44 days. Analysis of the worms indicated that the aromatic compounds, if present, were below the detection limit (less than 2 ng/g wet tissue).

The average weight of *Nereis virens* from a heavily oiled area in Portland, Maine was 0.48 ± 0.30 g in September and 0.37 ± 0.25 g in June. Worms from a nearby "clean" area averaged 2.3 ± 1.4 g in September and 2.5 ± 1.8 g in June.

The MFO activity of *Nereis succinea* from control trays averaged 100 units/ mg protein while worms exposed to crude oil and benz(a)anthracene averaged 200 and 550 enzyme units/ mg protein, respectively. The MFO activity of *Nereis virens* fed clams contaminated with benz(a)anthracene was 1440 ± 540 (n=7) enzyme units/mg protein while for worms fed uncontaminated clams the activity was 410 ± 230 (n=7) enzyme units/mg protein. The concentration of cytochrome P-450 was 143 picomoles of P-450/mg microsomal protein in worms fed clams containing benz(a)anthracene and 62 picomoles P-450/mg microsomal protein in worms fed uncontaminated clams. MFO activity of *Nereis virens* from a heavily oiled area in Portland, Maine averaged 1.93 ± 1.03 enzyme units/mg protein in June and 0.56 ± 0.20 in Spetember. Worms from the "clean" area averaged 1.33 ± 0.77 in June and 0.44 ± 0.27 in September.

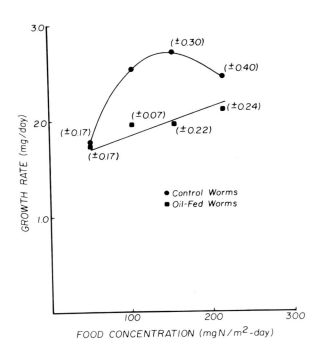

FIGURE 1:   Effects of Crude Oil on Growth of Nereis
succinea.   *Worms in sediment trays were given food (2.6%
nitrogen) with or without crude oil.   After 44 days the worms
were removed from the sediment and weighed (ash free dry
weights) to determine growth rates.   The worms (5 to 8) from
each tray were pooled for growth determinations.   The growth
occurring at each food concentration was averaged (n=2 or 3).
The standard error for each point is given in parenthesis,
except for oil treated worms at a food concentration of 104 mg
nitrogen/m²-day where one tray was lost.*

The concentration of benzo(a)pyrene in the heavily oiled
sediment and "clean" area were 1.8 and 0.02 μg/g sediment,
respectively.   The concentration of benzo(a)pyrene in *Nereis
virens* from the heavily oiled area and from the "clean" area
was 76.4 ng/g and 0.40 ng/g, respectively.   Analysis of other
polycylic aromatic hydrocarbons is presently under
investigation.

DISCUSSION

Our studies indicated decreased growth rates by *Nereis succinea* after being fed food contaminated by crude oil. Heip and Herman (1979), studied growth rates of *Nereis diversicolor* in the field. They found that individuals grew at a linear rate during their juvenile stages followed by slower growth during older stages. Growth rates were 2 to 4 mg/day for juveniles which is comparable to the growth rates we obtained for unexposed *Nereis succinea* at higher food concentrations. Thus, the reduced growth we obtained with oil-exposed worms could occur in natural populations. The lower weight of *Nereis virens* from heavily oiled sediment in Portland, Maine, relative to worms from a non-polluted area was assumed to be related to the presence of the oil in the sediment.

The observed effect of oil on growth could be due to either a slower feeding rate or to lower assimilation of food. We speculate that the effect is due to slower feeding rates because similar experiments with crustaceans, which are known to have sensitive chemosensory receptors showed a slower feeding rate when oil was in the food (Atema and Steih, 1974; Percy, 1976).

Studies reported by Rossi and Anderson (1978) showed that growth rates of juvenile *Neanthes arenaceodentata* were effected by water extracts of fuel oil. There was a decrease in growth as hydrocarbon concentrations increased from 60 to 182 µg/liter of total napthalenes. Quite high concentrations of water extracts of fuel oil were required to affect growth. The differences between our experiments and those of Rossi and Anderson (1978) included: (1) exposure through food versus water; (2) different species; (3) exposure to heavier fractions of oil in the case of crude oil in the food, whereas lower weight aromatic hydrocarbons are found in water extracts of fuel oil. Ecological studies of oil polluted areas have shown that *Nereis succinea* tend to remain in such polluted areas (Grassle and Grassle, 1974), despite effects on growth rates.

The growth of many deposit feeding polychaetes and other detritivors appear to be limited by the level of nitrogen which is often low in the available detritus (Tenore, *et al.*, 1979). Only with high levels of nitrogen does growth level off, as observed in our experiments. Investigations concerned with the effect of oil or other pollutants on benthic animals should consider different food concentrations with emphasis on the concentrations of utilizable nitrogen.

The explanation for the resistance of certain polychaete species to the toxic effects of oil may be due to the activity of a mixed function oxygenase (MFO) system. In our experiments *Nereis virens* and *Nereis succinea* after being fed food contaminated with crude oil or benz(a)anthracene showed increases in cytochrome P-450 and MFO activity.

MFO activity in *Nereis virens* from both the oil polluted and control areas were significantly different at the two times collected (June and September). A seasonal study of MFO activity using the livers of mosquitofish showed MFO activity changing during the year (Chambers and Yarbrough, 1979). In crabs MFO changes were correlated with changes in the molt cycle (Singer and Lee, 1977). We speculate that changes in the stage of gonad development may be related to the observed MFO activity changes in *Nereis virens*.

Possibly related to these changes are the observations of Tripp and Fries (unpublished data) that *Nereis virens* from the oil polluted areas in Maine do not show gamete formation. We speculate that oil interferes with the production of hormones which are synthesized by cytochrome P-450 dependent systems.

SUMMARY

Exposure to the polychaete, *Nereis succinea*, from coastal Georgia, to crude oil in the food resulted in a decrease in growth rate and an increase in mixed function oxygenase (MFO) activity relative to unexposed controls. At a food concentration of 156 mg nitrogen/m^2 - day the growth rate of worms fed oil-contaminated food and control worms were 1.91 ± 0.22 and 2.73 ± 0.30 mg/day, respectively. We speculate that oil effected the feeding rate of *Nereis succinea*. *Nereis virens* from an oil-polluted area of Maine had a lower weight and slightly higher MFO activity than worms from a "clean" site. An increase of MFO and cytochrome P-450, a component of the MFO system, was found in intestinal homogenates from *Nereis virens* fed food contaminated with benz(a)anthracene, a polycyclic aromatic hydrocarbon.

ACKNOWEDGMENTS

These studies were supported by the National Science Foundation, Office for the International Decade of Ocean Exploration, Grant No. OCE77-24516. Hydroxy benzo(a)pyrenes were provided by A. R. Patel through the National Cancer Institute Carcinogenesis Research Program. Benzo(a)pyrene analysis of some sediment samples were carried out by C. S. Giam (Texas A & M). Preliminary results on histological changes in *Nereis virens* were provided by M. R. Tripp and C. Fries (University of Delaware).

LITERATURE CITED

Armstrong, H. W., K. Fucik, J. W. Anderson, and J. W. Neff. 1979. Effects of oil field brine effluent on sediments and benthic organisms in Trinity Bay, Texas. Mar. Environ. Res. 2: 55–69.

Atema, J. and L. Steih. 1974. Effects of crude oil on the feeding behavior of the lobster *Homarus americanus*. Environ. Pollut. 6: 77–86.

Baker, J. M. 1976. Investigation of refinery effluent effects through field surveys. In: Marine Ecology and Oil Pollution, pp. 201–205, ed. by J. M. Baker, New York: John Wiley and Sons.

Bickers, D. R., L. C. Harber, A. Kappas, and A. P. Alvares. 1972. Polychlorinated biphenyls: comparative effects of high and low chlorine containing Aroclors on hepatic mixed function oxidase. Res. Commun. Pathol. Paarmacol., 3: 505–512.

Chambers, J. E. and J. D. Yarbrough. 1979. A seasonal study of microsomal mixed-function oxidase components in insecticide-resistant and susceptible mosquitofish, *Gambusia affinis*. Toxicol. Appl. Pharmacol. 48: 497–507.

Chasse, C. 1978. The ecological impact on near shores by the Amoco Cadiz oil spill. Mar. Pollut. Bull. 9: 298–301.

Gelboin, H. V. 1972. Studies on the mechanisms of microsomal hydroxylase induction and its role in carcinogen action. Rev. Can. Biol. 31: 31–60.

Giger, W. and M. Blumer. 1974. Polycyclic aromatic hydro-carbons in the environment: isolation and character-ization by chromatography, visible, ultraviolet, and mass spectrometry. Anal. Chem. 46: 1663–1671.

Grassle, J. P. and J. F. Grassle. 1974. Opportunistic life histories and genetic systems in marine benthic poly-chaetes. J. Mar. Res. 32: 253-284.

Gruger, E. H., T. Hruby, and N. L. Karrick. 1976. Sublethal effects of structurally related tetrachlor-, pentachloro-, and hexachlorobiphenyl on juvenile coho salmon. Environ. Sci. Technol. 10: 1033-1037.

Heip, C. and R. Herman. 1979. Production of *Nereis diversi-color* O. F. Müller (Polychaeta) in a shallow brackish water pond. Estuarine and Coastal Mar. Sci. 8: 297-305.

Lee, R. F. 1976. Metabolism of petroleum hydrocarbons in marine sediments. In: Sources, Effects, and Sinks of Hydrocarbons in the Aquatic Environemtn, pp. 333-344. Washington, D.C.: American Institute of Biological Sciences.

Lee, R. F., S. C. Singer, K. R. Tenore, W. S. Gardner and R. M. Philpot. 1979. Detoxification system in polychaete worms: importance in the degradation of sediment hydro-carbons. In: Marine Pollution: functional responses, pp. 23-37, ed. by W. Vernberg, A. Calabrese, F. P. Thurberg and J. Vernberg. New York: Academic Press.

Levell, D. 1976. The effect of Kuwait crude oil and the dis-persant BP 1100X on the lugworm, *Arenicola marina* L. In: Marine Ecology and Oil Pollution, pp. 131-158, ed. by J. M. Baker. New York: John Wiley and Sons.

Lowry, O. H., N. J. Rosenbrough, L. A. Farr and R. J. Randall. 1951. Protein measurement with the Folin phenol reagent. J. Biol. Chem. 193: 265-275.

Malins, D. C. 1977. Biotransformation of petroleum hydro-carbons in marine organisms indigenous to the Artic and subartic. In: Fate and Effects of Petroleum Hydrocarbons in Marine Ecosystems and Organisms, pp. 47-59, ed. by D. Wolfe. New York: Pergamon Press.

Nebert, D. W. and H. V. Gelboin. 1968. Substrate-inducible microsomal aryl hydroxylase in mammalian cell culture. I. Assay and properties of inducible enzyme. J. Biol. Chem. 243: 6242-6249.

Omura, T. and R. Sato. 1964. The carbon monoxide-binding binding pigment of liver microsomes. J. Biol. Chem. 239: 2370-2378.

Payne, J. F. 1976. Field evaluation of benzo(a)pyrene hydroxylase induction as a monitor for marine pollution. Science 191: 945-946.

Percy, J. A. 1976. Responses of marine crustaceans to crude oil and oil tainted food. Environ. Pollut. 10: 155-162.

Philpot, R. M., M. O. James and J. R. Bend. 1976. Metabolism of benzo(a)pyrene and other xenobiotics by microsmal mixed-function oxidases in marine species. In: Sources, Effects and Sinks of Hydrocarbons in the Aquatic Environment, pp. 184-189. Washington, D.C.: American Institute of Biological Sciences.

Prouse, N. J. and D. C. Gordon. 1976. Interactions between the deposit feeding polychaete *Arenicola marina* and oiled sediment. In: Sources, Effects and Sinks of Hydrocarbons in the Aquatic Environment, pp. 407-422. Washington, D.C.: American Institute of Biological Sciences.

Reish, D. J. 1971. Effect of pollution abatement in Los Angeles harbours. Mar. Pollut. Bull. 2: 71-74.

Rhoads, D. C. 1967. Biogenic reworking of intertidal and subtidal sediments in Barnstable Harbor and Buzzards Bay, Mass. J. Geol. 75: 61-76.

Rossi, S. and J. W. Anderson. 1978. Effects of #2 fuel oil water soluble fractions on growth and reproduction in *Neanthes arenaceodentata*. Water, Air and Soil Pollut. 9: 155-170.

Sanders, H. L., J. F. Grassle and G. R. Hampson. 1972. The West Falmouth Oil Spill. I. Biology. Woods Hole Oceanographic Institution, Technical Report No. 72-20.

Singer, S. C. and R. F. Lee. 1977. Mixed function oxygenase activity in blue crab, *Callinectes sapidus*: tissue distribution and corrleation with changes during molting and development. Biol. Bull. 153: 377-386.

Spies, R. B., P. H. Davis and D. H. Stuermer. 1979. The infaunal benthos of petroleum-contaminated sediments: study of a community at a natural oil seep. In: Ecological Impact of Oil Spills, pp. 735-755. Washington, D.C.: American Institute of Biological Sciences.

Tenore, K. R., J. Teitjen and J. Lee. 1977. Effect of meio-
    fauna on incorporation of aged eelgrass, *Zostera marina*,
    detritus by the polychaete *Nephthys incisa*. J. Fish. Res.
    Bd. Canada. 34: 563-567.

Tenore, K. R., R. B. Hanson, B. E. Dornseif and C. N.
    Wiederhold. 1979. The effect of organic nitrogen supp-
    lement on the utilization of different sources of
    detritus, Limnol. Oceanogr. 24: 350-355.

Wattenberg, L. W., J. L. Leong and P. J. Strand. 1962. Benz-
    pyrene hydroxylase activity in the gastrointestinal tract.
    Cancer Res. 22: 1120-1125.

COMPARISON OF SEVERAL PHYSIOLOGICAL MONITORING TECHNIQUES
AS APPLIED TO THE BLUE MUSSEL, *MYTILUS EDULIS*,
ALONG A GRADIENT OF POLLUTANT STRESS IN
NARRAGANSETT BAY, RHODE ISLAND

Donald K. Phelps
Walter Galloway

U.S. Environmental Protection Agency
Environmental Research Laboratory
Narragansett, Rhode Island

Frederick P. Thurberg
Edith Gould
Margaret A. Dawson

U.S. Department of Commerce
National Oceanic and Atmospheric Administration
National Marine Fisheries Service
Milford Laboratory
Milford, Connecticut

INTRODUCTION

Narragansett Bay, Rhode Island, has characteristics that make it attractive as a field laboratory. At any one point in time the physical characteristics of its bottom waters are relatively constant from its mouth to its head: temperature, for example, varies little more than 2-3°C, although on an annual basis it may undergo an excursion from -1° to 26°C, essentially from arctic to tropical conditions. Salinity remains relatively constant throughout the year, within a 4°/oo range.

The industrial history of the Bay area is a particularly useful factor in field-monitoring work. The first large-scale industry established in the United States was the Slater Mill, located on the Blackstone River, a major

tributary to the Bay (Phelps and Galloway, 1980). Pollutant input has increased since the time of that initial
textile effluent in 1793, from a variety of industries and
from a heavily populated metropolitan area comprising
several cities. Man's activities have thus produced a
pollution gradient in the Bay that overrides physical
variation as a controlling factor in the distribution of
resident marine species, especially in the upper reaches
of the Providence River. Elevated levels of metals and some
organic chemicals have been found in Bay sediments, in the
water column, and in the tissues of animals taken from Bay
waters (Farrington and Quinn, 1973; Phelps and Myers, 1977;
Phelps and Galloway, unpublished data). Stress in marine
animals exposed to even low levels of some of these pollutants
has been repeatedly demonstrated, as for example petroleum
hydrocarbons (Jeffries, 1972; Anderson, 1979), heavy metals
(Calabrese *et al.*, 1977), and pesticides (Miller and Kinter,
1977; Christiansen and Costlow, in press).

Because the waters of this ocean estuary present a
gradient of pollutant stress from its contaminated head to
its relatively clean mouth, the Bay thus provides an excellent
opportunity to examine, in a field laboratory, some effects
on marine life of pollution arising from man's activities.
Such a study was performed in 1977, using the criterion
scope for growth, a multiparametric index of whole-animal
respiration and feeding rates reflecting gross and net
growth efficiencies (Widdows *et al.*, in press). Four stations
were established on a transect along the pollution gradient in
Narragansett Bay, from a highly polluted site (Station 1) to a
relatively clean site (Station 4), shown in Figure 1. Blue
mussels (*Mytilus edulis*) were exposed for 1 month at each of
these four stations and their scope for growth measured in
August and in October. Scope for growth data reflected the
stress gradient in both months, and the O:N ratio alone was a
good stress indicator in August, but not after spawning, in
October (Widdows *et al.*, 1979). At the same time, the concentration of several heavy metals in mussel tissues was
determined; data for lead, cadmium, and iron showed no
particular pattern, but nickel reflected the stress gradient
(Fig. 2), and analysis of the water column gave similar
results, inversely proportional to scope for growth. A
subsequent laboratory study with mussels, involving a 96 hr-
exposure to 1000 ppb Ni failed to establish significant
physiological stress, as determined by gill respiration
(Thurberg, unpublished data). Nickel is strongly related,
however, to the degree of overall pollution in our field
stress gradient. Petroleum hydrocarbons in mussel tissues
also followed the stress gradient (Fig. 3).

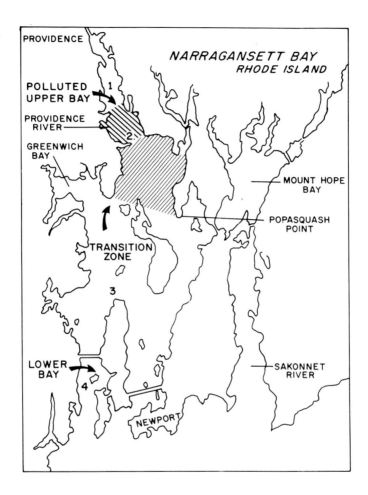

FIGURE 1. Map of Narragansett Bay showing the sites where mussels were held from April to September, 1979. Station 1 is a heavily polluted station, where the animals all died very quickly; Station 2, the next most polluted, and Station 4, a relatively clean area near the mouth of the Bay, were used for the field exposures in this study.

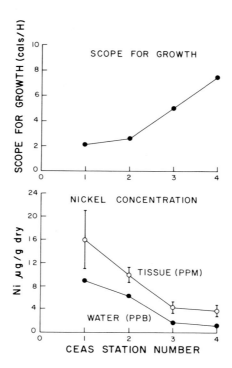

FIGURE 2. *Nickel concentrations in* M. edulis *tissues (open circles) and water column (solid circles) and scope for growth in mussels taken at the same time from 4 stations in Narragansett Bay, R. I. (from Phelps and Galloway, 1980). See Figure 1 for station sites.*

    At this point, it seemed desirable to obtain data for more than a single month's exposure, and to examine the relative effectiveness of other biological stress indicators. After a feasibility study in the summer of 1978, therefore, we undertook the work reported here. Because the mussels set out at the heavily polluted Station 1 did not survive through the summer, we used Station 2 as the polluted site and Station 4, the cleanest site, for comparison, setting the animals out in April, 1979, and sampling at monthly intervals from May to September.
    The frequent sampling and the large number of animals involved dictated our choice of stress criteria, physiological and biochemical measurements that are relatively simple and rapid to perform: gill-tissue respiration, serum-ion concentrations, and energy-related enzyme activities in

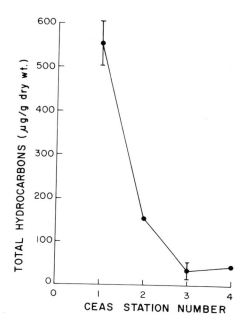

FIGURE 3.   Total petroleum hydrocarbons in tissues of M. edulis held at stations in the Narragansett Bay pollution gradient (from Phelps and Galloway, 1980).

gills and posterior adductor muscle. We also report on another suggested stress criterion, the adenylate energy charge, performed in 1978 on mussels held at Stations 2 and 4.

METHODS AND MATERIALS

Mussels were distributed and held along the Bay transect in submerged plastic baskets as described earlier by Phelps and Galloway (1980).

Twenty-five mussels from Station 2 and Station 4 were brought on ice to Milford, Connecticut, and kept in running seawater (22±2 ppt; 19±2°C) for 12 to 14 hours. Twenty animals of approximately equal size from each station were then prepared for respiration, serum ion, and biochemical studies. Hemolymph was sampled from the posterior adductor muscle via syringe and 22-gauge needle, then centrifuged to remove debris. Osmolality was read on an Advanced 3L

osmometer[1], and serum sodium, potassium, and calcium were
determined with a Coleman flame photometer. Gills from one
side of each mussel were removed and placed in a 15-ml War-
burg-type flask, one flask per mussel. Each flask contained
5 ml seawater from the holding tank. Oxygen consumption was
monitored over a 4-hr period at 20°C in a Gilson Differential
Respirometer. Oxygen-consumption rates were calculated as
microliters of oxygen consumed per hr per g dry weight of
gill tissue ($\mu\ell$ hr^{-1} g^{-1}), corrected to microliters of dry
gas at standard temperature and pressure.

For biochemical examination, pooled gill-tissue homogenates were prepared immediately after excision of the
tissues, then frozen-stored at -80°C until testing; posterior
adductor muscle was cleaned and packaged in air-tight plastic
bags, then held at -80°C until testing. Tissues were
homogenized in all-glass equipment as follows: fresh gill
tissue, 1:4 w/v, in cold doubly glass-distilled water;
PAM[2], 1:9, w/v, in cold 1% Triton, a non-ionic surfactant
(W-1339, Ruger Chemical Company, Irvington-on-Hudson, N.Y.).
Centrifugation was at 4°C and 40,000 g for both tissue
homogenates, 45 min for gill (after thawing frozen preparations in RT tap water) and 60 min for PAM. The clear
supernates, E^{5x} gill and E^{10x} PAM, were used for enzyme
analyses.

Enzyme assays

All solutions were made in doubly glass-distilled
water. Biochemicals were obtained from Sigma Chemical
Company, St. Louis, Mo. Activities of all enzymes were
determined by standard fixed-wavelength spectrophotometry
using pyridine nucleotide coenzymes either directly or by a
coupled-assay procedure; assay temperature was 25°C.

[1]*Use of trade names does not constitute endorsement by the
NMFS, NOAA, or by the EPA.*

[2]*Abbreviations used: PAM, posterior adductor muscle;
NADH, reduced form of nicotinamide adenine dinculeotide;
NADP, nicotinamide adenine dinucleotide phosphate; EDTA,
ethylenediamine tetraacetic acid, dipotassium salt, dihydrate;
cis-OA, cis-oxaloacetic acid; ADP, adenosine diphosphate;
PEP, phospho(enol)pyruvate; Tris, tris(hydroxymethyl)-
aminomethane; G6P, D-glucose-6-phosphate, monosodium
salt; AAT, aspartate aminotransferase, E.C. 2.6.1.1; G6PDH,
glucose-6-phosphate dehydrogenase, E.C. 1.1.1.49; LDH,
lactate dehydrogenase, E.C. 1.1.1.27; MDH, malate dehydrogenase, E.C. 1.1.1.37; PK, pyruvate kinase, E.C. 2.7.1.40.*

Unit of activity in each case was μmoles NADH oxidized or NADP reduced per min per mg biuret protein. Assay concentrations (mM) in a 3.00-ml reaction volume were: MDH=buffer pH 9.0 glycine (90) and EDTA (0.9), NADH (0.15), cis-OA (1.0), and 0.10 ml E^{100x} (1:19, v/v with water, for E^{5x} gill; 1:9 v/v with water, for E^{10x} PAM); PK=glycylglycine buffer pH 8.0 (60), NADH (0.15), $MgCl_2$ (8.3), KCl (7.5) ADP (0.31), PEP (0.53), 175 units LDH, and 0.10 ml E^{5x} gill or E^{10x} PAM; G6PDH=Tris buffer pH 8.0 (82), NADP (0.3), $MgCl_2$ (2.5), G6P (1.0), and 0.10 ml E^{5x} gill; LDH=glycylglycine buffer pH 7.5 (70), NADH (0.15), Na pyruvate (3.3), and 0.30 ml E^{10x} PAM; AAT=K phosphate buffer pH 7.5 (90), L-aspartate, potassium salt (200), NADH (0.35), α-ketoglutarate, potassium salt (6.7), 20 units MDH, and 0.20 ml E^{10x} PAM. Determination of statistical significance was made using the Student's t test.

RESULTS AND DISCUSSION

Gill-Tissue Respiration

     Monthly sampling of mussels held along a pollution gradient showed generally increasing oxygen-consumption rates in gills of animals taken from the polluted site, Station 2; gill-tissue respiration is a widely-used criterion of physiological stress in bivalves (Thurberg, *et al.*, 1974). Oxygen-consumption rates for animals collected at Station 2 in May were depressed as compared to Station 4 animals, but by June these rates were about equal, and from that time on animals from Station 2 had significantly elevated ($P<0.05$) oxygen-consumption values (Fig. 4), although in September at a somewhat lesser level, as was also seen in the biochemical data. These elevated rates indicate increased metabolic activity that is probably the result of the high pollution levels found at Station 2. The depressed rates of oxygen consumption in May during the early period of exposure to Station 2 waters may have been due to decreased activity, shell closure, and anerobic metabolism, all in an initial effort to "close out" unfavorable environmental conditions.
     These elevated summer gill-respiration rates corroborate the results of the earlier scope for growth study (Widdows *et al.*, in press). In that study, mussels were held at stations along the Narragansett Bay pollution gradient for one month before testing. In both August and October tests, scope for growth was significantly depressed at the polluted sites (Fig. 2). This means that an abnormally high amount of available energy was used for maintenance and little was left

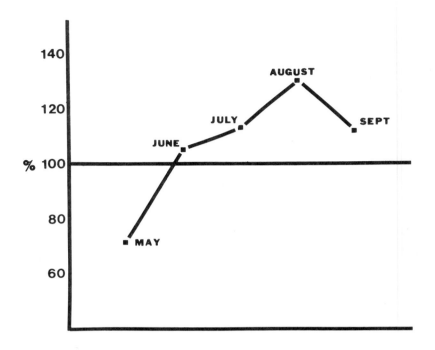

FIGURE 4. Gill-tissue oxygen consumption of M. edulis
*sampled monthly at two stations in Narragansett Bay, R.I.
The horizonal line in the middle of the figure represents
data for animals from the relatively clean Station 4, taken
as unity, and the points connected by lines represent data
for animals taken from the polluted Station 2, presented as
percent of values for the clean station.*

over for growth. In the present study, the high gill-tissue
oxygen-consumption measurements for animals taken July-
September at the pollutant site indicate the same condition in
mussels held along the gradient for the entire summer season.
    Both of these physiological indices of metabolic stress
provide similar information, and both have advantages and
disadvantages that must be weighed before selection for a
monitoring program. Scope for growth gives a much more
complete metabolic picture, including feeding rates, oxygen-
consumption rates, and oxygen:nitrogen ratios - all incorporated in a single index (Bayne *et al.*, 1980). A disadvantage
of this measure is the requirement for a time-consuming suite

of measurements and a considerable variety of equipment, which limits the number of animals that can be sampled and measured for any given point in time. On the other hand, gill-tissue oxygen consumption, although it gives only one part of the metabolic picture, can be performed on many animals in a short period of time, and can also be easily measured in a shipboard laboratory. For many studies this simple indicator of metabolic activity is sufficient, especially where numerous stations and many animals must be sampled to obtain a statistically significant sample size. Such a large sample size is often necessary to filter out the effects of the many environmental factors that influence metabolic activity (Thurberg, 1980).

Gill Enzyme Activity

Monthly sampling of *Mytilus* gills for enzyme analysis during the 1979 field study showed that glycolytic activity (as measured by PK) in control mussels at Station 4 dipped from a high in April and May, when the animals first spawn, to a low in June and July and then rose again, reaching the levels of late April by early September (Fig. 5). Bayne and Thompson (1970) observed that the general tissue metabolism of *Mytilus* is depressed in early summer, after spawning, and both gill glycolysis and gill respiration in this field study agree with their findings. In animals from the polluted Station 2, gill glycolysis followed a similar course but at significantly lower levels, both at the end of spawning in early May ($P<0.01$) and June ($P<0.005$) and during renewed glycolytic activity in early September ($P<0.01$). The markedly lower gill glycolysis in polluted mussels in May, and the diminished capacity in September to sustain the summer's glycolytic surge, both confirm the gill respiratory data.

Despite wide sample variability in May and June, biosynthetic activity in the gill (as measured by the pentose shunt enzyme G6PDH) was observed to rise steadily in clean *Mytilus* from April to September (Table 1). In animals from the polluted station, however, data indicate somewhat lower pentose shunt activity in June, possibly the result of an abnormally high ammonia concentration in the water at that station at that time (Table 2), the only hydrographic parameter being monitored that was markedly different from values at the clean station.

We favor this explanation in the light of the following observations: In August of the previous year, ammonia values at the polluted station rose to 18 and 20 μmoles l^{-1}, accompanied by a decrease in dissolved oxygen levels.

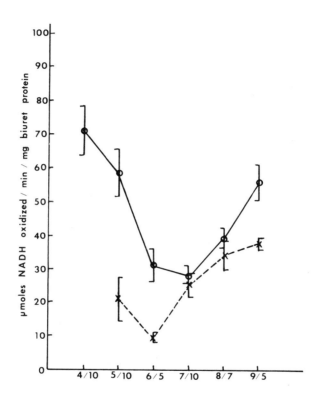

FIGURE 5. *Pyruvate kinase activity in gill tissue of*
M. edulis *sampled monthly at No. 2, polluted (x----x), and*
at No. 4, clean (o----o), stations in Narragansett Bay, R.I.
Each point represents the arithmetic mean for 6-9 sample
pools, 2-3 animals per pool, and the bars represent standard
error. Differences between the two stations were highly
significant in May, June, and September.

TABLE 1. G6PDH Activity (μmoles NADP reduced min^{-1}) in gills of *Mytilus edulis* taken at monthly intervals in 1979 from clean or polluted stations in Narragansett Bay. Each sample pool comprised either 2 or 3 animals.

		April 10	May 10	June 5	July 10	Aug. 7	Sept. 5
Station 4 (clean)	(N) pooled samples	(8)	(6)	(6)	(9)	(9)	(8)
	$\bar{x}$	37.5	61.9	85.2	91.9	95.1	106.1
	S.E.	1.5	18.7	23.8	13.3	9.6	13.5
	% sample variability, 100 (S.E./$\bar{x}$)	4.0	30.2	27.9	14.5	10.1	12.7
Station 2 (polluted)	(N) pooled samples		(6)	(6)	(8)	(6)	(8)
	$\bar{x}$		73.2	46.3	122.3	117.7	108.3
	S.E		21.7	13.6	18.0	12.9	5.2
	% sample variability, 100 (S.E./$\bar{x}$)		29.6	29.4	14.7	11.0	4.8

TABLE 2., Hydrographic parameters at Station 2 (polluted) and 4 (clean) in Narragansett Bay during the summer of 1979.

		June 4	July 9	Aug 6	Sept 4
Btm. Temp.:	Sta. 2	14.4	19.5	22.2	21.8
($^\circ$C)	4	13.8	18.6	22.6	21.8
Btm. Sal.:	Sta. 2	26.5	28.9	30	27.4
(o/oo)	4	29.5	30.6	28	30.2
D.O.:	Sta. 2	5.08	8.01	1.19	4.25
(mg l^{-1})	4	9.18	8.04	5.82	6.55
NH_3:	Sta. 2	15.65	1.37	----	8.50
(μmoles l^{-1})	4	3.12	2.25	----	2.50

Although for our preliminary work we sampled on only 3 dates that year (July, August, October), we found that mussels taken in August from that same polluted station showed significant stress, both in elevated gill-tissue respiration (P<0.01) and in elevated activity of adductor muscle AAT (P<0.001) and gill-tissue MDH (P<0.005), both enzymes that have served as criteria of sublethal metal stress in tissues of marine crustaceans (Gould *et al.*, 1976; Gould, 1980). Because of the possible implication of high ammonia levels in the 1978 observations of stress in mussels at the polluted site, therefore, we exposed healthy *Mytilus* (from the population set out in the field in April of 1979) for 22 days to 20 μmoles NH_3 l^{-1} (as NH_4Cl) in basins of aerated water that was changed daily, and sampled control and exposed animals at 1, 8, 15, and 22 days. G6PDH activity was lower in the gills of exposed mussels at 22 days (P<0.05), an indication of lower biosynthetic rates, and the same animals had depressed hemolymph potassium (P<0.05). On the other hand, gill respiratory rates for these animals did not differ from the controls; we conclude, therefore, that a high ammonia level alone is not sufficient to produce the physio-logical stress signals observed in August 1978. But the lower gill G6PDH activity in the ammonia-exposed mussels of this brief laboratory study, together with this year's field observation linking decreased gill G6PDH and high ammonia concentrations, suggest that further work in this area could prove fruitful.

Gill MDH activity was not significantly different in the polluted animals during the 1979 field study.

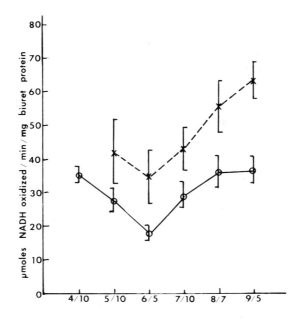

FIGURE 6.   Lactate dehydrogenase activity in posterior adductor muscle of M. edulis sampled monthly at No. 2, polluted (x----x), and No. 4, clean (o----o), stations in Narragansett Bay, R.I.   Each point represents the arithmetic mean for 8-15 samples, and the bars represent standard error. Differences between the stations were highly significant in August and September.

Adductor Muscle Enzyme Activity

In the posterior adductor muscle, the only statistically significant change observed was an increasingly elevated LDH activity (Fig. 6) in animals from the polluted station in August (P<0.05) and September (P<0.005), a possible indication that the sublethally stressed animals were drawing upon energy reserves in the muscle to support a growing metabolic demand in the increasingly stressed gill, the latter evinced by the weakening PK in September (Fig. 5). This interpretation would support the observations of Dunning and Major (1974), who, working with *Mytilus* exposed to sublethal levels of oil, found glycogen stores of the posterior adductor muscle to be depleted, although normally this animal retains muscle glycogen for use during gametogenesis, and will not draw upon this energy reserve even during prolonged starvation.

When gill PK and G6PDH and adductor muscle LDH in the
polluted animals are calculated as percent of values for
animals from the clean stations, two noteworthy patterns
emerge (Fig. 7). First, the gill tissue enzymes follow
the same course as do the similarly plotted gill oxygen-
consumption data (Fig. 4): an even greater depression in the
polluted animals than in control animals after spring
spawning, followed by an overcompensation that gradually
lessens. The second pattern is the picture of inversely
proportional activity of adductor muscle LDH, in comparison
with the gill tissue enzymes; in the pollutant-stressed
animals, energy reserves of the posterior adductor muscle
are apparently mobilized to support the weakened gill
metabolism. The "sparing" of adductor muscle glycogen
stored for gametogenesis, therefore, does not operate
effectively in pollutant-stressed mussels.

Hemolymph Ions

Although some differences in hemolymph ion concentration
were observed (Table 3), they did not reflect the pollutant
stress gradient indicated by the other measures of physiological condition tested. Mussels are reported to have no
osmoregulatory ability, very little ion-regulation, and a slow
adaptation to any change of salinity (Shumway, 1977). In this
study, the mussels were taken from two stations of similar
salinities (Table 2), then placed in a significantly lower
salinity overnight before testing (see Methods and Materials).
For any future work of a similar nature, it may be advisable
to bleed the mussels immediately after their removal from the
stations in the pollution gradient. In the light of our
results here, however, hemolymph ion concentrations did not
reflect field stress.

Adenylate Energy Charge

The AEC data for mussels field-coll. ·ted at Stations 2
and 4 did not support the picture of stre..ful field conditions indicated by other physiological criteria. The
data did appear to reflect a seasonal change in condition of
all field-exposed animals between April and August, the two
sampling dates. The ATP: AMP ratio seemed more promising as
a stress indicator, but the sample number was insufficient to
prove a statistically significant difference between the two
groups. From the results of this attempt to apply this
energy-intensive and time-consuming technique to field
studies, we do not consider the AEC a practical monitoring
tool.

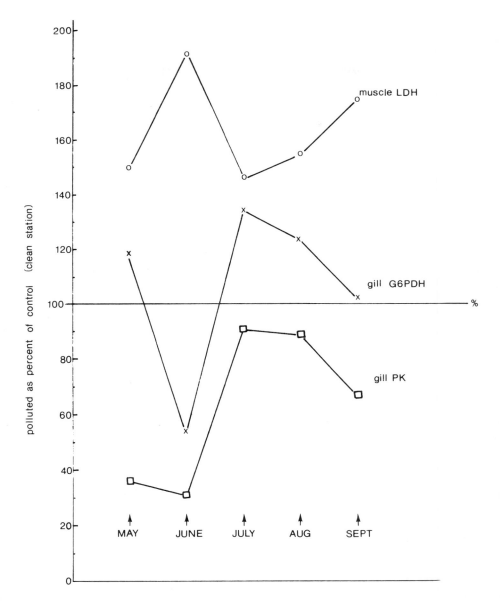

FIGURE 7.   *Posterior adductor muscle LDH and gill PK and G6PDH activities in* M. edulis *from polluted Station 2, as percent of corresponding activities in animals from clean Station 4.*

TABLE 3. Hemolymph ion concentration and osmolatities of mussels taken from station 2 (polluted) and Station 4 (clean) during the summer of 1979.

		Na mEq l^{-1}	K mEq l^{-1}	Ca mEq l^{-1}	Osmolality mOsm
May	St. 2	248 ± 2.6	9.53 ± 0.51	10.57 ± 0.81	723 ± 1.7
	St. 4	253 ± 2.5	9.09 ± 0.63	9.76 ± 0.24	735 ± 5.9
June	St. 2	313 ± 4.6	10.33 ± 0.64	12.54 ± 0.28[a]	827 ± 14.5
	St. 4	379 ± 3.7[a]	11.25 ± 0.60	17.11 ± 0.58	861 ± 10.0
July	St. 2	344 ± 3.5	10.17 ± 0.40	16.05 ± 0.32	763 ± 4.4
	St. 4	335 ± 2.2	11.10 ± 0.71	16.17 ± 0.29	765 ± 2.0
August	St. 2	309 ± 1.6[a]	11.43 ± 0.50[a]	15.86 ± 0.29	792 ± 4.1
	St. 4	353 ± 5.0	9.48 ± 0.25	16.06 ± 0.53	796 ± 3.9
September	St. 2	313 ± 3.9[a]	9.03 ± 0.27	12.52 ± 0.30[a]	775 ± 0.5[a]
	St. 4	293 ± 2.3	9.91 ± 0.51	11.31 ± 0.25	764 ± 1.2

All values are mean of 15–20 samples ± standard error.
[a] significantly different from St. 4; $p < .05$.

Pea Crabs

We also took note of the pinnotherid crabs (*Pinnotheres maculatus*) living as commensals within the mantle cavity of the mussels collected from Stations 2 and 4. In June, half of the 20 mussels examined at each station contained a single crab. During July, August, and September, however, mussels from the polluted Station 2 contained only one-third to one-half as many crabs as mussels from the cleanest station (Table 4). These limited data suggest that the crabs either died at the polluted station or left the mussels in an attempt to escape unfavorable environmental conditions. They also support the theory that a lack of pinnotherid commensals in certain areas may be directly related to the level of pollution (J. B. Pearce, pers. comm.).

Tissue Residue Analysis

Another monitoring tool is the chemical analysis of body tissues for specific pollutants (Goldberg *et al.*, 1978). Such analyses were performed on *Mytilus* tissues collected during the summers of 1977 through 1979. Significant findings were that of the metals tested, nickel concentrations reflected the pollution gradient in the Bay (Fig. 2), as did petroleum hydrocarbons (Fig. 3), and that the metal levels remained remarkably stable over this three-year period of observation (Phelps and Galloway, unpublished data).

TABLE 4.   Pea crabs (*Pinnotheres maculatus*) found within the mantle cavity of mussels collected from Stations 2 (polluted) and 4 (clean).

	Number found at Sta. 2	Number found at Sta. 4	Sta. 2 as % Sta. 4
June 4	10	10	100
July 9	7	14	50
Aug. 6	2	6	33
Sept. 4	5	10	50

SUMMARY

Sublethal pollutant stress in mussels was clearly detected in a field-monitoring situation using these criteria of physiological response: scope for growth (Widdows *et al.*, 1980), tissue residue analysis (Phelps and Galloway, 1980), and gill respiration and glycolytic rates in either gill or posterior adductor muscle (this report). We have drawn the following assessment of their relative usefulness in this field study of pollutant stress:
1. Scope for growth and gill-tissue oxygen consumption are both valuable indices of metabolic stress in mussels. Scope for growth is the more comprehensive index and the more labor-intensive. It has the advantage of relating an index of condition directly to such vital whole-body functions as growth and reproduction. Gill respiration is a simpler and faster measure that can be used with larger numbers of animals in a short period of time with limited laboratory space available, as at sea. It does not, however, provide direct information on feeding rates or nitrogen metabolism.
2. Examination of enzyme activity in mussel gill and posterior adductor muscle confirmed the results of gill-tissue oxygen consumption and the scope for growth. Such measurements of subcellular metabolic activity also provide information on the flow of and capacity for energy expenditure, and often give earlier and more sensitive indication of metabolic stress, picking up significant differences before whole-tissue or whole-animal response is detectable.
3. Hemolymph ion and adenylate energy charge measurements did not detect pollutant stress in mussels held at the polluted field station of this study. We therefore conclude that these physiological indices are not appropriate for monitoring biological response to pollutant stress in this animal, at least, under field conditions.
4. The relative abundance of pinnotherid crab commensals we consider to be an interesting potential indicator of relative pollution in the field.
5. Analysis of animal tissues for specific toxicants has been demonstrated to be a reliable indicator of relative degree of pollution in the field (Goldberg *et al.*, 1978; Phelps and Galloway, 1980). Levels of nickel in *Mytilus* tissues for each of the four stations were remarkably stable over a three-year testing period, and both nickel and petroleum hydrocarbons reflected the stress gradient in Narragansett Bay. Relative levels of contaminants in tissues alone, however, are very difficult to interpret and have limited

application. This study presents the necessary coupling of
tissue residue analyses to biological effects, a blending of
chemistry and biology that is a powerful tool for field
monitoring.

ACKNOWLEDGEMENTS

We thank Drs. Gerald E. Zaroogian, Angela Ivanovici, and
John H. Gentile for sharing with us the preliminary results
of their AEC work with *M. edulis*. A fuller report on this
subject, including some laboratory studies, is in preparation.
We thank George Morrison for providing the data shown in
Table 2.

LITERATURE CITED

Anderson, J. W.   1979.   An assessment of knowledge concerning
    the fate and effects of petroleum hydrocarbons in the
    marine environment. In: Marine Pollution: Functional
    Responses (W. B. Vernberg, F. P. Thurberg, A. Calabrese,
    and F. J. Vernberg, eds). Academic Press: N.Y., pp. 3-21

Bayne, B. L., D. A. Brown, F. Harrison, and P. P. Yevich.
    1980. Mussel health. In: The International Mussel
    Watch. National Academy of Sciences: Washington, D.C.,
    pp. 163-235.

Bayne, B. L., M. N. Moore, J. Widdows, D. R. Livingstone, and
    P. Salkeld. 1979. Measurement of the responses of
    individuals to environmental stress and pollution:
    studies with bivalve molluscs. Phil. Trans. R. Soc.
    Lond. B   286:   563-581.

Bayne, B. L., and R. J. Thompson. 1970. Some physiological
    consequences of keeping *Mytilus edulis* in the laboratory.
    Helgoländer wiss. Meeresunters. 20:   526-581.

Calabrese, A., F. P. Thurberg, and E. Gould. 1977. Effects
    of cadmium, mercury, and silver on marine animals. Mar.
    Fish. Rev. 39(4):   5-11.

Christiansen, M. E., and J. D. Costlow, Jr. In press.
    Persistence of the insect growth regulator Dimilin
    in brackish water: a laboratory evaluation using larvae
    of an estuarine crab as indicator. Helgoländer wiss.
    Meeresunters. 33.

Dunning, A., and C. W. Major. 1974. The effect of cold
    seawater extracts of oil fractions upon the blue mussel,
    Mytilus edulis. In: Pollution and Physiology of Marine
    Organisms, F. J. Vernberg and W. B. Vernberg, eds.
    Academic Press: N.Y., pp. 349-366.

Farrington, J. W., and J. G. Quinn. 1973. Petroleum
    hydrocarbons in Narragansett Bay. I. Survey of sediments
    and clams (*Mercenaria mercenaria*). Estuarine Coastal
    Mar. Sci. 1: 71-79.

Goldberg, E. D., V. T. Bowen, J. W. Farrington, G. Harvey,
    J. H. Martin, P. L. Parker, R. W. Risebrough, W.
    Robertson, E. Schneider, and E. Gamble. 1978. The
    mussel watch. Environmental Conservation 5(2): 101-125.

Gould, E. 1980. Low-salinity stress in the American lobster,
    Homarus americanus, after chronic sublethal exposure to
    cadmium: biochemical effects. Helgoländer wiss.
    Meeresunters. 33: 174-184.

Gould, E., R. S. Collier, J. J. Karolus, and S. Givens. 1976.
    Heart transaminase in the rock crab, *Cancer irroratus,*
    exposed to cadmium salts. Bull. Environ. Contam.
    Toxicol. 15: 636-643.

Jeffries, H. P. 1972. A stress syndrome in the hard clam
    Mercenaria mercenaria. J. Invert. Pathol. 20: 242-251.

Miller, D. S., and W. B. Kinter. 1977. DDT inhibits nutrient
    absorption and osmoregulatory function in *Fundulus
    heteroclitus*. In: Physiological Responses of Marine
    Biota to Pollutants (F. J. Vernberg, A. Calabrese, F. P.
    Thurberg, W. B. Vernberg, eds.). Academic Press: N.Y.,
    pp. 63-74.

Phelps, D. K., and W. B. Galloway. 1980. A report on the
    coastal environmental assessment station (CEAS) program.
    Rapp. P.-v. Réun. Cons. Int. Explor. Mer 179: 76-81.

Phelps, D. K., and A. C. Myers. 1977. Ecological con-
    siderations in site assessment for dredging and spoiling
    activities. In: Proceedings of the Second U.S.-Japan
    Meeting on Management of Bottom Sediments Containing
    Toxic Substances, October 1976. EPA Ecological Research
    Series, EPA-600-3-77-083.

Shumway, S. E. 1977. Effect of salinity fluctuation on the osmotic pressure and Na^+, Ca^{2+}, and Mg^{2+} ion concentrations in the hemolymph of bivalve molluscs. Mar. Biol. 41: 153-177.

Thurberg, F. P. 1980. The use of physiological techniques in monitoring pollution: a consideration of its problems and current research. Rapp. P.-v. Réun. Cons. Int. Explor. Mer 179: 82-87.

Thurberg, F. P., A. Calabrese, and M. A. Dawson. 1974. Effects of silver on oxygen consumption of bivalves at various salinities. In: Pollution and Physiology of Marine Organisms (F. J. Vernberg and W. B. Vernberg, eds.). Academic Press, N.Y., pp. 67-78.

Widdows, J. L., D. K. Phelps, and W. B. Galloway. In press. Measurement of physiological condition of mussels transplanted along a pollution gradient in Narragansett Bay. Mar. Environ. Poll.

MERCURY IN MUSSELS
OF BELLINGHAM BAY, WASHINGTON (U.S.A.):
THE OCCURRENCE OF MERCURY-BINDING PROTEINS

G. Roesijadi
A. S. Drum
J. R. Bridge

BATTELLE
Pacific Northwest Laboratories
Marine Research Laboratory
538B Washington Harbor Road
Sequim, Washington

INTRODUCTION

Mercury is presently regarded as a trace element of
concern from the standpoint of marine pollution due to its
high toxicity and increases in anthropogenic inputs to the
environment (National Academy of Sciences, 1978, Young *et al.*,
1979). Mercury has been reported to occur at elevated levels
in certain coastal environments (Crecelius *et al.*, 1975;
Gardner *et al.*, 1978; Skei, 1978), and marine animals inhabiting mercury-contaminated areas have been shown to possess
relatively high tissue burdens of the metal (de Wolf, 1975;
Davies and Pirie, 1978; Gardner *et al.*, 1978). Although the
toxicity of mercury to marine animals has been extensively
studied (e.g. Corner and Sparrow, 1956; Corner and Rigler,
1958; Calabrese *et al.*, 1977), relatively little is known
about the mechanisms by which this element is sequestered and
stored within tissues. Such mechanisms may enable animals to
tolerate potentially toxic metals and may be involved in
processes associated with the ability of animals to survive in
mercury-contaminated habitats.

Extensive studies have demonstrated the existence of the
low molecular weight, metal-binding protein, metallothionein,
in higher animals ranging from fish to humans (Cherian and

Goyer, 1978; Kojima and Kagi, 1978). This protein has a high affinity for certain trace metals, including mercury, and it has been suggested that the protein may serve a protective function by binding intracellular excesses of the metals and thus preventing their binding to sensitive cellular sites (Winge *et al.*, 1973; Brown *et al.*, 1977; Brown and Parsons, 1978). Recent studies (e.g., Casterline and Yip, 1975; Noel-Lambot, 1976) have shown that metallothionein-like proteins exist in the tissues of certain marine invertebrates.

The distribution of mercury in gills and digestive glands of the marine mussel, *Mytilus edulis*, is described in this paper. Mercury is known to concentrate in bivalve gills and digestive glands (Wrench, 1978). Mussels exposed to mercury in the laboratory are compared to mussels from Bellingham Bay, Washington, an area with a known industrial source for mercury contamination (Lee, 1971; Crecelius *et al.*, 1975). Mussels from Bellingham Bay have been shown to possess relatively high tissue mercury concentrations (Lee, 1971; Rasmussen and Williams, 1975) when compared to individuals collected from areas uncontaminated by mercury.

MATERIALS AND METHODS

Animals

Mussels to be used for laboratory exposures to mercury were collected at low tide from Sequim Bay, Washington during October, 1978 and held in flowing, unfiltered sea water of 8°C and 32°/$_{oo}$ salinity prior to initiation of experiments.

For field studies, mussels were collected in December, 1978 from designated sites in Bellingham Bay, Washington. Mussels from Sequim Bay were used as reference controls. For storage, mussels were frozen in liquid nitrogen at the collection sites, then transported on dry ice to the laboratory where they were placed in a -65°C freezer.

Laboratory Experiments

Exposures to 1 μg/l mercury (nominal concentration) for 90 days were conducted in fiberglas aquaria containing flowing sea water from Sequim Bay, Washington of 8°C, 32°/$_{oo}$, and 8.5 ppm dissolved oxygen at 1.5 ℓ/min. Mercury as $HgCl_2$ dissolved in double-deionized water (using mixed bed anion-cation exchange resins) adjusted to pH 2 with HCl was added at 2.5 mℓ/min. Similar pH 2 water, but without mercury, was added to the control tank. The resulting pH of exposure sea water

(7.85 ± 0.03) was within the natural variability of pH for Sequim Bay during the experimental period. Weekly measurements were made for total mercury, temperature, salinity, pH, and dissolved oxygen.

Exposure sea water was analyzed for total mercury by reduction with sodium borohydride to convert Hg^{2+} to Hg^0. Mercury vapors were concentrated by amalgamation on a two-stage gold foil trap prior to sweeping into a Laboratory Data Control (Rivera Beach, Florida) flameless mercury analyzer. The detection apparatus is essentially similar to that described by Fitzgerald and Gill (1979). The measured mercury concentration in the exposure tank was 0.81 ± 0.11 (S.D.) $\mu g/\ell$. Mercury concentrations in control sea water were found to be below detection (<10 ng/ℓ) using the sodium borohydride method. Therefore, control sea water samples were oxidized as described in Bothner and Robertson (1975), then reduced with stannous chloride for release of mercury vapors prior to analysis as described above. Total mercury levels in control sea water were 7.2 ± 1.1 (S.D.) ng/ℓ and consistent with current estimates for natural levels of mercury in sea water (Baker, 1977; Matsunaga et al., 1975; Mukerji and Kester, 1979).

Following exposure to 1 $\mu g/\ell$ mercury for 90 days, groups of exposed and control mussels were placed in clean, flowing sea water for 24 hours, to allow cleansing, then either frozen whole at -65°C or transferred to static aquaria for additional exposure to 1 $\mu g/\ell$ mercury as $^{203}HgCl_2$ or 0.05 $\mu moles/\ell$ ($U^{-14}C$)-cysteine for 24 hours. Soluble fractions of gills and digestive glands of the frozen mussels were analyzed by gel chromatography for incorporation of mercury into soluble constituents; specifically into low molecular weight, metallothionein-like proteins. Gills and digestive glands of ^{203}Hg- or ^{14}C-treated mussels were used to test the influence of pre-exposure to mercury on the incorporation of ^{203}Hg and ^{14}C of cysteine, a major component of metallothionein, into metallothionein-like proteins.

Gills and digestive glands were prepared for gel chromatography by homogenizing in two volumes of 0.75 M sucrose containing 1% 2-mercaptoethanol and 0.1 mM phenylmethylsulfonyl fluoride (The latter was omitted in radioactive samples), then centrifuged at 35,000x g for 15 min. Supernatants were heated at 70°C for 10 min (except radioactive samples which were heated to 70°C then held at that temperature for 1 min.), then recentrifuged for 35,000x g for 15 min. Final supernatants were stored at -65°C prior to separation on a 1.6 x 85 cm column of Sephadex G-75 fine gel (Pharmacia Fine Chemicals, Piscataway, N.J.) with 0.1 N NH_4HCO_3 as eluting buffer. Fractions were analyzed for mercury by digestion in HNO_3 and excess $KMnO_4$, followed by addition of $NH_2OH \cdot HCl$ reduction with stannous chloride and mercury detection as described earlier. ^{203}Hg and ^{14}C in fractions were determined by using Aquasol II scintillation cocktail (New England Nuclear, Boston, Ma.) and a LS-150 liquid scintillation spectrometer (Beckman Instrument, Inc.; Fullerton, Ca.). Values for each fraction were expressed as a function of the wet weight of original tissue sample.

Field Study

Sampling stations for Bellingham Bay mussels are shown in Figure 1 and represent stations along the shoreline at in-creasing distances from the known mercury source, the waste discharge of a chloralkali plant located on the northeast shoreline of Bellingham Bay. General current patterns direct the flow of water toward the sampling stations in a clockwise direction, with localized eddies occurring in the area of interest (Bothner, 1973). Exact locations of the stations were determined by the availability of mussels.

Whole mussels, gills, and digestive glands were prepared for mercury analyses by wet digestion in hot, concentrated nitric acid, followed by addition of excess $KMnO_4$. Prior to analysis, $NH_2OH \cdot HCl$ was added to the digested samples which were then reduced with stannous chloride and analyzed for mercury as described above. Ten individual whole mussels were analyzed. For gills and digestive glands, five samples, each of which consisted of the pooled organs of five individuals, were analyzed.

Soluble extracts of gills and digestive glands were pre-pared as already described and analyzed by gel chromatography on Sephadex G-75 fine gel. Fractions were analyzed for ab-sorbance at 254 nm, mercury, zinc, copper, and cadmium. The levels of the latter three elements were determined in chroma-tographic fractions since they are known constituents of metallothionein. Mercury in fractions was determined as described above. Zinc, copper, and cadmium were measured by direct aspiration into the flame of an atomic absorption spectrophotometer (Instrumentation Laboratories IL251, Wilmington, Ma.).

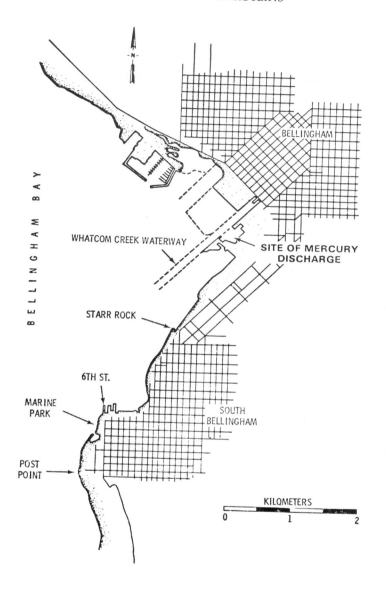

Figure 1: Locations of collection sites for mussels in Bellingham Bay, Washington. Post Point, Marine Park, 6th Street Boat Ramp, and Starr Rock are the collection sites. From Nautical Chart No. 18424 of the National Atmospheric and Oceanic Administration.

RESULTS

Laboratory Experiments

The concentrations of mercury in the gills, digestive
gland, and whole animal for the 90-day mercury exposure are
compared to those for unexposed mussels in Table 1. Concentrations of mercury in the exposed group represented a mag-
nification of greater than 1000x the concentration in the
exposure sea water. Mercury was concentrated in gills and
digestive gland when compared to the whole animal in mercury-
exposed mussels. In unexposed mussels, mercury was concentrated in digestive glands but not gills. Concentrations of
mercury in the latter mussels were extremely low ($\sim$1000x) when
compared to the mercury-exposed group.

*TABLE 1: Mercury concentrations ($\bar{x} \pm 1$ S.E. $\times 10^{-3}$
micromoles Hg/g wet weight) in the whole animal, gills, and
digestive gland of mussels exposed to 1 $\mu g/\ell$ (5.0 $\times 10^{-3}$
micromole/ℓ) mercury for 90 days and mussels unexposed to
mercury.*

Treatment	Whole Animal	Gills	Digestive gland
90-day exposure	*43.8±13.5*	*169.5±57.8*	*67.3±18.4*
Unexposed	*0.045±0.005*	*0.34±0.001*	*0.085±0.005*

Chromatographic profiles for mercury and 254 nm absorbance
of gills and digestive gland of mercury-exposed mussels are
shown in Figure 2. In both gills and digestive glands,
mercury eluted as three peaks associated with the void volume
(peak I: >70,000 daltons molecular weight), 11,000 daltons
(peak II), and substances eluting below the working range of
G-75 (peak III: <3,000 daltons). Peak I corresponds with
high molecular weight proteins; peak II with the metallothionein-like cadmium-binding protein of *Mytilus edulis* (Noel-
Lambot, 1976; Talbot and Magee, 1978), and peak III with the
low molecular weight pool substances (e.g., peptides, amino
acids, and nucleotides). Absorbance peaks which corresponded
with the mercury peaks were observed in both gill and digestive gland. Relatively large amounts (43% in gills and 45% in

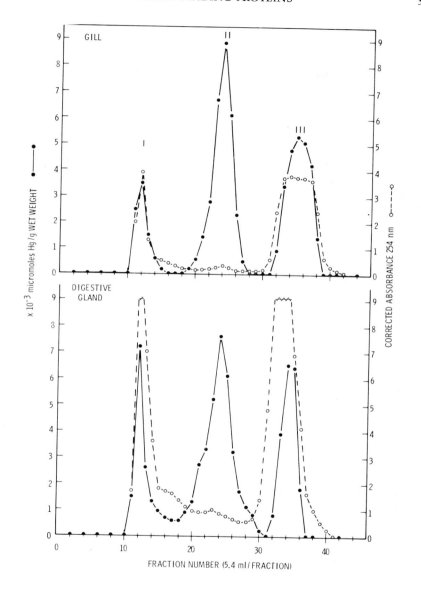

FIGURE 2. *Sephadex G-75 chromatographic profiles of soluble extracts from gills and digestive gland of mussels exposed to 1 μg/ℓ mercury for 90 days.*

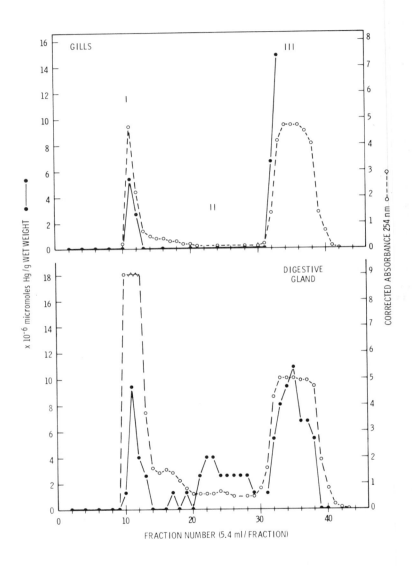

Figure 3: *Sephadex G-75 chromatographic profiles of soluble extracts from gills and digestive glands of mussels previously unexposed to mercury. Peak III was cut off due to malfunction of fraction collector.*

digestive glands) of mercury in the soluble fraction was associated with peak II which is referred to in this paper as the low molecular weight mercury-binding protein. Lower amounts of mercury were associated with Peak I, high molecular weight proteins (11% in gill and 16% in digestive gland), and with peak III, the low molecular weight pool (37% in gill and 28% in digestive gland).

The low concentrations of mercury in tissues of unexposed mussels (Table 1) were consistent with mercury elution profiles (Figure 3) which had peak concentrations up to three orders of magnitude lower than those of exposed mussels. Mercury in soluble fractions of unexposed gills occurred on peaks I and III, with higher amounts associated with the latter. No detectable mercury was observed in the region of peak II, the low molecular weight mercury-binding protein of gills. In digestive gland, small amounts of mercury were detected in the region of peak II, as well as peaks I and III. Correspondingly, 254 nm absorbance was detected in the region of peak II of digestive glands, but not of gills.

When 90-day mercury-exposed (i.e., pre-exposed, see Materials and Methods section) and control mussels were treated with $^{203}HgCl_2$ for 24 hours, gills were found to contain the highest ^{203}Hg concentrations (72 and 60%, respectively, of the whole body ^{203}Hg) and were identified as the primary organ for the short-term incorporation of inorganic mercury (Table 2). Digestive glands also exhibited relatively high ^{203}Hg concentrations when compared to other organs (values for other organs not shown).

TABLE 2. ^{203}Hg (dpm $^{203}Hg/g$ wet weight) in the whole animal, gills, and digestive gland of mussels exposed to $^{203}HgCl_2$ (1 µg $^{203}Hg/\ell$ for 24 hours, Pre-exposed mussels were exposed to 1 µg/ℓ mercury for 90 days prior to the ^{203}Hg exposure.

Treatment	Whole Animal	Gill	Digestive Gland
Pre-exposed	561,771	2,465,272	445,627
Control	747,215	3,098,979	809,738

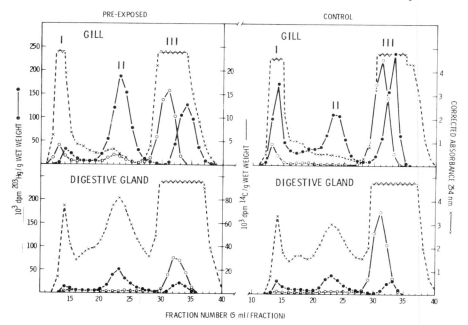

FIGURE 4. Sephadex G-75 chromatographic profiles of soluble extracts from gills and digestive glands of pre-exposed and control mussels which were treated with [203]*Hg or* [14]*C-cysteine for 24 hours.*

[203]Hg profiles of gill tissue (Figure 4) indicated considerable differences due to prior treatment in the subcellular distribution of [203]Hg among high molecular weight proteins (peak I), mercury-binding proteins (peak II), and the low molecular weight pool (peak III). In gills of pre-exposed mussels, [203]Hg in the soluble fraction was associated mainly with the mercury-binding protein (39% of the total soluble [203]Hg) and, to a lesser extent (28%), to components in the low molecular weight pool. Only a small amount (6%) was associated with the high molecular weight proteins. This was not the case in the control mussels in which 22% of the soluble [203]Hg was present on the high molecular weight proteins and only 12% on the mercury-binding protein. Less pronounced differences in [203]Hg profiles were observed between pre-exposed and control digestive glands. Most of the ^{14}C of cysteine was recovered from the low molecular weight pool for all organs. However, the relatively large degree of incorporation of ^{14}C into the mercury-binding proteins of gills of pre-exposed mussels provided evidence for the *de novo* synthesis of the proteins in those organs during the short-

term experiment. Other organs, including gills of controls, exhibited low ^{14}C activity in the region of the low molecular weight, mercury-binding protein.

Field Studies

Gills, digestive gland, and whole mussels from Sequim and Bellingham Bays were analyzed for total mercury (Table 3). Post Point, Marine Park, Sixth Street Boat Ramp, and Star Rock represent sampling stations in Bellingham Bay, ranked in order of decreasing distance from the mercury discharge. All samples from Sequim Bay were lower in mercury than those from Bellingham Bay ($p < 0.001$, one-way analysis of variance; $p < 0.05$, Student-Newman-Keul's multiple comparison among means). Digestive glands contained the highest concentrations of mercury regardless of collection site; levels of mercury in digestive glands did not differ among collection sites in Bellingham Bay. Gills, however, exhibited differences in concentration among stations in Bellingham Bay; levels generally increased with increasing proximity to the discharge area.

TABLE 3. *Concentrations of mercury in the whole animal, gills and digestive gland of mussels of Sequim and Bellingham Bays ($\bar{x} \pm S.E. \times 10^{-3}$ micromoles Hg/g wet weight).*

Collection Site	Whole Animal	Gills	Digestive Gland
Sequim Bay	0.045±0.003	0.035±0.002	0.085±0.003
Bellingham Bay			
Post Point	0.175±0.013	0.095±0.004	0.259±0.045
Marine Park	0.150±0.011	0.115±0.007	0.279±0.009
6th St. Boat Ramp	0.189±0.009	0.170±0.009	0.274±0.016
Starr Rock	0.165±0.018	0.155±0.007	0.269±0.016

Chromatographic profiles for gills and digestive gland of Sequim Bay mussels unexposed to mercury were shown in the previous section. Gel elution profiles for Bellingham Bay mussels were qualitatively similar to those shown for Sequim Bay mussels. However, mercury concentrations associated with the respective peaks were higher, and mercury was also detected in the region of peak II of gills of Bellingham Bay mussels. The values are summarized in Tables 4 and 5. Prominent zinc peaks were also detected in the region of the mercury-binding proteins of all the digestive glands, along with traces of cadmium and copper in some samples (data for those metals are not shown here).

The amounts of mercury present on the mercury-binding proteins appeared to be related to the mercury content of tissues. For example, mercury associated with such substances in gills increased from 0.002×10^{-3} μmoles/g to 0.015×10^{-3} μmoles (Table 4) in samples whose total mercury contents also exhibited increases (Table 3). In digestive gland, mercury levels on mercury-binding proteins of Sequim Bay mussels were lower than those from Bellingham Bay, but those among Bellingham Bay samples showed no apparent differences (Table 5). A similar pattern was observed with the total mercury concentrations in digestive gland of Sequim and Bellingham Bay mussels (Table 3).

With the exception of a single Starr Rock digestive gland sample, the highest concentrations of mercury among constituents of the soluble fraction occurred in the low molecular weight pool. Amounts of mercury on high molecular weight proteins were relatively low. There were no apparent patterns in the distribution of mercury associated with either the low molecular weight pool or high molecular weight proteins with respect to the mercury content of the tissues.

TABLE 4. *Mercury concentrations on chromatographic peaks of gills of Sequim and Bellingham Bay mussels separated on Sephadex G-75 (x10^{-3} micromoles Hg/g wet weight).*

Collection Site	High Molecular Wt Proteins	Mercury-binding Proteins	Low Molecular Wt Pool
Sequim Bay	0.008	Not detectable	>0.022 (1)
Bellingham Bay			
Post Point	0.009	0.002	0.073
Marine Park	0.036	0.003	0.065
6th St Boat Ramp	0.012	0.012	0.052
Starr Rock	0.013	0.015	0.049

(1) *Low molecular weight pool peak of Sequim Bay samples was cut off due to malfunction of fraction collection, see Figure 3.*

TABLE 5. *Mercury concentrations on chromatographic peaks of digestive glands of Sequim and Bellingham Bay mussels separated on Sephadex G-75 (x 10^{-3} micromoles Hg/g wet weight).*

Collection Site	High Molecular Wt Proteins	Mercury-binding Proteins	Low Molecular Wt Pool
Sequim Bay	0.015	0.022	0.053
Bellingham Bay			
Post Point	0.027	0.063	0.164
Marine Park	0.047	0.050	0.141
6th St. Boat Ramp	0.031	0.055	0.102
Starr Rock	0.036	0.059	0.051

DISCUSSION

Laboratory experiments described in this paper have demonstrated the existence of a low molecular weight, mercury-binding protein in the marine mussel *Mytilus edulis*. The apparent molecular weight of this protein (11,000 daltons) was similar to that reported for the cadmium-binding protein of the mussel (Noel-Lambot, 1976; Talbot and Magee, 1978) and similar to that reported for metallothionein in other species (Nordberg, 1978; Olafson *et al.*, 1979) when estimated on Sephadex gels. Gills and digestive glands, which are organs known to concentrate mercury, were capable of binding relatively large quantities of mercury on those proteins (45% of the soluble mercury of the respective organs).

In both mercury pre-exposed mussels and control mussels which were subjected to a subsequent short-term treatment with ^{203}Hg, gills accounted for about 70% of the ^{203}Hg taken up by the whole animal. A major fraction of the gill ^{203}Hg was associated with the mercury-binding protein, especially in pre-exposed animals (39% of the total soluble ^{203}Hg as opposed to 12% in controls). Since ^{14}C incorporation into the mercury-binding proteins indicated enhanced *de novo* synthesis of the proteins in pre-exposed gills, induction of the proteins by the mercury pre-exposure may be the mechanism by which the mercury-binding capacity of the pre-exposed gills was enhanced. If metallothionein-like proteins are involved in the sequestration and detoxification of trace metals as suggested by previous investigators (Cherian and Goyer, 1978; Kojima and Kagi, 1978), induction of the protein in the gill and their subsequent binding to mercury may play an important role in protecting the gills and other organs. It is known that mercury taken up by gills are transported and redistributed to other organs (Cunningham and Tripp, 1975; Wrench, 1978), possibly via hemocytes as is known to occur with ferritin (Coombs and George, 1978). The induction of mercury-binding proteins at the site of uptake (i.e., gills) may be of advantage to the organism since subsequent transport of the metals to organs for elimination or storage would be facilitated if the metals are initially bound to substances whose role is to sequester and detoxify the metals. Elevated levels of toxic metals on high molecular weight proteins such as that described for control gills have been associated with cytotoxic effects of the metal (Brown *et al.*, 1977; Brown and Parsons, 1978). Induction of the mercury-binding protein in gills may have enabled those proteins to selectively sequester mercury which would otherwise have bound to the high molecular weight proteins and possibly other subcellular sites.

Induction of low molecular weight, copper-binding proteins in gills of copper-exposed mussels has also been reported recently (Viarengo et al., in press), indicating that induction of gill metal-binding protein may be a generalized response to metal exposure.

The role of the digestive gland in the short-term (24 hours) incorporation of mercury was not as significant as that of gills with respect to mercury content as indicated by the considerably lower levels of ^{203}Hg incorporation. However, under conditions of a continued 90-day exposure, considerable mercury was incorporated into the digestive gland, both into the whole organ and the mercury-binding protein. Whether the mercury in the digestive gland was transported from the gill or taken up directly via the digestive tract is still to be determined. The low level of ^{14}C incorporation into the mercury-binding protein of digestive gland suggests a lack of induction by mercury exposure in that organ. Mercury in digestive gland may have bound to existing metallothionein-like proteins, the presence of which was indicated by a zinc peak coincident with that of mercury in the region of the mercury-binding protein.

In conjunction with the laboratory experiments discussed above, a field-oriented study was conducted to determine if similar mercury-binding mechanisms were operative in natural populations of mussels which inhabit mercury-contaminated areas. Bellingham Bay was chosen as a study site since it is known to be contaminated with mercury due to operation of a chloralkali plant which discharges mercury wastes into the bay. Discharge rates were estimated to be as high as 4.5 to 9 kg/day between 1965 and 1970 (Crecelius et al., 1975). The discharge was reduced to 0.23 kg/day in 1970; then to 0.05 kg/day, the present rate, in 1974 (Department of Ecology, State of Washington). From the time that the reduction in mercury discharge was instituted, mercury levels in sediment, which were reported to be as high as 100 ppm adjacent to the outfall in 1970, have declined with a half-life of ∿1.3 years (Crecelius et al., 1975). Mercury levels in Bellingham Bay mussels also declined following the reduction in discharge (Table 6). The present levels of about 0.169×10^{-3} micromoles/g for mussels in Bellingham Bay are about three times higher than those which occur in mussels of uncontaminated areas of Puget Sound (Rasmussen and Williams, 1975; Roesijadi et al., this volume).

Examination of the subcellular distribution of mercury in Bellingham Bay mussels indicated an enhanced binding to molecules eluting in the region of mercury-binding proteins when compared to Sequim Bay mussels. It appears that the amount of mercury bound to the mercury-binding protein is

TABLE 6. Levels of mercury in whole mussels from Bellingham Bay, Washington from 1970 to 1978 ($\bar{x} \pm 1$ S.E. x 10^{-3} micromoles Hg/g wet weight).

Date	Hg Concentration	Source of Information
May, 1970	1.3, 1.1	Lee (1971)
March-May 1973	0.598±0.015	Rasmussen and Williams (1975)
December, 1978	0.169±0.015	Roesijadi et al. (this volume)

reflective of the total mercury concentration in tissues. Thus, the amount of mercury on the protein correlated with the total mercury content of tissues, although that fraction was not the most significant mercury pool in the soluble tissue extracts. The major fraction of soluble mercury was associated with the low molecular weight pool substances, the identity of which is presently not known. The differences in the relative importance of the mercury-binding protein in laboratory and field-exposed mussels may be related to the amounts of mercury accumulated by tissues. It is possible that a threshold level in tissue must be reached before high levels of mercury occur on the mercury-binding protein, especially in tissue such as that of gills in which induction by mercury was demonstrated in laboratory experiments.

Mussels collected from mercury-contaminated areas have been reported to possess tissue burdens of mercury ranging from about 0.050×10^{-3} to 2.49×10^{-3} micromoles/g mercury (de Wolf, 1975; Eganhouse and Young, 1976, Davis and Pirie, 1978), depending on the location of collection. That range of concentrations is similar to that observed between our Sequim and Bellingham Bay samples and laboratory-exposed samples. Continued experimentation utilizing both laboratory and field approaches would be extremely useful in contributing to our understanding of the responses of marine animals to mercury pollution.

SUMMARY

Laboratory experiments demonstrated the existence of metallothionein-like, low molecular weight, mercury-binding proteins in the marine mussel *Mytilus edulis*. Relatively large quantities of mercury were associated with such proteins in gills and digestive gland, the organs of interest in the present study. ^{14}C-incorporation indicated induction of the protein in gills, but not in digestive gland. Mercury in digestive gland may have bound to existing metal-binding proteins. Short-term incorporation of mercury occurred primarily in gills. The induction of mercury-binding proteins in gills may have facilitated detoxification of mercury at the site of uptake.

Mercury in mussels of Bellingham Bay was shown to have decreased from 1970 to 1978, the collection date for the present study. Mercury levels were low but approximately three times higher than those from uncontaminated areas. Mercury associated with the mercury-binding protein of gills and digestive glands of Bellingham Bay mussels was low and reflected the concentrations measured in the whole tissues.

ACKNOWLEDGMENTS

This study was supported by the U. S. Department of Energy under Contract No. EY-76-C-06-1830. We would like to thank Richard Burkhalter and Bruce Johnson of the Washington State Department of Ecology for their time and assistance in making available the records of that department.

LITERATURE CITED

Baker, C. W. 1977. Mercury in surface waters of seas around the United Kingdom. Nature 270: 230-232.

Bothner, M. H. and D. E. Robertson. 1975. Mercury contamination of seawater samples stored in polyethylene containers. Analyt. Chem. 47: 592-595.

Brown, D. A., C. A. Bawden, K. W. Chatel, and T. R. Parsons. 1977. The wildlife community of Iona Island jetty, Vancouver, B.C., and heavy-metal pollution effects. Environ. Conserv. 4: 213-216.

Brown, D. A. and T. R. Parsons. 1978. Relationship between cytoplasmic distribution of mercury and toxic effects to zooplankton and chum salmon (*Oncorhynchus keta*) exposed to mercury in a controlled ecosystem. J. Fish. Res. Board Can. 35: 880-884.

Calabrese, A., F. P. Thurberg, and E. Gould. 1977. Effects of cadmium, mercury and silver on marine animals. Mar. Fish. Rev. 39: 5-11.

Casterline, Jr., J. L. and G. Yip. 1975. The distribution and binding of cadmium in oyster, soybean, and rat liver and kidney. Arch. Environ. Contam. Toxicol. 3: 319-329.

Cherian, M. G. and R. A. Goyer. 1978. Metallothioneins and their role in the metabolism and toxicity of trace metals. Life Sci. 23: 1-10.

Coombs, T. L. and S. G. George. 1978. Mechanisms of immobilization and detoxification of metals in marine organisms. In: Physiology and behavior of marine organisms. pp. 179-197. Ed. by D. S. McLusky and A. J. Berry. Pergamon Press, New York.

Corner, E. D. S. and B. W. Sparrow. 1956. The modes of actions of toxic agents. I. Observations on the poisoning of certain crustaceans by copper and mercury. J. Mar. Biol. Ass. U.K. 35: 531-548.

Corner, E. D. S. and F. H. Rigler. 1958. The modes of action of toxic agents. III. Mercuric chloride and N-amylmercuric chloride on crustaceans. J. Mar. Biol. Ass. U.K. 37: 85-96.

Creclius, E. A., M. H. Bothner, R. Carpenter. 1975. Geochemistries of arsenic, antimony, mercury, and related elements in sediments of Puget Sound. Environ. Sci. Tech. 9: 325-333.

Cunningham, P. A. and M. R. Tripp. 1975. Accumulation, tissue distribution and elimination of $^{203}HgCl_2$ and $CH_3^{203}HgCl$ in the tissues of the American oyster, *Crasostrea virginica*. Mar. Biol. 31: 321-334.

Davies, I. M. and J. M. Pirie. 1978. The mussell *Mytilus edulis* as a bio-assay organism for mercury in sea water. Mar. Pollut. Bull. 9: 128-132.

de Wolf, P. 1975. Mercury content of mussels from West European coasts. Mar. Pollut. Bull. 6: 61-63.

Eganhouse, R. D. and D. R. Young. 1976. Mercury in tissues of mussels of Southern California. Mar. Pollut. Bull. 7: 145-147.

Fitzgerald, W. F. and G. A. Gill. 1979. Subnanogram determination of mercury by two-stage gold amalgamation and gas phase detection applied to atmospheric analyses. Anal. Chem. 50: 1714-1720.

Gardner, W. S., D. R. Kendall, R. R. Odom, H. L. Windom, J. A. Stephens. 1978. The distribution of methyl mercury in a contaminated salt marsh ecosystem. Environ. Pollut. 15: 243-251.

Kojima, Y. and J. H. R. Kagi. 1978. Metallothionein. Trends Biochem. Sci. 3: 90-93.

Lee, R. A. 1971. Mercury in Washington State, 22 pp. Olympia, Washington: Washington State Department of Ecology.

Matsunaga, K. S. Konishi, M. Nishimura. 1979. Possible errors caused prior to measurement of mercury in natural waters with special reference to mercury. Environ. Sci. Tech. 13: 63-65.

Mukherji, P. and D. R. Kester. 1979. Mercury distribution in the Gulf Stream. Science 204: 64-66.

National Academy of Sciences. 1978. An assessment of mercury in the environment. 184 pp. Washington, D.C. National Academy of Sciences.

Noel-Lambot, F. 1976. Distribution of cadmium, zinc and copper in the mussel *Mytilus edulis*. Existence of cadmium-binding proteins similar to metallothioneins. Experientia 32: 324-326.

Nordberg, M. 1978. Studies on metallothionein and cadmium. Environ. Res. 15: 381-404.

Olafson, R. W., R. G. Sim, and K. G. Boto. 1979. Isolation and chemical characterization of the heavy metal-binding protein and metallothionein from marine invertebrates. Comp. Biochem. Physiol. 62B: 407-416.

Rasmussen, L. F. and D. C. Williams. 1975. The occurrance
    and distribution of mercury in marine organisms in
    Bellingham Bay. Northwest Sci. 49: 87-94.

Skei, J. M. 1978. Serious mercury contamination of sediments
    in a Norwegian semi-enclosed bay. Mar. Pollut. Bull.
    9: 191-193.

Talbot, V. and R. J. Magee. 1978. Naturally-occurring
    heavy metal-binding proteins in invertebrates. Arch.
    Environ. Contam. Toxicol. 7: 73-81.

Viarengo, A., A. Pertica, G. Mancinelli, S. Palmero, and
    M. Drunesu. (in press). Rapid induction of copper
    binding proteins in the gills of metal-exposed mussels.
    In: Proc. 1st Meeting European Soc. Comp. Biochem-
    Physiol. Biochem., pp. 81-83. Ed. by R. Gilles.
    Pergamon Press.

Winge, D., J. Krasno, and A. V. Colucci. 1973. Cadmium
    accumulation in rat liver: correlation between bound
    metal and pathology. In: Trace element metabolism in
    animals - 2. pp. 500-502. Ed. by W. G. Hoekstra, J. W.
    Suttie, H. E. Ganther, and W. Mertz. University Park
    Press, Baltimore.

Wrench, J. J. 1978. Biochemical correlates of dissolved
    mercury uptake by the oyster, *Ostrea edulis*. Mar.
    Biol. 47: 79-86.

Young, D., C. K. Bertine, D. Brown, E. Crecelius, J. Marint,
    F. Morel, and G. Roesijadi. 1979. Trace metals. In:
    Scientific Problems Relating to Ocean Pollution, pp.
    130-152. Ed. by E. D. Goldberg. National Oceanic and
    Atmospheric Administration, U. S. Department of Commerce.

MONITORING SEA SCALLOPS IN THE OFFSHORE WATERS
OF NEW ENGLAND AND THE MID-ATLANTIC STATES:
ENZYME ACTIVITY IN PHASIC ADDUCTOR MUSCLE

Edith Gould

National Oceanic and Atmospheric Administration
National Marine Fisheries Service
Northeast Fisheries Center
Milford Laboratory
Milford, Connecticut

INTRODUCTION

A marine field-monitoring program, Ocean Pulse (OP), was
initiated two years ago by the Northeast Fisheries Center
(NOAA, National Marine Fisheries Service) to assess the health
of marine animal stocks in the offshore waters between Cape
Hatteras and the Canadian border, and to perform seasonal
monitoring of selected species. Animal specimens are taken
seasonally from a group of stations in these waters (Fig. 1)
and examined by a variety of methods, as no single criterion
can adequately serve. Relatively clean stations have been
paired with those that are known to be polluted (Table 1), on
the basis of similarity in hydrographic parameters, as well as
general sediment structure where such information is available.
The multidisciplinary approach includes physiology, biochemistry, chemistry, microbiology, genetics, pathology, and
benthic ecology. In this report, some baseline biochemical
data are presented for a single tissue, the phasic portion of
the adductor muscle, in a single species, *Placopecten
magellanicus,* as an example of one kind of biological monitoring. The purpose is to determine the range of variation
in some energy-related biochemical patterns in this relatively
sessile animal. Establishing the range of normalcy should
enable the detection of any significant change that could be
interpreted as a response to environmental stress.

TABLE 1. Paired stations, based on bottom temperatures and depths.

Test Sample	Test Date	Concentrations (sw:effluent)
Hyperion - present final effluent	12/19/78 1/24/79 3/29/79	500:1, 108:1, 23:1, 5:1 108:1, 39:1, 14:1, 5:1 103:1, 64:1, 45:1, 35:1, 24:1, 19:1, 15:1, 14:1
JWPCP - present final effluent	11/28/78 1/ 9/79 2/20/79	500:1, 108:1, 23:1, 5:1 108:1, 39:1, 14:1, 5:1 108:1, 39:1, 14:1, 5:1
JWPCP - projected final effluent	11/28/78 1/ 9/79 2/20/79	500:1, 108:1, 23:1, 5:1 108:1, 39:1, 14:1, 5:1 108:1, 39:1, 14:1, 5:1
ORCOSAN - present final effluent	12/ 5/78 2/ 6/79 3/13/79 4/ 9/79	500:1, 108:1, 23:1, 5:1 108:1, 39:1, 14:1, 5:1 103:1, 64:1, 45:1, 35:1, 24:1, 19:1, 15:1, 14:1 300:1, 75:1, 60:1, 43:1, 35:1, 26:1, 19:1, 16:1
ORCOSAN - projected final effluent	12/:5/78 2/ 6/79	500:1, 108:1, 23:1, 5:1 108:1, 39:1, 14:1, 5:1
Hyperion - sludge	2/13/79 3/ 6/79	500:1, 300:1, 160:1, 130:1, 108:1, 80:1, 65:1, 23:1 1000:1, 700:1, 600:1, 500:1, 300:1, 160:1, 130:1, 108:1

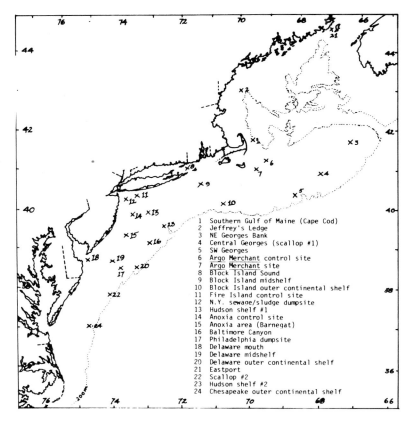

FIGURE 1. Ocean Pulse stations in the offshore waters of New England and the mid-Atlantic states.

    A commercially important animal, the sea scallop can live
for as long as 20 years and is found in widespread populations
in the waters of the northwest Atlantic continental shelf,
from Newfoundland to the Virginia Capes (Posgay, 1957). It is
a swimming bivalve whose movements are vigorous but random,
with no directed population movements or seasonal migration
(Baird, 1954; Serchuk *et al.*, 1979). In a long-term tagging
study, Posgay (1963) found that after 4 to 5 years, 85% of
more than 2000 specimens were recovered within two miles of
the site where they were originally set out. This relatively
limited sphere of activity makes the scallop a particularly
useful species for monitoring offshore waters. Moreover, the
adductor muscle is a large, easily-sampled, discrete tissue
that holds up well, biochemically speaking, under good conditions of frozen storage (-80°C), and has been observed to
reflect polluted field conditions (Mearns and Young, 1977),

as well as experimentally induced sublethal metal stress
(Gould, unpublished data). Because the scallop adductor
muscle is composed of two distinct kinds of tissue (the
striated phasic and the smooth catch muscle), which differ in
function as well as in structure (deZwaan *et al.*, in press),
sampling is carefully restricted to the larger (ca. 80%)
phasic portion (Fig. 2).

Our sampling efforts for the sea scallop during OP
cruises have been considerably expanded by the long-term, con-
tinuing cooperation of the NEFC Resource Survey Investigation
at Woods Hole, Massachusetts, whose members take scallop
adductor muscle samples for us at random sites during their
seasonal groundfish and shellfish survey cruises. It is
these survey samples, especially, that have provided the wide
variety of bottom temperatures and depths for our baseline
data. Sample number per station varies from 6 to 18.

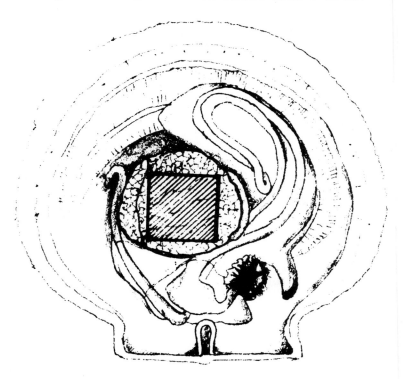

FIGURE 2. *The sea scallop,* Placopecten magellanicus, *on
the half shell.* *The hatched square in the middle of the
adductor muscle (phasic portion) represents the tissue sample
taken from each animal.*

As for the biochemical parameters, we started with the energy-related enzymes that have been found to respond to stress in the experimental animals we have used in heavy-metal research (lobster, rock crab, winter flounder): key glycolytic enzymes and the major transaminase. Experimental exposures of the scallop to sublethal heavy-metal stress, recently begun in our laboratory here, will help us to discover whether these enzymes (and others presently under study) will serve as stress indicators in the scallop, also. In earlier work with other animals, we have found malate dehydrogenase (E.C. 1.1.1.37, MDH) to be a good indicator of general stress, especially in crustaceans (Gould, 1980); pyruvate kinase (E.C. 2.7.1.40, PK) is the terminal regulator of anaerobic glycolysis in the Embden-Myerhof pathway, operating unidirectionally toward glucose breakdown, or energy expenditure; and aspartate aminotransferase (E.C. 2.5.1.1, AAT) is one aspect of nitrogen metabolism that has served to indicate stress in crustaceans (Biesinger and Christensen, 1972; Gould *et al.*, 1976). Normal ranges of activity are presented in terms of the depth and bottom temperature of waters from which the animals were taken. To these three metabolic criteria has recently been added octopine dehydrogenase (E.C. 1.5.1.11, ODH), an enzyme that provides oxidizing potential during energy production in cephalopod molluscs (Fields and Hochachka, 1975) and other marine molluscs (Regnouf and van Thoai, 1970; Gäde, 1980), especially swimming bivalves (deZwaan *et al.*, in press). Because we have only begun to complete ODH data, however, that enzyme is not reported here.

Biochemical work poses several technical problems not applicable to most of the other disciplines; frozen-storage history, for example, of tissues packaged at sea and transported to the laboratory for analysis is no small variable, and can affect enzyme stability as well as the relative concentrations of metabolites. Uniformity of extraction procedures and stability of some of the reagent biochemicals are other factors to be closely watched. These are areas of concern in any experimental study, of course, but they become critical in a monitoring effort that spans years. It is mandatory, therefore, to have specific and consistent methodology, and a rigorous system of controls and standards.

METHODS

The phasic portion of the scallop adductor muscle (here-
inafter referred to simply as the adductor muscle) was sampled
by cutting a square portion from the middle of the tissue
(Fig. 2), then placed in a small plastic pouch with as much
air excluded as possible, wrapped tightly in masking tape, and
frozen at the lowest possible temperature (we have a sea-going
freezer held at $-40°C$). Samples were kept frozen in dry ice
during transport to the analyzing laboratory, then placed in a
freezer at $-80°C$.

Each muscle homogenate was prepared in an all-glass homogenizer, 1:9, w/v, with cold 1% Triton (a non-ionic surfactant) and a small amount of 25-μm glass powder to fac-
ilitate grinding. Centrifugation was at $4°$ and 40,000 x $\underline{g}$ for
60 min; the undiluted supernatants served as the crude enzyme
preparations (E10x).

Enzyme activities were determined with a twin-beam
recording spectrophotometer, by following the change in absorbance at 340 nm. Slopes were drawn from the fastest
portion of the curve, and activities expressed as μmoles NADH
oxidized min^{-1} mg biuret $protein^{-1}$. Data were analyzed statistically by the Student's $\underline{t}$ test; sample number ranged from
5 to 18 animals per station for each sampling date.

Assay concentrations ($m\underline{M}$) were: $\underline{AAT}$ - pH 8.0 Tris buffer
(103), L-asparate (200), NADH (0.40), 2-oxoglutarate (6.7),
62.5 units MDH, and 0.1 ml E10x; $\underline{MDH}$ - pH 9.0 glycine (90) and
EDTA.K_2 (9) buffer, oxaloacetate (1.0), NADH (0.15), and 0.10
ml E100x; $\underline{PK}$ - pH 7.5 triethanolamine buffer (58.3), $MgCl_2$
(8.3), KCl (75), NADH (0.15), adenosine diphosphate (0.23),
phospho(enol)pyruvate (0.8), 140 units LDH, and 0.10 ml E10x.
Reaction mixtures were first tested with standard purified
enzyme suspensions before use with the scallop muscle preparations.

RESULTS AND DISCUSSION

Baseline data gathered thus far are shown in Figures 3
(AAT), 4 (MDH), and 5 (PK), with the activities plotted
against depth (upper graph) and bottom temperature (lower
graph). Each point represents an arithmetic mean for 5-18
animals taken from a single station on a single day. A
single small sample (5) of immature animals (<5.5 cm) compared
to 9 adults (8.0-11.0 cm) from a station with similar hydrographic parameters, had higher transaminase activity

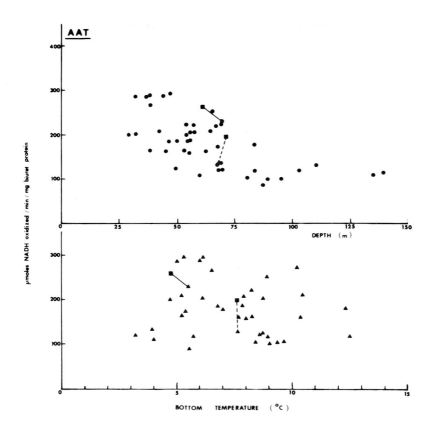

FIGURE 3. Asparate aminotransferase activity in scallop adductor muscle, in terms of depth and bottom temperature of waters from which the animals were taken. Each point represents the arithmetic mean of 5-18 samples from a single station on a single day. The lines connecting paired points represent data sets from scallops collected in May (pre-drilling, solid line) and October (after drilling had begun, dashed line) of 1978 from the Baltimore Canyon drilling site (square) and a nearby control station (dot). The same conventions are used for Figures 4 and 5.

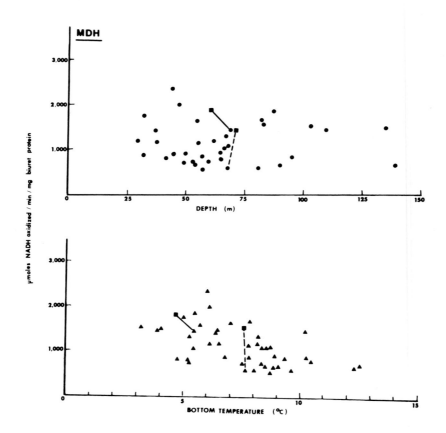

FIGURE 4. Malate dehydrogenase activity in scallop
adductor muscle.

(P <0.01) in the immature animals and somewhat lower glycolysis. Except for the latter samples, animal size has
ranged from 7.0 to 14.5 cm.

No pattern has been seen as yet for either sex or season.
Because most of the data reported here were obtained from
animals taken either in the spring or in the fall months, however, one would not expect to see a seasonal pattern. A
proper study of this important variable should involve at
least monthly sampling of a single population.

In the overall patterns for adult scallops, AAT activity
in Figure 3 seems to decrease slightly with increasing depth
(and possibly MDH with increasing temperature, Figure 4), but
the dominant observation is the great amount of apparently

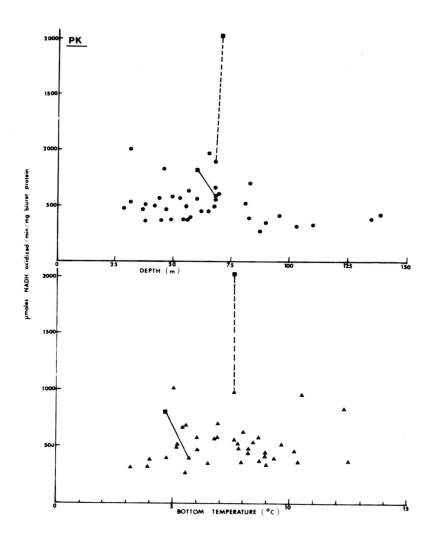

FIGURE 5. Pyruvate kinase activity in scallop adductor muscle.

normal scatter, which is not necessarily attributable to the technical variables mentioned above. We have found, for example, significantly different transaminase activities (P <0.005) between groups of animals taken on the same day from "clean" sites separated by only 8 miles, having the same

depth and bottom temperature, and the tissue samples prepared
and analyzed on the same day. Each data set, however, fell
within the broad range thus far established for each enzyme;
although one set of animals may have metabolic rates statistically different from the other, the values are not abnormally different.

What does attract the attention in monitoring work is an
abnormal statistical difference as exemplified by the upper
pair of connected points in Figure 5. In each of the graphs
in Figures 3-5, the lines connecting two points represent data
sets from scallops collected in May and October of 1978 from
the Baltimore Canyon drilling site (square) and from a nearby
control station (circle), before and after drilling had begun.
The PK data for post-drilling are the points connected by the
upper (dashed) line, and indicate a flare of glycolytic
activity in the animals collected near the drilling site
shortly after drilling was initiated. Sample numbers for
upper and lower points were 17 and 16, respectively (P <0.001).
These data, originally reported by Calabrese et al. (1979),
reflected a transitory phenomenon, as values showed a return
to the normal range in animals taken from the Baltimore
Canyon site during the following summer. Such a burst of
glycolysis could be explained simply by a period of uncommonly
vigorous activity; scallops are capable of speedy and sustained movement under the stimulus of noise or vibration
(Baird, 1954; Posgay, personal communication). We are presently gathering data for scallops taken from Georges Bank at
sites near which oil-drilling activity is under consideration.

Our weak point at the moment is the lack of experimental
data. The scallop's intolerance of water temperature above
20°C (Serchuk et al., 1979) restricts chronic-exposure
studies, which use flow-through systems, to the winter months.
Initial 2-month exposures to sublethal concentrations
(10 mg l^{-1}) of cadmium and of silver have just been completed,
however, and short-term experiments on the effects of starvation, hypoxia, and exercise are presently underway, to
gather data using the energy-related biochemical criteria
described here. Knowledge of the effects of natural stress
and of pollutant stress, together with the growing body of
baseline data, will help us to distinguish and interpret
significant abnormalities in the field.

SUMMARY

The sea scallop, *Placopecten magellanicus*, is a particularly attractive subject for environmental monitoring of
the waters of North America's continental shelf because it is
relatively sessile, of commercial importance, and found in
large populations in the offshore waters of New England and
the mid-Atlantic states. For the past two years, we have
collected information on the activity of energy-related
enzymes in scallop adductor muscle, in an attempt to establish
normal baseline ranges in terms of biological and hydrographic
variables. The enzymes chosen for monitoring are those that
have been shown to vary significantly in the tissues of sublethally stressed marine animals. Thus far there have been no
significant changes consistently attributable to sex or
season of the year, although there are differences in enzymatic activities between mature and immature animals. Some
indication can be seen of decreasing transaminase activity
with increasing depth and, to a lesser extent, of decreasing
MDH activity with increasing bottom temperature.

LITERATURE CITED

Baird, F. T., Jr. 1954. Migration of the deep sea scallop
      (Pecten magellanicus). Fisheries Circular No. 14. Dept.
      Sea and Shore Fisheries, Augusta, Maine.

Biesinger, K. E., and G. M. Christensen. 1972. Effects of
      various metals on survival, growth, and reproduction of
      Daphnia magna. J. Fish. Res. Bd. Canada. 29: 1691-1700.

Calabrese, A., F. P. Thurberg, E. Gould, and J. T. Graikoski.
      1979. Ocean Pulse: Some physiological, biochemical, and
      bacteriological activities - year 2 (1978-1979).   ICES
      document C.M.1979/E:62, Marine Environmental Quality
      Committee.

Fields, J. H. A., and P. W. Hochachka. 1975. Octopine de-
      hydrogenase in squid mantle. Comp. Biochem. Physiol.
      52B: 158.

Gäde, G. 1980. Biological role of octopine formation in
      marine molluscs. Mar. Biol. Letters 1: 121-135.

Gould, E. 1980. Low-salinity stress in the lobster, *Homarus americanus,* after chronic sublethal exposure to cadmium: Biochemical effects. Helgol. wiss. Meeresunters. 33: 174-184.

Gould, E, R. S. Collier, J. J. Karolus, and S. Givens. 1976. Heart transaminase in the rock crab, *Cancer irroratus,* exposed to cadmium salts. Bull. Environ. Contam. Toxicol. 15: 635-643.

Mearns, A. J., and D. R. Young. 1977. Chromium in the marine environment. In: Pollutant Effects on Marine Organisms, (C. S. Giam, editor). Lexington Books (D. C. Heath Co.: Lexington, Massachusetts), pp. 43-44.

Posgay, J. A. 1957. The range of the sea scallop. The Nautilus 71: 55-57.

Posgay, J. A. 1963. Tagging as a technique in population studies of the sea scallop. Int. Comm. Northwest Atl. Fish. Spec. Publ. No. 4.

Regnouf, F., and N. van Thoai. 1970. Octopine and lactate dehydrogenases in mollusc muscles. Comp. Biochem. Physiol. 32: 411-416.

Serchuk, F. M., P. W. Woods, J. A. Posgay, and B. E. Brown. 1979. Assessment and status of sea scallop (*Placopecten magellanicus*) populations off the northeast coast of the United States. Proc. Nat. Shellfish. Assn. 69: 161-191.

Zwaan, A. de, R. J. Thompson, and D. R. Livingstone. In Press. Physiological and biochemical aspects of the valve snap and valve closure responses in the giant scallop *Placopecten magellanicus*. II. Biochemistry. J. Comp. Physiol.

EFFECTS OF MUNICIPAL WASTEWATER ON
FERTILIZATION, SURVIVAL, AND DEVELOPMENT OF
THE SEA URCHIN, *STRONGYLOCENTROTUS PURPURATUS*

Philip S. Oshida
Teresa K. Goochey
Alan J. Mearns

Southern California Coastal Water Research Project
Long Beach, California

INTRODUCTION

Nearly one billion gallons (3.78×10^9 liters) of municipal wastewater are discharged daily into coastal waters of the
Southern California Bight. Most of the wastewaters are unchlorinated and are injected into the ocean through deep ocean
outfalls (60 m depth, several km offshore) fitted with multi-
port diffusers which initially dilute the wastewaters by
factors of 80 to 200 (Hendricks, 1977). An important purpose
of this dilution is to mitigate possible toxicity to marine
life in the water column. However, marine organisms have
rarely been used to assess the toxicities of these effluents.
Instead, freshwater fish have been, and still are being used
as 96-hour bioassay organisms (Kopperdahl, 1976) for the
determination of wastewater toxicity. Because of this
reliance on freshwater fish, it is uncertain whether or not
water quality near the discharges is suitable for survival and
growth of sensitive stages of marine organisms. Woelke
(1972), Cardwell (1979), Cardwell *et al*. (1977, 1977b, and
1979), Stober *et al* (1977 and 1978) and Kobayashi (1971, 1972,
and 1973) have demonstrated such toxicity can indeed occur in
the sea near industrial and domestic discharges, and their
results suggest that echinoderm and oyster eggs and larvae are
quite sensitive bioassay test organisms.
Conversely, the freshwater fish used in the wastewater
tests were not very sensitive and often demonstrated 100 percent survival in 50 to 100 percent effluent for 96 hours.

Since fish had such low mortality in high concentrations of
wastewater, the lowering of toxicity due to improvements in
effluent quality, such as alterations in processing or source
control, may go undetected.

To reduce this uncertainty, the Southern California
Coastal Water Research Project began a program to develop a
marine bioassay system for local receiving water monitoring.
Bioassays were conducted to determine the concentrations of
wastewater that were not harmful to sensitive life stages of
the purple sea urchin, *Strongylocentrotus purpuratus*, to compare the sensitivites of *S. purpuratus* and freshwater fish
used in effluent bioassays, and to test the toxicities of
present and projected effluents.

To address the subject of wastewater toxicity in the
ocean, we examined the effects of municipal wastewater and
digested sludge on fertilization, survival, and development
of purple sea urchin gametes, embryos, and larvae. These
early life stages are both sensitive to marine pollutants and
critical for normal growth of the urchins (Kobayashi, 1971).

MATERIALS AND METHODS

The experimental procedures were adapted from the
bioassay techniques of Kobayashi (1971) on sea urchins,
Stober *et al.* (1977) on sand dollars, and Woelke (1972) on
oysters. Urchins were collected intertidally, held in a
laboratory cold-bath and induced to spawn. The eggs and
sperm were exposed, separately, to the test concentrations of
effluent before being added together in the test solution.
The eggs were subsampled after a short time and microscopically examined for the presence or absence of the fertilization membrane (fertilization), and later for the percentage of normally developed 48-hour larvae.

This study investigated the toxicities of six effluents:
City of Los Angeles' Hyperion Treatment Plant (Hyperion)
    1.   Present final effluent (5-mile)
    2.   Primary and secondary digested sludge (7-mile)
County of Los Angeles' Joint Water Pollution Control
Plant (JWPCP)
    3.   Present final effluent
    4.   Projected final effluent: four parts fine-screened
        primary and two parts Eunox secondary effluent
County Sanitation Districts of Orange County, Plants 1
and 2 (ORCOSAN)
    5.   Present final effluent

6.   Projected final effluent: one part present
primary effluent and two parts secondary effluent

The average concentrations of general constituents,
trace metals, and chlorinated hydrocarbons, as well as average
flow rates for all the present effluents and Hyperion sludge
are summarized by Schafer (1980).   Young (1978) has identified
the EPA "Priority Pollutants" for these wastewaters.   All
effluent samples collected by the treatment plant personnel
were 24 to 36 hour composites.   Glass bottles that had been
acid-washed and kilned were used for collection to insure
against metal or hydrocarbon contamination of the effluent
samples.   The Hyperion sludge samples were one-hour composites which had been collected in an acid-washed plastic
bucket.

For the experiment, all effluents and sludge were diluted
with filtered (3 µm) natural seawater.   A list of experimental
dilutions is given in Table 1.   A 900ml portion of each
dilution was poured into each of two one-liter beakers and a
50 ml portion was poured into each of two polypropylene cups.
The beakers were used for egg exposure and development, and
the cups for sperm exposure.   The remaining portions of the
effluent dilutions were sampled and measured for dissolved
oxygen (Winkler titration), pH (Beckman pH meter) and salinity
(refractometer).   The beakers and cups were placed in a 12°C
water bath and the gametes were added.

Salinity was monitored to determine variability in gamete
and egg-development due to salinity.   The salinity dilutions,
which ranged from 65 percent seawater (21 ppt, parts per
thousand) to 90 percent (29 ppt), were made by adding deionized water to natural seawater (33 ppt).   These dilutions
were set up and measured in the same manner as the effluent
dilutions.

Four replicate beakers and four replicate cups containing
filtered natural seawater were set up as controls in a manner
similar to the effluent and salinity dilutions and measured
for dissolved oxygen, pH, and salinity.

The urchins used in this study were collected intertidally by hand from rocks at Malaga Cove, Palos Verdes
Peninsula, California.   They were packed with brown algae
(Egregia sp.) to prevent desiccation and transported to the
laboratory where they were held in recirculating seawater
aquariums at 12°C and fed brown algae, Egregia sp.   Urchins
were induced to spawn by injecting 0.5 ml of 0.5 M potassium
chloride into the coelom (Hinegardner, 1967).   Spawning female
urchins were inverted, aboral side downwards, onto 100ml Pyrex
beakers filled with 12°C seawater.   The diameter of the beaker
was smaller than the diameter of the urchin test and allowed
the urchin to remain "perched" on the beakers with the
gonadopores immersed in seawater.   Eggs were collected only

TABLE 1.   Effluent and sludge dilutions used in each experiment.

Test Sample	Test Date	Concentrations (sw:effluent)
Hyperion - present final effluent	12/19/78 1/24/79 3/29/79	500:1, 108:1, 23:1, 5:1 108:1, 39:1, 14:1, 5:1 103:1, 64:1, 45:1, 35:1, 24:1, 19:1, 15:1, 14:1
JWPCP - present final effluent	11/28/78 1/ 9/79 2/20/79	500:1, 108:1, 23:1, 5:1 108:1, 39:1, 14:1, 5:1 108:1, 39:1, 14:1, 5:1
JWPCP - projected final effluent	11/28/78 1/ 9/79 2/20/79	500:1, 108:1, 23:1, 5:1 108:1, 39:1, 14:1, 5:1 108:1, 39:1, 14:1, 5:1
ORCOSAN - present final effluent	12/ 5/78 2/ 6/79 3/13/79 4/ 9/79	500:1, 108:1, 23:1, 5:1 108:1, 39:1, 14:1, 5:1 103:1, 64:1, 45:1, 35:1, 24:1, 19:1, 15:1, 14:1 300:1, 75:1, 60:1, 43:1, 35:1, 26:1, 19:1, 16:1
ORCOSAN - projected final effluent	12/ 5/78 2/ 6/79	500:1, 108:1, 23:1, 5:1 108:1, 39:1, 14:1, 5:1
Hyperion - sludge	2/13/79 3/ 6/79	500:1, 300:1, 160:1, 130:1, 108:1, 80:1, 65:1, 23:1 1000:1, 700:1, 600:1, 500:1, 300:1, 160:1, 130:1, 108:1

during the first 15 minutes of spawning to minimize the use of immature eggs that might be shed with prolonged spawning. Eggs collected from an average of six females were combined, passed through gauze to remove spines and debris, and rinsed twice with seawater (Hinegardner, 1967). They were then mixed with seawater to achieve a uniform density and a volumetric sample was taken to count the numbers of eggs released. Using a technique similar to Woelke's (1972), approximately 31,500 eggs were then added to each beaker of test sample with an automatic pipet and maintained for exactly 30 minutes prior to the addition of the sperm to the egg-exposure beakers.

Spawning male sea urchins were kept out of water with their oral side on wet paper towels to collect sperm under "dry" conditions (Hinegardner, 1967), so as not to activate the sperm. Sperm of several males were removed with a pasteur pipet to a test tube and kept below 5°C. Prior to the addition of sperm to the test sample cups, nine drops of sperm were added to 45 ml of seawater in a 100ml beaker and thoroughly stirred. This addition of the sperm to seawater "activated" the sperm. Fifteen minutes after we had added the eggs to the egg-exposure beakers, 1.2 ml of sperm solution were added to each of the sperm-exposure cups. At this time, the eggs and sperm were being simultaneously and separately exposed to the test samples.

Fifteen minutes after the sperm had been added to the sperm-exposure cups, which was also 30 minutes after the eggs had been added to the egg-exposure beakers, the sperm solutions were added to their respective egg-exposure beakers to allow for fertilization. Fifteen minutes later the eggs in each beaker were mixed to achieve a uniformly dense solution, and a 10 ml sample was taken from each beaker with an automatic pipet. Each sample was placed in a vial, pre-served with ten percent borax-buffered formalin, and saved for future microscopic examination.

The eggs remaining in the beakers were stirred, with electric stirrers, at about 60 rmp, aerated, and kept at 12°C. The result was a nearly homogeneous solution with very little sedimentation, even with the digested sludge. Sampling at 48 hours was accomplished in the same manner as the sampling at 15 minutes. Egg survival was carefully monitored in each experiment.

The eggs and larvae which had been sampled and preserved were examined microscopically (100X) on a Sedgewick-Rafter counting chamber. For the 15-minute samples, "fertilization" of the eggs was defined by the presence of an obvious fertilization membrane around each egg; for the 48 hour samples, "normal" development was defined by the presence of

the archenteron at the gastrula stage. The toxicity of a
test sample was defined as the inhibitory effects of the test
sample on fertilization, and the further development of sea
urchin eggs (Kobayashi, 1972).

In the statistical analysis of the data, comparisons were
restricted to those between the control and each of the
treatments (individual effluent, sludge, or salinity
dilutions). The control and treatment data were compared by
calculating the T-statistic with respect to the difference
between the two independent sample means. The treatments
that differed significantly at $p \leq 0.05$ were identified.

RESULTS

Municipal wastewater significantly reduced fertilization
of sea urchin eggs at effluent concentrations generally higher
than one to seven percent effluent (Table 2). There was a
trend of decreased percentage of fertilization with increased
concentrations of effluent that paralleled reductions in the
percentages of normal development of the 48-hour larvae (Table
3). The concentrations of effluent that did not elicit toxic
responses were found to be similar in repeated experiments;
usually only changed due to different experimental concentration levels. Hyperion digested-sludge (7-mile) concentrations at, and above 0.2 percent, significantly reduced
fertilization of the eggs.

Urchin gametes and fertilized eggs exposed to the
control solutions of uncontaminated natural seawater showed
very little variability, with a mean percentage of fertilization of 94.7 percent (n = 40, standard error, S.E. was
0.84 percent) and a mean percentage of normal development
of 92.1 percent (n = 48, S.E.: 0.80 percent).

The salinity tests that were conducted concurrently with
effluent and sludge tests generally showed that fertilization
and normal 48-hour development were reduced at salinity
levels lower than 28 ppt. The results of salinity tests
showed that the effects of lowered salinity were negligible
at "present" effluent concentrations that caused harmful
effects to the urchins. The salinities in the lowest
"present" effluent concentrations that caused reduction in
fertilization and development were at or higher than 30 ppt in
all but one instance (Hyperion test, 12/19/78), while in the
salinity tests, the salinities tested that caused significantly harmful effects were at or below 28 ppt.

The "projected" effluent from ORCOSAN caused significant
reductions in normal development at 20 percent effluent, which
had salinity of about 26 ppt. The toxicity in these 5:1

TABLE 2. A summary of the highest effluent concentrations that had no effect on fertilization of sea urchin, *Strongylocentrotus purpuratus*, eggs, and the lowest concentrations that significantly (p≤ 0.05) reduced fertilization.

Test Sample	Test Date	Highest concentration of effluent tested that caused no effect on fertilization		Lowest concentration of effluent tested that caused significantly reduced fertilization	
		Dilution	% Effluent	Dilution	% Effluent
Hyperion - present final effluent	12/19/78	23:1	4.3	5:1	20
	1/24/79	39:1	2.6	14:1	7.2
	3/20/78**	14:1	7.2	—	—
JWPCP - present final effluent	1/ 9/79*	—	—	108:1	0.9
	2/20/79*	—	—	108:1	0.9
JWPCP - projected final effluent	1/ 9/79*	—	—	108:1	0.9
	2/20/79	39:1	2.6	14:1	7.2
ORCOSAN - present final effluent	2/ 6/79	108:1	0.9	39:1	2.6
	3/13/79*	—	—	103:1	1.0
	4/ 9/79	43:1	2.3	35:1	2.9
ORCOSAN - projected final effluent	2/ 6/79	39:1	2.6	14:1	7.2
Hyperion - sludge	2/13/79*	—	—	500:1	0.2
	3/ 6/79	700:1	0.14	600:1	0.17

*All concentrations of effluent tested caused significantly reduced fertilization.

**No experimental concentrations of effluent caused significantly reduced fertilization.

dilutions of effluent may not have been attributable to projected effluent, but, in fact, may have reflected lowered salinity effects.

The pH, dissolved oxygen, and salinity levels in the controls, effluent dilutions, and sludge dilutions showed only minimal changes during the 48-hour experimental periods. The pH mean values were 7.90 (n=11, S.E.: 0.011), 7.78 (n=17, S.E.: 0.041), and 7.89 (n=72, S.E.: 0.008) for control, sludge and effluent solutions, at the start of the experiment with mean changes in pH after 48 hours of 0.15 (S.E.: 0.013), 0.11 (S.E.: 0.020), and 0.13 (S.E.: 0.007) units respectively. The same samples were also analyzed for dissolved oxygen and salinity, and the mean dissolved oxygen values at the start of the experiments were 7.56 (S.E.: 0.049), 6.69 (S.E.: 0.364), and 7.27 (S.E.: 0.070) mg/l, with mean changes in dissolved oxygen over 48 hours of 0.51 (S.E.: 0.095), 0.52 (S.E.: 0.062), and 0.97 (S.E.: 0.075) mg/l, for control, sludge and effluent solutions, respectively. The mean salinities at the start of the experiments were 32.5 (S.E.: 0.144), 32.6 (S.E.: 0.445), and 30.9 (S.E.: 0.267) ppt with mean changes after 48 hours of 0.33 (S.E.: 0.128), 0.19 (S.E: 0.101) and 0.27 (S.E.: 0.050) ppt for control sludge and effluent solutions, respectively. The lower mean salinity for the effluent solutions was due to the lower seawater dilutions used with these freshwater effluents.

DISCUSSION

This sea urchin fertilization and larval bioassay appears to be an accurate, reliable, and feasible test for measuring the toxicities of seawater dilutions of municipal wastewaters and sludges. These results are consistent with information found by other researchers. Cardwell *et al.* (1979) conducted 48-hour toxicity tests on fertilized eggs of the Pacific oyster, *Crassostrea gigas,* with chlorinated and unchlorinated sewage-treatment plant effluents from Olympia, Washington; both effluents were found to be toxic to developing oysters at ten percent effluent, but innocuous at one percent effluent. Stober *et al.* (1977) conducted a success of fertilization test with only the sperm of the sand dollar, *Dendraster excentricus,* exposed to both chlorinated and unchlorinated West Point, Seattle, Washington effluent prior to fertilization and showed that fertilization was reduced by 50 percent in 4.4 percent chlorinated effluents. In the Stober *et al.* (1977) test, the eggs were not exposed to effluent prior to fertilization, thereby measuring only the effects of

TABLE 3. A summary of the highest effluent concentration tested that had no effect on 48-hour sea urchin, *Strongylocentrotus purpuratus*, development and the lowest concentrations that significantly (p ≤ 0.05) reduced normal development.

Test Sample	Test Date	Highest concentration of effluent that caused no effect on development		Lowest concentration of effluent that caused significantly reduced normal development	
		Dilution	% Effluent	Dilution	% Effluent
Hyperion - present final effluent	12/19/78	108:1	0.9	23:1	4.3
	1/24/79	108:1	0.9	39:1	2.6
	3/20/79**	14:1	7.2	-	-
JWPCP - present final effluent	11/28/78	23:1	4.3	5:1	20
	1/ 9/79*	-	-	108:1	0.9
	2/20/79*	-	-	108:1	0.9
JWPCP - projected final effluent	11/28/78	23:1	4.3	5:1	20
	1/ 9/79	108:1	0.9	39:1	2.6
	2/20/79	108:1	0.9	39:1	2.6
ORCOSAN - present final effluent	12/ 5/78	23:1	4.3	5:1	20
	1/ 9/79	108:1	0.9	39:1	2.6
	3/13/79*	-	-	103:1	1.0
	4/ 9/79	300:1	0.3	75:1	1.3
ORCOSAN - projected final effluent	12/ 5/78	23:1	4.3	5:1	20
	2/ 6/79	14:1	7.2	5:1	20
Hyperion - sludge	2/13/79*	-	-	500:1	0.2
	3/ 6/79*	-	-	1000:1	0.1

*All concentrations of effluent caused significantly reduced normal development.

**No concentrations of effluent caused significantly reduced normal development.

the effluent on the ability of the sperm to fertilize the eggs. The sea urchin experiment described here showed very similar results; the significantly harmful effluent concentrations being above one to seven percent effluent.

It is now possible to demonstrate how the toxic effects of wastewater might be distributed in the ocean by relating the effluent concentrations that caused toxic reactions in sea urchins to the measured and theoretical concentrations of wastewater around sewage outfalls. Hendricks (1977) made *in situ* measurements of the initial dilution zones of Hyperion outfall, JWPCP outfall, ORCOSAN outfall, and San Diego City's Point Loma outfall in October 1976. His goals were to compare various methods of predicting initial dilution zones with the actual measured dilutions produced by the existing outfall systems. Hendricks' results indicated that the "minimum initial dilutions" associated with the four outfalls ranged from a low of 100 to 1 seawater to effluent, to a high of 290 to 1, for the well-stratified conditions existing at the time of measurement. Using the 100 to 1 level as a hypothetical "minimum initial dilution", the Hyperion and ORCOSAN effluents would not be toxic to sea urchin fertilization and 48-hour development, after initial dilution. The JWPCP present effluent would still be toxic to sea urchins in this test at the 100 to 1 dilution, but may not be toxic at the 135 to 1 dilution that Hendricks (1977) had measured for this outfall in 1976.

The Hyperion digested-sludge discharge is still very toxic as the dilutions at which there were no effects on sea urchin fertilization or development were in the 600-1000 to 1 range. The "minimum average dilution" for the sludge outfall is about 100-150 to 1 (Hendricks, personal communication). Therefore, the sludge discharge and effluent discharge from the Hyperion Treatment Plant should be toxic and non-toxic, respectively, after initial dilution.

Since California has no standard marine bioassay for the measurement of the toxicity of wastewaters discharged into the ocean, the toxicities of such effluents have been, and still are, measured using freshwater fish exposed to freshwater dilutions of effluents (Kopperdahl, 1976). The results of the dischargers' freshwater fish bioassays and this study's sea urchin bioassays appear on Table 4. It appeared that the 48-hour sea urchin fertilization test was five to ten times more sensitive to wastewater effluent toxicity than the freshwater fish test.

The sea urchin test has several advantages when studying wastewater toxicity in the marine environment. It utilizes a marine organism in seawater solutions to measure toxicity and is sensitive to alterations in effluent toxicity that

TABLE 4. Summary of 96-hour LC50 results for freshwater fish, *Pimephales promelas*, (fathead minnows), bioassays with wastewaters (1/78-5/79) and effluent concentrations that caused 50 precent reduction in fertilization of sea urchin, *Strongylocentrotus purpuratus*, eggs.

Present Effluents Tested	Freshwater Bioassay with Fathead Minnows 96-hr LC50 (% effluent)			Marine Bioassay with Sea Urchin Gametes 50% reduction in fertilization occurred between and (% effluent)		
	n	mean	standard error	n		
Hyperion Final	16	98.6[A]	1.01	2	2.6	20
ORCOSAN Final	16	77.1[B]	3.29	3	0.97	7.1
JWPCP Final	9	26.6[C]	2.64	2	0.92	2.6

[A] Hyperion Treatment Plant and Santa Monica Bay 1978 Annual Summary Report and personal communication with Phil Chang.

[B] County Sanitation Districts of Orange County, California: Annual Report 1978, Operations and Marine Monitoring, and personal communication with Monica Farris.

[C] County Sanitation Districts of Los Angeles County, Joint Water Pollution Control Plant Annual Summary, 1979.
County Sanitation Districts of Los Angeles County, Monthly Monitoring Reports (January 1978 to December 1978).

occur when changes in plant operations or source control are implemented. Investigators can measure the toxicities of hypothetical effluents by adding and reducing specific contaminants to wastewaters or by mixing different proportions of primary, secondary, or tertiary treated wastewater.

Further studies will be conducted with the sea urchin fertilization and development test. Seawater sampling to measure toxicity in and near the zones of initial dilution will be conducted using a tracer (e.g. ammonia) in the effluent to confirm dilution measurements. The toxicity of these samples will be measured and compared to similar samples diluted in the laboratory to determine how the laboratory dilutions relate to actual *in situ* dilutions of wastewater and seawater. Seawater samples will also be taken at various depths in coastal regions to measure the natural variability in seawater toxicity, as well as to check for the influence of storm run-off and upwelling. Future tests will compare the sea urchin to other marine species to measure and calibrate interspecific differences in sensitivity to complex effluents and isolated contaminants.

SUMMARY

1. Municipal wastewaters reduced fertilization of urchin eggs at concentrations generally higher than one to seven percent effluent.

2. Using a 100 to 1 seawater to effluent ratio as a hypothetical "minimum initial dilution" of effluents and sludge injected into the ocean, two of three effluents tested did not affect fertilization outside the zone of initial dilution.

3. Digested sludge, as discharged into Santa Monica Bay, California, was very toxic within and near the zone of initial dilution.

4. This urchin fertilization test was 5-10 times more sensitive than the 96-hour freshwater fish bioassay now being used by the dischargers.

ACKNOWLEDGEMENTS

The authors are indebted to Jean L. Wright for expert technical assistance. This project was supported in part by the University of Southern California Marine and Freshwater Biomedical Center, USPH Grant No. 1 P30 ES0 19 65-01 5RC. Southern California Coastal Water Research Project Publication Number 162.

LITERATURE CITED

Cardwell, R. D., S. Olsen, M. I. Carr, and E. W. Sanborn. 1979. Causes of oyster larvae mortality in south Puget Sound. Nat. Oceanic Atmos. Admin. Tech. Memo., ERL MESA-39, 73 pp.

Cardwell, R. D., C. E. Woelke, M. I. Carr, and E. W. Sanborn. 1977a. Evaluation of water quality in Puget Sound and Hood Canal in 1976. Nat. Oceanic Atmos. Admin. Tech. Memo., ERL MESA-21. 36 pp.

Cardwell, R. D. 1977a. Appraisal of a reference toxicant for estimating the quality of oyster larvae. Bull. Environ. Contam. Toxicol. 18: 719-725.

Cardwell, R. D. 1977b. Evaluation of the efficacy of sulfite pulp mill pollution abatement using oyster larvae. In: Aquatic toxicology and hazard evaluation, ASTM STP 634 pp. Ed. by F. L. Mayer and J. L. Hamelink. American Society for Testing and Materials.

Cardwell, R. D. 1978. Variation of toxicity tests of bivalve mollusc larvae as a function of termination technique. Bull. Environ. Contam. Toxicol. 20: 128-134.

Cardwell, R. D. 1979. Toxic substance and water quality effects on larval marine organisms. Washington Dept. of Fisheries, Tech. Rpt. No. 45, 71 pp.

County Sanitation Districts of Los Angeles County, Joint Water Pollution Control Plant Annual Summary. 1979.

County Sanitation Districts of Los Angeles County, Monthly Reports. January 1978 to December 1978.

County Sanitation Districts of Orange County, California, Annual Report 1978.

Hendricks, T. J. 1977. *In situ* measurements of initial dilution. In: Southern California Coastal Water Research Project, Annual Report 1977. El Segundo, CA.

Hinegardner, R. T. 1967. Echinoderms. In: Methods of Developmental Biology. Eds: F. W. Wilt and N. K. Wessels. Thomas Y. Crowell Company, New York.

Hyperion Treatment Plant and Santa Monica Bay 1978 Annual Summary Report-NPDES Permit No. CA0109991 (File No. 1492).

Kobayashi, N. 1971. Fertilized sea urchin eggs as an indicatory material for marine pollution bioassay, preliminary experiments. Publ. Seto Mar. Biol. Lab. 18: 379-406.

Kobayashi, N. 1972. Marine pollution bioassay by using sea urchin eggs in the Inland Sea of Japan (The Seto-Naikai). Publ. Seto Mar. Biol. Lab. 19: 359-381.

Kobayashi, N. 1973. Studies on the effects of some agents on fertilized sea urchin eggs, as part of the bases for marine pollution bioassay. I. Publ. Seto Mar. Biol. Lab. 21: 109-114.

Kopperdahl, F. R. 1976. Guidelines for performing static acute toxicity fish bioassays in municipal and industrial wastewaters. California State Water Resources Control Board report. 65 pp.

Schafer, H. A. 1980. Characteristics of municipal wastewater discharges. In: Southern California Coastal Water Research Project Annual Report 1979-80. Ed. by W. Bascom. Long Beach, California.

Stober, Q. J., P. A. Dinnel, M. A. Wert, D. H. DiJulio, and R. E. Nakatani. 1977. Toxicity of West Point effluent to marine indicator organisms, part II. University of Washington, College of Fisheries, Final Report FRI-UW-7737.

Woelke, C. E. 1972. Development of a receiving water quality bioassay criterion based on the 48-hour Pacific oyster (*Crassostrea gigas*) embryo. Washington Department of Fisheries, Technical Report No. 9. 93 pp.

Young, D. R. 1978. Priority pollutants in municipal wastewaters. In: Southern California Coastal Water Research Project, Annual Report 1978. Ed. by W. Bascom. El Segundo, California.

Section IV

PHYSIOLOGICAL MONITORING

PHYSIOLOGICAL EFFECTS OF SOUTH LOUISIANA
CRUDE OIL ON LARVAE OF THE AMERICAN LOBSTER
(HOMARUS AMERICANUS)

Judith M. Capuzzo
Bruce A. Lancaster

Woods Hole Oceanographic Institution
Woods Hole, Massachusetts

INTRODUCTION

Interest in oil and gas exploration in the coastal regions
of the northwest Atlantic has increased in recent years in
efforts to find petroleum resources. Increased exploitation,
however, poses many unanswered questions concerning potential
toxic effects of oil from drilling and transport operations
on a commercially important species and the resulting economic
impact on established fisheries.

The American lobster (*Homarus americanus*) is found off the
northeastern coast of the United States and off Atlantic
Canada and is of particular ecological and economic importance. The life cycle of the lobster includes both planktonic
and benthic stages (Herrick, 1896, 1911) and, thus, the
lobster may be exposed to a wide range of stress conditions
as a result of oil contamination in the sea. Short-term exposure of planktonic larval stages to oil dispersed in surface
waters could result in reduced survival of larval lobsters,
increased susceptibility to other environmental stresses and
changes in the rates of growth and development. Wells and
Sprague (1976) found stage I lobster larvae to be more sensitive to Venezuelan crude oil than later larval stages and
observed changes in development time to the first post-larval
stage, the occurrence of intermediate larvae and reductions in
food consumption as a result of oil exposure. Forns (1977)
reported a threshold sensitivity between 0.1 and 1.0 ppm South
Louisiana crude oil for planktonic larval lobsters with

reduced feeding and swimming activity and prolonged development time being observed at the higher exposure concentration.

An understanding of the effects of crude oil on development, growth and energetics of larval lobsters is needed to
assess the impact of this marine pollutant on lobster populations. The objectives of this study were: (1) to assess
the physiological effects of short-term exposure of sublethal levels of crude oil on all larval stages of the American lobster; (2) to determine the effects of sublethal exposure on the molt to the postlarval form; and (3) to determine any subsequent changes in biochemical composition
associated with crude oil exposure.

MATERIALS AND METHODS

Female egg-bearing lobsters were obtained from local fish-
ermen and maintained in flowing seawater (30-31 °/oo salinity)
at ambient temperatures on a diet of mussels and squid. The
eggs developed normally and began to hatch when seawater temperatures reached 18-20°C. After hatching, stage I larvae
were transferred to fiberglass plankton-kreisels, described by
Hughes *et al.* (1974) and maintained in flowing seawater on a
diet of live or frozen brine shrimp (*Artemia salina*) until
used in bioassays.

Each larval stage (I-IV) was exposed to 0.25 ppm South
Louisiana crude oil for 96 h at 20°C in a continuous flow bioassay system described by Capuzzo *et al.* (1976). This bioassay system comprises 24-500 ml fleakers[R] (Corning Glass)
modified by the addition of an intake tube and an outflow port
at the 400 ml mark, covered with Nitex[R] screening to allow
continuous flow; seawater-crude oil mixtures were supplied to
the assay chambers by peristaltic pumping and temperature and
flow rates were maintained at constant levels. Seawater or
seawater-crude oil mixtures were delivered from 13-liter glass
carboys; prior to addition to the 13-liter carboys, oil-seawater dispersions (250 ppm) were mixed by gentle stirring for
24 h at 10°C (Anderson *et al.*, 1974a). Aliquots of seawater-
crude oil mixtures were withdrawn by pipet from below the surface of the stock solution and added to the carboys directly
and diluted at the designated concentration. Carboys were
replaced every 6 h and the turnover time in each assay chamber
was 1 h. Hydrocarbon concentrations of oil-seawater dispersions supplying the assay chambers were assayed by gas
chromatography. Water samples were extracted with hexanes and
concentrated using a Bucchi rotary evaporator; gas chromatograms were compared with a sample of South Louisiana crude oil

(Figs. 1 and 2). A comparison of the two chromatograms
indicates that the whole oil fraction rather than just the
water soluble components is present in the seawater-crude oil
mixture. This would appear to be a more realistic present-
ation of oil to the bioassay organisms than presenting only
the water soluble fraction. There is some enhancement
relative to the composition of crude oil at possibly the
naphthalene peak and at the C_{13} peak; identification of the
naphthalene peak is based on comparing its retention time
against a standard. The concentration of total hydrocarbons
during the bioassay experiments was 0.25 ± 0.05 ppm (mean of 6
determinations ± 1 S.D.).

For each larval stage, 5 larvae were added to an assay
unit, maintained on a diet of *Artemia* nauplii to reduce
cannibalism, and monitored for survival, respiration rates,
ammonia excretion rates and O:N ratios at selected times
during the exposure period and compared with control organisms
maintained under identical conditions. Respiration rates of
individual larvae were measured using both microrespirometers
and a Gilson differential respirometer according to the
techniques described by Capuzzo and Lancaster (1979). Equilibrium between the gas and liquid phase of O_2 diffusion in
the respirometer flasks was enhanced by gently shaking the
respirometer flasks; this had no effect on the survival of
larval lobsters. At the end of each set of oxygen uptake
measurements, the seawater in the respirometer flasks was
analyzed for NH_4^+-N by the method of Solorzano (1969) in
order that an *in situ* estimate of ammonia excretion rates and
the O:N ratio (atomic ratio of oxygen consumed to NH_4^+-N excreted) could be made; ammonia levels were compared with control blanks.

At the end of the exposure period, animals were weighed,
dried at 70°C for 24 h and reweighed, and assayed for biochemical composition. The relative percentages of protein,
lipid, carbohydrate, chitin and ash were determined for each
stage of oil-exposed and control lobsters. Protein, carbohydrate, chitin and ash were determined by the methods
described by Raymont *et al*. (1964) using dry tissues; lipid
content was analyzed by the method of Marsh and Weinstein
(1966).

To evaluate the effect of crude oil exposure on the molt
to the postlarval form, stage IV larvae were exposed to 0.25
ppm South Louisiana crude oil for 96 h and respiration rates,
ammonia excretion rates and O:N ratios were monitored daily.
After the 96-h exposure period (Day 4), lobsters were transferred to clean seawater and maintained for 1 week to assess
success of molting to the postlarval form, physiological
changes associated with molting and degree of recovery from

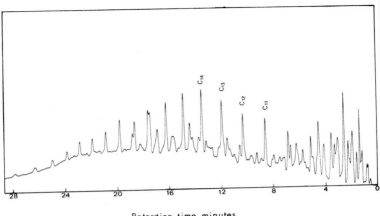

Retention time, minutes

FIGURE 1.   Gas chromatogram of South Louisiana crude oil. Column: Stainless steel, 14 ft., 3% OV-25 on 100/120 Supelcoport® (Supelco, Inc.); Detector: FID; Carrier Gas: N₂; Detector Temperature: 200°C; Injector Temperature: 300°C; Column Temperature: 70°-300°C, programmed at 10°C/minute.

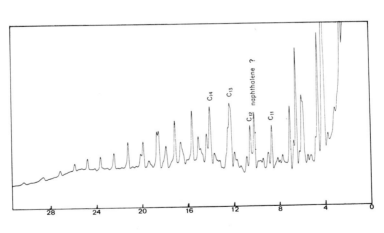

Retention time, minutes

FIGURE 2.   Gas chromatogram of hexane extract of seawater-crude oil mixture; details of analysis as in Figure 1.

oil exposure. During the post-exposure period the various physiological parameters described above were measured at 96 h (Day 8) and 168 h (Day 11). At the end of the post-exposure period, animals were assayed for biochemical composition.

Differences among physiological parameters measured and biochemical composition of the oil-exposed and control lobsters were assessed by analysis of variance (Sokal and Rohlf, 1969).

RESULTS

Survival of larval lobsters exposed to 0.25 ppm South Louisiana crude oil is presented in Table 1. There was no significant difference in survival of lobsters from the control and oil-exposed groups for each larval stage. Stage IV larvae, however, had significantly higher (P<0.01) survival rates in both groups than the other three larval stages; the mortality of stages I, II and III larvae is largely attributed to cannibalism.

Respiration rates measured at 24 h and 72 h during the exposure period were not significantly different from one another for each larval stage in the control group (Fig. 3).

Table 1. Survival of larval lobsters exposed to 0.25 ppm South Louisiana crude oil for 96 h at 20°C

Stage	Control (%)[a]	Oil-exposed (%)[a]
I	43.2 (1.2)	48.9 (1.5)
II	41.3 (1.5)	46.6 (1.3)
III	50.0 (1.8)	51.1 (1.6)
IV	92.7 (1.5)	96.0 (1.5)

[a] Mean value of 12 replicate assays ($\pm$ 1 S. E.).

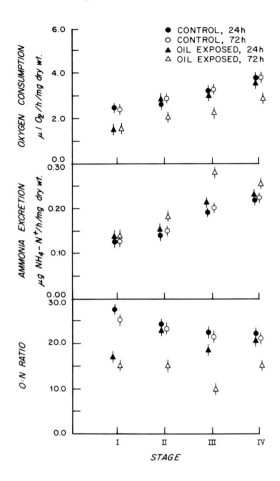

FIGURE 3. Respiration rates, ammonia excretion rates and O:N ratios of larval lobsters exposed to 0.25 ppm South Louisiana crude oil for 96 h at 20°C; each point is the mean value of 8 determinations ± 1 S.E.

The weight-specific respiration rates increased with each successive larval stage as previously reported by Capuzzo and Lancaster (1979); these measurements reflect the level of metabolism during feeding for each larval stage.

With the exception of stage I larvae, there was no signigicant difference in respiration rates at 24 h for any larval stage between the control and oil-exposed groups; however, significant reductions (P<0.01) were observed at 72 h.

Significant reductions in respiration rates of stage I larvae were observed at 24 h and 72 h (Table 2).

Ammonia excretion rates of larval lobsters measured at 24 h and 72 h during the exposure period are presented in Figure 3. There was no significant difference between the two sets of measurements for any of the larval stages in the control group or for stage I and IV lobsters in the oil-exposed group. Oil-exposed stage II and III lobsters had significantly higher (P<0.01) NH_4^+-N excretion rates at 72 h (Table 3).

The O:N ratios of control and oil-exposed lobsters are presented in Figure 3. For control lobsters the O:N ratios ranged from 20.6 to 27.3 and are within the range previously reported for larval lobsters (Capuzzo and Lancaster, 1979). The O:N ratio of oil-exposed lobsters at 24 h ranged from 17.0 to 22.4; however, a greater reduction in O:N ratio was observed at 72 h (range = 9.6-14.9).

In previous work in our laboratory we found that the weight-specific respiration and ammonia excretion rates of lobsters are highest in stage IV larvae and decrease with the molt to the postlarval form (stage V). The results in the experiments with stage IV and V lobsters are in agreement with those findings but the decline in metabolic rates for control lobsters takes place just prior to molting to stage V (Fig. 4). All animals successfully molted to stage V and molting was synchronous for all lobsters on day 5. Metabolic rates were consistent for control lobsters during the post-exposure period; the O:N ratio of stage IV and V lobsters averaged 24.5.

For oil-exposed lobsters respiration rates were significantly lower (P<0.01) during the first three days of exposure and the O:N ratios were significantly reduced; ammonia excretion rates were not significantly different from the rates of control organisms except on day 1 (Fig. 4). At day 4 there was a slight increase in both metabolic parameters relative to control organisms but a reduced O:N ratio was determined. All organisms successfully molted to stage V, but respiration rates and ammonia excretion rates were significantly reduced (P<0.01) during the post-exposure period (Table 4). The O:N ratio of oil-exposed lobsters during the exposure period averaged 16.8 and during the post-exposure period averaged 25.5.

The dry weight and biochemical composition of all larval and postlarval stages from the control and oil-exposed groups are presented in Table 5. There was no significant difference in dry weight of any lobster stage between control and oil-exposed groups. Protein was the major biochemical constituent in all larval stages and the first postlarval stage of both control and oil-exposed groups. For control lobsters there

Table 2.   Anaylsis of variance or respiration rates measured for each larval stage of the American lobster at 24 h and 72 h during exposure to 0.25 ppm South Louisiana crude oil.

Source of variation	S.S.	d.f.	M.S.	F-ratio
Stage I				
24 h - between	4.00	1	4.00	-
within	0.23	14	0.02	-
total	4.23	15	-	248.9[a]
72 h - between	2.56	1	2.56	-
within	0.23	14	0.02	-
total	2.79	15	-	159.3[a]
Stage II				
24 h - between	0.04	1	0.04	-
within	0.23	14	0.02	-
total	0.27	15	-	2.5[b]
72 h - between	2.56	1	2.56	-
within	0.36	14	0.03	-
total	2.92	15	-	99.6[a]
Stage III				
24 h - between	0.16	1	0.16	-
within	0.36	14	0.03	-
total	0.52	15	-	6.2[b]
72 h - between	4.84	1	4.84	-
within	0.36	14	0.03	-
total	5.20	15	-	188.2[a]
Stage IV				
24 h - between	0.04	1	0.04	-
within	0.59	14	0.04	-
total	0.63	15	-	0.9[b]
72 h - between	2.56	1	2.56	-
within	0.56	14	0.04	-
total	3.12	15	-	64.0[a]

[a] $P < 0.01$

[b] Not significant

Table 3.   Analysis of variance of ammonia excretion rates for each larval stage of the American lobster at 24 h and 72 h during exposure to 0.25 ppm South Louisiana crude oil.

Source of variation	S.S.	d.f.	M.S.	F-ratio
Stage I				
24 h - between	-	1	-	-
within	-	14	-	-
total	-	15	-	-
72 h - between	0.0004	1	0.0004	-
within	0.0009	14	-	
total	0.0013	15	-	6.2^a
Stage II				
24 h - between	0.0004	1	0.0004	-
within	0.0009	14	-	-
total	0.0013	15	-	6.2^a
72 h - between	0.0016	1	0.0016	-
within	0.0009	14	-	-
total	0.0025	15	-	24.9^b
Stage III				
24 h - between	0.0036	1	0.0036	-
within	0.0009	14	-	-
total	0.0045	15	-	56.0^b
72 h - between	0.0144	1	0.0144	-
within	0.0022	14	0.0001	-
total	0.0166	15	-	89.6^b
Stage IV				
24 h - between	0.0004	1	0.0004	-
within	0.0022	14	0.0001	-
total	0.0026	15	-	2.5^a
72 h - between	0.0010	1	0.0010	-
within	0.0024	14	0.0001	-
total	0.0034	15	-	6.1^a

[a] Not significant

[b] $P<0.01$

Table 4. Analysis of variance of respiration rates and ammonia excretion rates of stage IV and V lobsters during 96 h exposure to 0.25 ppm South Louisiana crude oil and the post-exposure period.

Source of variation	S.S.	d.f.	M.S.	F-ratio
Respiration rates				
Day 1 - between	0.64	1	0.64	-
within	0.36	14	0.03	-
total	1.00	15	-	24.9^a
Day 2 - between	7.84	1	7.84	-
within	0.36	14	0.03	-
total	8.20	15	-	305.0^a
Day 3 - between	5.76	1	5.76	-
within	0.36	14	0.03	-
total	6.12	15	-	224.0^a
Day 4 - between	1.00	1	1.00	-
within	0.36	14	0.03	-
total	1.36	15	-	38.9^a
Day 8 - between	4.84	1	4.84	-
within	0.36	14	0.03	-
total	5.20	1	-	188.2^a
Day 11- between	2.56	1	2.56	-
within	0.36	14	0.03	-
total	2.92	15	-	99.6^a
Ammonia excretion rates				
Day 1 - between	0.0064	1	0.0064	-
within	0.0036	14	0.0002	-
total	0.0100	15	-	24.9^a
Day 2 - between	0.0004	1	0.0004	-
within	0.0009	14	-	-
total	0.0013	15	-	6.2^b
Day 3 - between	0.0007	1	0.0007	-
within	0.0025	14	0.0001	-
total	0.0032	15	-	4.1^b
Day 4 - between	0.0036	1	0.0036	-
within	0.0009	14	-	-
total	0.0045	15	-	56.0^a
Day 8 - between	0.0144	1	0.0144	-
within	0.0009	14	-	-
total	0.0153	15	-	224.0^a
Day 11- between	0.0100	1	0.0100	-
within	0.0009	14	-	-
total	0.0109	15	-	155.6^a

[a] P<0.01
[b] Not significant

Table 5. Dry weight and biochemical composition of the four larval stages and the first postlarval stage of[a] the American lobster from control and oil-exposed groups[a]

Component	Group	I	II	Stage III	IV	V[b]
dry weight, mg[c]	control	1.0+0.1	1.9+0.1	2.9+0.1	7.2+0.2	9.0+0.2
	oil	1.0+0.1	2.0+0.1	2.8+0.1	7.2+0.3	8.9+0.3
% ash[d]	control	18.4+0.2	19.0+0.4	21.2+0.5	24.5+1.0	23.0+0.5
	oil	18.5+0.3	18.9+0.3	20.5+0.5	22.5+1.0	22.5+0.5
% protein[d]	control	68.8+0.2	65.0+0.8	63.5+1.0	62.0+0.6	62.5+0.8
	oil	72.9+1.0	71.0+0.5	66.3+0.7	64.6+0.6	64.2+0.6
% carbohydrate[d]	control	2.0+0.1	1.9+0.1	1.5+0.1	1.4+0.1	1.4+0.1
	oil	1.9+0.1	1.9+0.1	1.5+0.1	1.4+0.1	1.4+0.1
% lipid[d]	control	5.9+0.2	5.3+0.5	5.4+0.5	4.2+0.3	3.2+0.2
	oil	4.5+0.2	3.9+0.4	4.0+0.5	3.7+0.2	3.1+0.2
protein:lipid	control	11.7	12.3	11.8	14.8	19.5
	oil	16.2	18.2	16.6	17.5	20.7

a Mean Values of 6 determinations $\pm$ 1 S.E.
b Sampled at the end of the post-exposure period
c Dried for 24 h at 60°C
d % dry weight basis

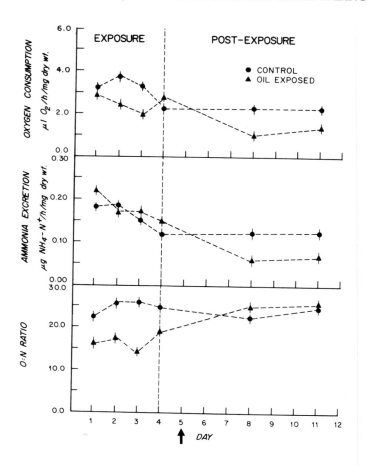

FIGURE 4. Respiration rates, ammonia excretion rates and O:N ratios of stage IV and V lobsters exposed to 0.25 ppm South Louisiana crude oil for 96 h at 20°C and maintained for 1 week following exposure in clean seawater; arrow indicates molt to stage V; each point is the mean of 8 determinations ± 1 S.E.

were significant decreases (P<0.01) in both carbohydrate and lipid content and significant increases in ash and chitin content in the later larval and postlarval stages (stages III, IV and V). Similar trends were observed among oil-exposed lobsters, but there were significant differences in protein and lipid content and the protein:lipid ratio for each larval stage between control and oil-exposed groups; among oil-exposed lobsters there were significantly higher (P<0.01) protein contents and significantly lower lipid contents of all larval stages. Stage V lobsters, sampled at the end of the post-exposure period, had a slightly higher protein content but no significant difference in lipid content than control lobsters.

DISCUSSION

The effects of crude oil and refined oils on the survival, metabolism and energetics of marine crustaceans have recently been investigated (Wells and Sprague, 1976; Edwards, 1978; Laughlin et al., 1978). Other investigators have also observed delayed development of crustaceans as a result of oil exposure (Wells, 1972; Katz, 1973; Lindén, 1976; Wells and Sprague, 1976; Caldwell et al., 1977). With chronic exposure of the mud crab Rhithropanopeus harrisii to water soluble fractions of No. 2 fuel oil, Laughlin et al. (1978) found the early zoeal stages to be the most sensitive and larval development was significantly delayed as a result of oil exposure. The growth rates of crab stages and the size distribution and sex ratio of animals at the end of the 6-month exposure period, however, were not significantly different between control and oil-exposed groups.

Edwards (1978) observed changes in respiration rate, growth rate and net carbon turnover in the sand shrimp, Crangon crangon, with chronic exposure to the water soluble components of North Sea Brent Field crude oil. Acute exposure (24-48 h) of shrimp to low concentrations (5% WSF = 1 ppm) resulted in a reduction in respiration rate but increases and subsequent decreases in respiration rate were observed at higher concentrations (>10% WSF = >2 ppm). Similar changes in respiratory activity were observed by Lee et al (1974b) in the shrimps Penaeus aztecus and Palaemonetes pugio and by Percy (1977) in the amphipod Onisimus (Boekisimus) affinis. Anderson (1977) suggested that these changes in respiratory activity might be related to a metabolic response to the naphthalene component of water soluble fractions.

In the present study, reductions in respiration rates and O:N ratios of all stages of the American lobster were observed with exposure to 0.25 ppm South Louisiana crude oil. This concentration is within the range measured in surface waters beneath recently deposited oil slicks (Grose and Mattson, 1977; Law, 1978; Mackie et al., 1978) and, because of the abundance of all larval stages in near surface waters (Scarratt, 1964, Lund and Stewart, 1970), would be encountered

by larval lobsters immediately following a spill. Stage I
lobsters were the most sensitive of all larval stages with
significant changes in respiration rate occurring within the
first 24 h of the exposure period; significant differences in
respiration rates of other larval stages did not become
apparent until later in the exposure period. Wells and
Sprague (1976) also found stage I larvae to be the most sensitive, using mortality and swimming activity as response
parameters.

The O:N ratio has been used in several studies (Conover
and Corner, 1968; Corner and Cowey, 1968: Bayne and Scullard,
1977; Capuzzo and Lancaster, 1979) as an index of substrate
utilization for energy production and has been shown to vary
with stage of development, diet and degree of physiological
stress. Low O:N ratios (8.0 or below) indicate that an
organism is deriving all of its energy supply from protein
catabolism; higher values reflect increased dependence on
lipid and/or carbohydrate catabolism. The O:N ratios measured
for control lobsters provide an indication that protein
catabolism is the principal source of energy for larval
lobsters but lipid and/or carbohydrate sources are utilized to
some extent. The reduction in O:N ratios of larval lobsters
exposed to crude oil is an indication of increased dependence
on protein catabolism for all larval stages, especially after
72 h exposure. The changes in protein and lipid content and
the protein: lipid ratio is a further indication that there is
a significant shift in both energy storage and utilization
as a result of crude oil exposure. Simultaneous increases in
the rate of protein catabolism and protein content with
reductions in lipid content of oil-exposed larval lobsters
suggest that although protein is being metabolized for immediate energy needs, less protein is being utilized for *de
novo* synthesis of lipid reserves.

A change in energetics as a result of oil exposure may be
most crucial as a larval organism approaches metamorphosis
and begins to shift its habitat preference from pelagic to
benthic; such an impact may affect both larval recruitment
and postlarval survival as energy reserves may not be
sufficient during this transition period. Stage IV lobsters
exposed to 0.25 ppm crude oil showed reductions in the O:N
ratio but successfully molted to the postlarval form. Recovery of lobsters was not immediate upon transfer to clean
seawater; reduced metabolic rates but normal O:N ratios were
observed during the post-exposure period. These findings may
be an indication that although respiration rates and ammonia
excretion rates are reduced relative to control values during
the post-exposure period, the normal pattern of energy
metabolism and utilization is slowly being restored. The

similar protein:lipid ratio of control and oil-exposed post-larval lobsters is further evidence for this conclusion.

Anderson et al. (1974b) suggested that the recovery response of marine crustaceans exposed to oil was related to the release of accumulated hydrocarbons. Lee (1977) discussed the accumulation and turnover of petroleum hydrocarbons in marine crustaceans and Malins (1977) described several mechanisms of bioconversion of hydrocarbons that have been identified in marine organisms. It is apparent that the physiological effects of oil exposure are modified by the ability of the organism to accumulate and detoxify various components of crude and refined oils. Some marine crustaceans possess a mechanism for hydroxylating certain aromatic hydrocarbons (Corner, 1978); this has been recently demonstrated for shrimp and crab larvae (Lee, 1975; Sanborn and Malins, 1977). However, the relationship of accumulation and release of hydrocarbons, the metabolic pathways associated with transformation of hydrocarbons, the toxicities of the transformed hydrocarbons, and the interference with protein and lipid metabolism of marine larval crustaceans, especially those as important as the American lobster, need to be further explored.

SUMMARY

The sublethal effects of exposure to 0.25 ppm South Louisiana crude oil on larvae of the American lobster were evaluated. Respiration rates, ammonia excretion rates and O:N ratios were monitored at 24 h and 72 h during a 96-h exposure period. Significant reductions in respiration rates and O:N ratios were detected for all larval stages after 72 h exposure, indicating an increased demand on protein catabolism as a result of oil exposure.

In further studies the effect of crude oil on the molt of stage IV larvae to the postlarval form (stage V) and the recovery potential of lobsters transferred to clean seawater were investigated. Significant differences in all of the physiological parameters were detected between control and oil-exposed lobsters during exposure and post-exposure periods. Energy metabolism of larval and juvenile lobsters is apparently affected by crude oil exposure in the laboratory and recovery is not immediate in the post-exposure period.

These changes in energetics may become particularly critical during metamorphosis of larval lobsters and the transitional period between pelagic and benthic habits. Understanding these responses of larval lobsters to acute oil

exposures will provide us with a better basis for predicting the overall impact of oil spills on lobster populations.

ACKNOWLEDGMENTS

This research was supported by U. S. Department of
Interior, Bureau of Land Management under Contract No. AA 551-
CT9-5. Woods Hole Oceanographic Institution Contribution No.
4543.

LITERATURE CITED

Anderson, J. W. 1977. Responses to sublethal levels of
    petroleum hydrocarbons: are they sensitive indicators and
    do they correlate with tissue contamination: In: Fate
    and Effects of Petroleum Hydrocarbons in Marine Organisms
    and Ecosystems. pp. 95-114. Ed. by D. A. Wolfe. New
    York: Pergamon Press.

Anderson, J. W., J. M. Neff, B. A. Cox, H. E. Tatem, and G. M.
    Hightower. 1974a. Characteristics of dispersions and
    water-soluble extracts of crude and refined oils and their
    toxicity to estuarine crustaceans and fish. Mar. Biol.
    27: 75-88.

Anderson, J. W., J. M. Neff, B. A. Cox, H. E. Tatem, and G. M.
    Hightower. 1974b. The effects of oil on estuarine
    animals: toxicity, uptake and depuration, respiration.
    In: Pollution and Physiology of Marine Organisms. pp.
    285-310. Ed. by F. J. Vernberg and W. B. Vernberg. New
    York: Academic Press.

Bayne, B. L. and C. Scullard. 1977. Rates of nitrogen ex-
    cretion by species of Mytilus (Bivalvia:Mollusca). J.
    Mar. Biol. Ass. U.K. 57: 355-369.

Caldwell, R. S., E. M. Calderone, and M. H. Mallon. 1977.
    Effects of seawater-soluble fraction of Cook Inlet crude
    oil and its major aromatic components on larval stages of
    the Dungeness crab, Cancer magister Dana. In: Fate and
    Effects of Petroleum Hydrocarbons in Marine Ecosystems.
    pp. 210-220. Ed. by D. A. Wolfe. New York: Pergamon
    Press.

Capuzzo, J. M. and B. A. Lancaster. 1979. Some physiological and biochemical considerations of larval development in the American lobster, *Homarus americanus* Milne Edwards. J. Exp. Mar. Biol. Ecol. 40: 53-62.

Capuzzo, J. M., S. A. Lawrence and J. A. Davidson. 1976. Combined toxicity of free chlorine, chloramine and temperature to stage I larvae of the American lobster *Homarus americanus*. Water Research 10: 1093-1099.

Conover, R. J. and E. D. S. Corner. 1968. Respiration and nitrogen excretion by some marine zooplankton in relation to their life cycles. J. Mar. Biol. Ass. U.K. 48: 49-75.

Corner, E. D. S. 1978. Pollution studies with marine plankton. Part 1. Petroleum hydrocarbons and related compounds. Adv. Mar. Biol. 15: 289-380.

Corner, E. D. S. and C. B. Cowey. 1968. Biochemical studies on the production of marine zooplankton. Biol. Rev. 43: 393-426.

Edwards, R. R. C. 1978. Effects of water-soluble oil fractions on metabolism, growth and carbon budget of the shrimp *Crangon crangon*. Mar. Biol. 46: 259-265.

Forns, J. M. 1977. The effects of crude oil on larvae of lobster *Homarus americanus*. Proceedings, Oil Spill Conference (Prevention, Behavior, Control, Cleanup). EPA/API/USCG. pp. 569-573.

Grose, P. L. and J. S. Mattson. 1977. The *Argo Merchant* oil spill, a preliminary scientific report. U.S. Dept. of Commerce, NOAA, Boulder, Colorado.

Herrick, F. H. 1896. The American lobster: a study of its habits and development. Bull. U.S. Fish. Comm. 15: 1-252.

Herrick, F. H. 1911. Natural history of the American lobster. Bull. U.S. Bur. Fish. 29: 149-408.

Hughes, J. R., R. A. Shleser and G. Tchobanoglous. 1974. A rearing tank for lobster larvae and other aquatic species. Prog. Fish Cult. 36: 129-132.

Katz, L. M. 1973. The effects of water-soluble fractions of crude oil on larvae of the decapod crustacean *Neopanope texana* (Say). Environ. Pollut. 5: 199-204.

Laughlin, R. B., Jr., L. G. L. Young, and J. M. Neff. 1978. A long-term study of the effects of water-soluble fractions of No. 2 fuel oil on the survival, development rate, and growth of the mud crab *Rhithropanopeus harrisii*. Mar. Biol. 47: 87-95.

Law, R. J. 1978. Petroleum hydrocarbon analyses conducted following the wreck of the supertanker *Amoco Cadiz*. Mar. Pollut. Bull. 9: 293-296.

Lee, R. F. 1975. Fate of petroleum hydrocarbons in marine zooplankton. In: Proc. Conf. on Prevention and Control of Oil Pollution. American Petroleum Institute, Washington, D.C. pp. 549-553.

Lee, R. F. 1977. Accumulation and turnover of petroleum hydrocarbons in marine organisms. In: Fate and Effects of Petroleum Hydrocarbons in Marine Organisms and Ecosystems. pp. 60-70. Ed. by D. A. Wolfe. New York: Pergamon Press.

Lee, W. Y., K. Winters, and J. A. C. Nicol. 1978. The biological effects of the water-soluble fractions of a No. 2 fuel oil on the planktonic shrimp, *Lucifer faxoni*. Environ. Pollut. 15: 167-183.

Lindén, O. 1976. Effects of oil on the amphipod *Gammarus oceanicus*. Environ. Pollut. 10: 239-245.

Lund, W. A. and L. L. Stewart. 1970. Abundance and distribution of larval lobsters, *Homarus americanus,* off the coast of southern New England. Proc. Nat'l. Shellfish. Assn. 60: 40-49.

Mackie, P. R., R. Hardy, E. I. Butler, P. M. Holligan, and M. F. Spooner. 1978. Early samples of oil in water and some analyses of zooplankton. Mar. Pollut. Bull. 9: 296-297.

Malins, D. C. 1977. Biotransformation of petroleum hydrocarbons in marine organisms indigenous to the Arctic and Subarctic. In: Fate and Effects of Petroelum Hydrocarbons in Marine Organisms and Ecosystems. pp. 47-59. Ed. by D. A. Wolfe. New York: Pergamon Press.

Marsh, J. B. and D. B. Weinstein. 1966. Simple charring method for determination of lipids. J. Lipid Res. 7: 574-576.

Percy, J. A. 1977. Effects of dispersed crude oil upon the respiratory metabolism of an Arctic marine amphipod, *Onisimus (Boekisimus) affinis*. In: Fate and Effects of Petroleum Hydrocarbons in Marine Organisms and Ecosystems. pp. 192-200. Ed. by D. A. Wolfe. New York: Pergamon Press.

Raymont, J. E. G., J. Austin and E. Linford. 1964. Biochemical studies on marine zooplankton. I. The biochemical composition of *Neomysis integer*. J. Cons. perm. int. Explor. Mer 28: 354-363.

Sanborn, H. R. and D. C. Malins. 1977. Toxicity and metabolism of naphthalene: a study with marine larval invertebrates. Proc. Soc. Exp. Biol. Med. 154: 151-155.

Scarratt, D. J. 1964. Abundance and distribution of lobster larvae (*Homarus americanus*) in Northumberland Strait. J. Fish. Res. Bd. Canada 21: 661-680.

Sokal, R. R. and F. J. Rohlf. 1969. Biometry. San Francisco: W. H. Freeman and Co. 776 pp.

Solorzano, L. 1969. Determination of ammonia in natural waters by the phenolhypochlorite method. Limnol. Oceanogr. 14: 799-801.

Wells, P. G. 1972. Influence of Venezuelan crude oil on lobster larvae. Mar. Pollut. Bull. 3: 105-106.

Wells, P. G. and J. B. Sprague. 1976. Effects of crude oil on American lobster (*Homarus americanus*) larvae in the laboratory. J. Fish. Res. Bd. Canada 33: 1604-1614.

EXCRETION OF AROMATIC HYDROCARBONS
AND THEIR METABOLITES
BY FRESHWATER AND SEAWATER DOLLY VARDEN CHAR

Robert E. Thomas

Department of Biological Science
Chico State University
Chico, California

Stanley D. Rice

Northwest and Alaska Fisheries Center
Auke Bay Laboratory
National Marine Fisheries Service
P. O. Box 155
Auke Bay, Alaska

INTRODUCTION

Aquatic animals are often contaminated by petroleum hydrocarbons after oil spills, and many field and laboratory
studies have been made to measure uptake and loss of petroleum
hydrocarbons from their tissues. Most of these studies have
been with fish and have focused primarily on the accumulation
of petroleum hydrocarbons labeled with isotopes and identification of the tissues with major accumulation of isotopes
(Lee *et al.*, 1972; Korn *et al.*, 1976, 1977; Roubal *et al.*,
1977, 1978; Melancon and Leach, 1978). Usually the metabolism
or biotransformation of these hydrocarbons has not been
measured, but a few recent studies have been concerned with
the ability of organisms to metabolize aromatic hydrocarbons
(Pedersen *et al.*, 1974; Payne, 1976; Gruger *et al.*, 1977a, b;
Varanasi *et al.*, 1978; Statham *et al.*, 1978; Collier *et al.*,
1978; Malins *et al.*, 1979). In these studies, hydrocarbons
were rapidly assimilated into fish tissues, then rapidly

depurated. However, metabolites of petroleum hydrocarbons
persist in some tissues longer than the parent hydrocarbons
(Lee et al., 1976; Roubal et al., 1977; Varanasi et al.,
1978).

Several factors probably affect the uptake, metabolism,
and discharge of petroleum hydrocarbons from fish tissues, but
only the effect of molecular size of aromatic hydrocarbons has
been studied. An increase in size of aromatic hydrocarbon
molecules (increased number of aromatic rings or substitution
on aromatic rings) has been known to increase toxicity (Rice
et al., 1977a; Caldwell et al., 1977; Rossi and Neff, 1978).
As the number of aromatic rings increases, the accumulation
and retention of parent compound and metabolites by coho
salmon, Oncorhynchus kisutch, also increases (Roubal et al.,
1977).

Environmental factors probably influence the ability of
fish to accumulate and rid themselves of petroleum hydrocarbons. At lower temperatures, coho salmon fry had greater
retention of napthalene in several tissues (Collier et al.,
1978) and pink salmon, Oncorhynchus gorbuscha, fry had lower
survival after exposure to aromatic hydrocarbons (Korn et al.,
1979; Thomas and Rice, 1979). Increased salinity during
exposure of several species of euryhaline fish to aromatic
hydrocarbons or water-soluble fractions of crude oil has also
caused lower survival (Moles et al., 1979; Levitan and Taylor,
1979), but the effect of salinity changes on petroleum-hydro-
carbon uptake, metabolism, distribution in tissues, or
elimination has not been studied.

In this study, we determined: 1) the major excretory
routes of ^{14}C napthalene, ^{14}C toluene, and their metabolites
from Dolly Varden char (Salvelinus malma); 2) the percentage
of carbon-14 metabolites of ^{14}C toluene and ^{14}C naphthalene
excreted by each major excretory pathway; 3) the relationship
between tissue retention and excretion of ^{14}C toluene, ^{14}C
naphthalene, their metabolites; and 4) whether environmental
salinity affected any of these parameters.

We chose toluene (one aromatic ring) and naphthalene (two
aromatic rings) because they are present in significant
quantities in the water-soluble fraction of crude oil (Rice
et al., 1979; Korn et al., 1979) and are representative of
mono- and diaromatic hydrocarbons, which are believed to be
major contributors to the toxicity of this fraction (Moore
and Dwyer, 1974; Anderson1 1977; Rice et al., 1977a). We
chose Dolly Varden char because this species is euryhaline
after smoltification. In southeastern Alaska, older juveniles
and adults migrate from overwintering lakes to seawater in
spring (Armstrong, 1965); and in summer, they are typically
found in seawater near river mouths where they are alternately

exposed to freshwater and seawater daily. Dolly Varden char are very mobile in the summer and may feed in seawater and in freshwater drainages that are not connected to the over-wintering lakes (Armstrong, 1965).

METHODS

During June and July, Dolly Varden char were beach-seined in seawater near Auke Bay, Alaska. Only juvenile fish (30-50 g each) were collected because they were small enough for the experiments and mature enough to be euryhaline. The fish were maintained for a minimum of 20 days in aquaria with running seawater (30 $^{\circ}/_{\circ\circ}$) or freshwater. Water temperature was 8.7°C ± 1°C. The Dolly Varden char were fed salmon roe daily. All feeding was stopped 72 h before exposure to ^{14}C toluene or ^{14}C naphthalene.

The urinary bladders of four groups of 10 fish each were catheterized. These fish were held in freshwater or in sea-water in a "split box" exposure chamber (Figure 1) that isolated various excretory pathways from each other. The concentration of excreted carbon-14 was measured by liquid-scintillation spectrometry from samples of water taken from the gill or cloacal areas and from samples of urine taken from a catheter. Tissue samples were taken from all fish after 24-h exposures. The percentage of the recovered carbon-14 in polar metabolites was determined in water, urine, and

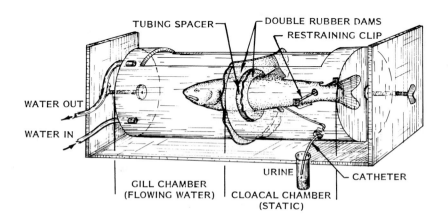

FIGURE 1. "Split-box" chamber to confine fish and separate avenues of excretion.

tissues. Differences in the carbon-14 activity of the various
samples from the four groups of fish were analyzed statistically by a one-way analysis of variance and Student's $\underline{t}$
test.

Preparation and Dosing of the Fish

On the day before the experiment, the fish were lightly
anesthetized with [1]MS-222 (0.125 g/l). A catheter was
inserted in the ureter and was held there with a silk
ligature looped around the urinary papillae. A ligature
through the caudal peduncle kept the fish in the chamber
(Fig. 1). While the fish were anesthetized, we constantly
irrigated water over the gills. The fish were then placed in
holding chambers and allowed to acclimate overnight. On the
morning of the experiment, the fish were again lightly
anesthetized with MS-222 and a gelatin capsule containing sublethal quantities (0.1-0.3 mg) of either ^{14}C toluene
(specific activity of 3.67 µCi/µmole) dissolved in corn oil
was inserted into the stomach. By administering the carbon-14
labeled hydrocarbon in a capsule, we could detect the capsule
if it were regurgitated. Two of the 60 fish regurgitated the
capsules, probably because the capsules were not properly
inserted in the stomach. It is possible that some of the
capsule solution could have been regurgitated later in the
experiment. If this did occur, we would have detected
abnormally high concentrations in the gill chamber or abnormally low concentrations in the tissues; however, neither
abnormal concentration occurred. After insertion of the
capsule, the fish were placed in the "split box" exposure
chamber with the anterior end through two rubber dams that
separated the gill and cloacal areas (Fig. 1). A small air
space between the rubber dams prevented any leakage between
the two chambers. Water flow through the gill chamber
(approximately 250 ml/min) was constantly monitored. Water
was not circulated in the cloacal chamber. The temperatures
in the gill chamber and the cloacal chamber were the same
(10.3°C ± 0.5 C°) because the whole "split box" chamber was
partially submerged in a temperature-control bath.

Measurement of Excreted Carbon-14

We stopped the water flow in the gill chamber for 15 min
and measure the amount of carbon-14 excreted by the gills
during the 15 min by quantifying the net accumulation of
carbon-14 in a water sample from the static gill chamber.
(From preliminary tests of oxygen depletion for varying
lengths of time for 60 g fish, we found that oxygen con-

centrations would not drop below 75% saturation in the gill chamber in 15 min.) The water samples from the gill chambers were collected at 2, 4, 6, 8, 12, and 24 h after placement of the capsules in the fish stomach. Carbon-14 excreted by the cloaca was measured by quantifying the accumulated carbon-14 in water samples taken from the cloacal chamber after 8 h and 24 h. Urine was collected with a catheter that emptied into a covered 12 or 25 ml conical-bottom centrifuge tube. The amount of carbon-14 excreted in the urine during the entire 24 h was quantified. Only a small amount of urine was collected during this period: less than 16 ml was collected from any one fish. Evaporation of labeled compounds from the urine was minimal because the collection tubes, which have a small exposed surface area, were covered; and the urine in them was maintained at 10°C by partial immersion in the water bath.

Tissue Sampling

After 24 h, the fish were killed with MS-222 and weighed. The following organs and tissues were removed and weighed: gill, heart, liver, stomach, intestine, gall bladder, brain-spinal cord, and muscle from the area below and lateral to the dorsal fin. After removal of these tissues, the head and reproductive tissues were removed and the remaining carcass weighed. The carcass weight was used to estimate total muscle mass. The immature reproductive organs were not sampled because they were small and variable in size. To determine the amount of carbon-14 in each tissue, we incinerated half of each sample in a Packard[1] sample oxidizer (model 306). The resulting $^{14}CO_2$ was dissolved in Carbosorb-Permafluor[1] for subsequent counting in a Beckman[1] liquid-scintillation system.

Measurement of Metabolites

To determine the percentage of carbon-14 present in the tissues as metabolites, we digested in papain the other half of the tissue sample and extracted the carbon-14 (Roubal *et al.*, 1977). This method separates the polar metabolites from the nonpolar parent hydrocarbons. Carbon-14 activity in aliquots of the nonpolar, hexane phase (parent hydro-carbons) and in aliquots of the aqueous, alkaline phase (polar metabolites) was measured with a Packard[1] Dimilume 30 scintillator.

Urine samples and water samples from the gill and cloacal chambers were extracted in the same manner as tissue samples, and the carbon-14 activity was determined for aliquots of the

hexane and aqueous phases. From the results, we were able to
determine the percentage of carbon-14 recovered as polar
metabolite.

RESULTS

Carbon-14 in Excretions

    Dolly Varden char in both freshwater and seawater
primarily excreted carbon-14 from the gills (Table 1): up to
20% of the administered activity was accounted for in the
gill chamber compared to maxima of 0.13% in the urine and 3.4%
in the cloacal chamber.
    We were unable to collect urine for the entire 24-hr
period from all experimental fish because of problems with
catheters. Urine samples were obtained for the total period
in only three to five fish per group. Volume of urine
collected from the fish during the 24-h period averaged 1.3
ml for fish kept in seawater and 15.7 ml for fish kept in
freshwater. Fish in seawater excrete less urine than fish in
freshwater because fish in seawater have decreased glomerular-
filtration rates. Less than 0.01% of the administered
carbon-14 was recovered in the urine of all groups except the
freshwater group exposed to ^{14}C toluene. In the freshwater
group exposed to ^{14}C toluene, 0.13% of the administered
activity was recovered in the urine (Table 1).
    The amount of carbon-14 recovered from the cloacal
chamber was small: about one-tenth of the carbon-14 excreted
from the gills (Table 1). Because the urinary-bladder
catheters failed and leaked in 5-7 fish per group, the carbon-
14 recovered from the cloacal chamber reflects not only gut
excretion but also small amounts of carbon-14 from the urine.
The quantities of carbon-14 recovered from the cloacal
chamber were about 10-1,000 times greater than quantities of
carbon-14 found in the urine, indicating that urine contamination was probably insignificant. Insoluble fecal
material was noted infrequently (less than 10% of the
chambers). The lack of fecal excretion probably resulted
from the nature of the animals' diet and lack of food for 72
h before the test. Because fecal matter was not usually found
in the cloacal chamber, no attempt was made to analyze the
feces independently. Carbon-14 recovered from the cloacal
chamber was probably somewhat less than the actual excretion
via this pathway because some volatile compounds were
probably lost from the static chamber. We believe these
losses were not large because the chamber was chilled and
the amount of carbon-14 in the 8-h sampling was proportion-

ately less than the amount of carbon-14 in the 24-h sampling (Table 1).

In general, fish exposed to ^{14}C toluene excreted more carbon-14 than fish exposed to ^{14}C naphthalene. About twice as much carbon-14 was recovered in the gill chamber after exposures to ^{14}C toluene as was recovered after exposure to ^{14}C naphthalene (Table 1). The gills excreted a peak concentration of carbon-14 from ^{14}C toluene after 6 to 8 h; however, they continued to excrete carbon-14 from ^{14}C naphthalene at a generally increasing rate over the entire 24-h test period (Fig. 2). Although the total amount of carbon-14 recovered in the urine was relatively small, more carbon-14 was recovered after exposure to ^{14}C toluene than after exposure to ^{14}C naphthalene. Fish excreted about three times more carbon-14 from the cloaca after exposure to ^{14}C toluene than after exposure to ^{14}C naphthalene.

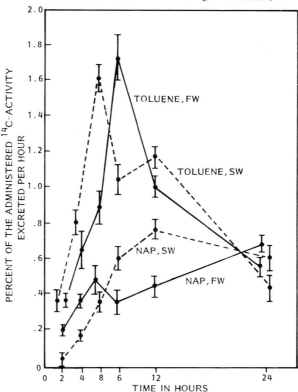

FIGURE 2. Gill excretion of ^{14}C toluene and ^{14}C naphthalene by Dolly Varden char in seawater and freshwater during first 24 h. Vertical bars indicate 95% confidence interval.

TABLE 1. Recovery of excreted carbon-14 after 24 h expressed as a percentage of administered. ^{14}C toluene or ^{14}C naphthalene had been administered orally to Dolly Varden char acclimated in seawater (SW) or freshwater (FW). Numbers in parenteses represent 8-h values.

Excreted carbon-14 as a percentage of administered

Excretory pathway	Toluene		Naphthalene	
	FW	SW	FW	SW
Gill chamber	20.3	19.2	10.8	12.6
Cloacal chamber	2.3 (0.4)	3.4 (1.6)	0.6 (0.1)	1.1 (0.4)
Urine	0.13	0.008	0.009	0.003
Total excreted in 24 h	22.7	22.6	11.4	13.7

Neither seawater nor freshwater environment had significant effect on excretion of carbon-14. Seawater fish excreted slightly less carbon-14 from ^{14}C toluene and slightly more carbon-14 from ^{14}C naphthalene into the gill chamber than freshwater fish. Seawater fish excreted slightly more carbon-14 from the cloaca than freshwater fish, whereas freshwater fish excreted more carbon-14 into the urine than seawater fish.

Metabolites in Excretions

The percentage of recovered carbon-14 excreted as metabolites depended on the excretory pathway, the toxicant, and the salinity of the environment. Less than half the excreted carbon-14 was in the metabolites (Table II). Although a large amount of carbon-14 was excreted by the gills, only a small amount was excreted as metabolites. In contrast, the percentage of carbon-14 recovered in metabolites from the cloacal chamber and urine was high (37-97%), but the total carbon-14 excreted by these pathways was small compared to the total carbon-14 excreted by the gills.

More carbon-14 metabolites from toluene were excreted into the gill and cloacal chambers than carbon-14 metabolites from naphthalene (Table II). In the urine, the percentages of carbon-14 from metabolites of ^{14}C toluene and ^{14}C naphthalene were about equal.

Freshwater and seawater environments also affected the excretion of metabolites (Table II). Fish in freshwater excreted more ^{14}C naphthalene metabolites by all three excretory pathways than fish in seawater. Freshwater fish excreted more toluene metabolites in the urine and cloacal chamber than seawater fish; however, freshwater fish excreted a smaller quantity of toluene metabolites into the gill chamber than seawater fish.

Carbon-14 in Tissues

The amount of carbon-14 retained in the tissues at 24 h was quite variable, depending on tissue, toxicant, and salinity of the environment (Table III). Stomachs and intestines had variable amounts of recovered carbon-14, apparently because these tissues contained large quantities of isotope that had not been absorbed into the blood. The amounts of carbon-14 retained in the stomach and intestine are useful only to account for the carbon-14 administered to the animals. The amount of carbon-14 in the gill tissues and in the heart varied considerably with individuals, probably because varying amounts of blood were left in the hearts and

TABLE 2. Percentage of recovered carbon-14 existing as metabolites in the excretions of Dolly Varden char acclimated in seawater (SW) and freshwater (FW) at 24-h after oral administration of ^{14}C toluene and ^{14}C naphthalene. Numbers in parentheses represent the 8-h values.

Percentage of recovered carbon-14 in excretions as metabolites

Excretory pathway	Toluene		Naphthalene	
	FW	SW	FW	SW
Gill chamber	12.7	28.4	10.2	6.5
Cloacal chamber	89.6 (86.5)	62.5 (44.3)	47.7 (28.7)	37.7 (24.7)
Urine	90.3	60.3	97.1	55.8
Total percentage of excreted carbon-14 appearing in metabolites	26.7	32.3	13.6	11.9

gills. The stomach contents often contaminated the gills when tissues were removed. Significant quantities of carbon-14 were found in the muscle because of the muscle's large mass. Other tissues, such as gall bladder and brain-spinal cord, contained higher concentrations of carbon-14 than muscle tissue, but because of their small mass compared to muscle, the total amount of isotope retained was small.

Fish exposed to ^{14}C toluene retained less carbon-14 in the tissues than fish exposed to ^{14}C naphthalene (Table II and III), except in the heart and gall bladder. For all tissues combined, 1.5 to 4 times more carbon-14 was retained after exposure to ^{14}C naphthalene than after exposure to ^{14}C toluene.

The seawater environment had little effect on tissue retention of carbon-14 (Table III). The total carbon-14 recovered in the tissues after administration of ^{14}C toluene was the same for freshwater and seawater fish, but seawater fish retained more carbon-14 after administration of ^{14}C naphthalene. The retention of carbon-14 by the tissues was not consistent; however, the muscle and brain-spinal cord (two of the more stable and important tissues) of seawater fish consistently retained more carbon-14 after administration of ^{14}C toluene and ^{14}C naphthalene than the muscle and brain-spinal cord of freshwater fish.

Carbon-14 Metabolites in Tissues

At 24 h, many tissues contained more metabolites than parent compounds (Table IV). As expected, the gall bladder had the highest concentration of the carbon-14 in metabolites although the total amount of carbon-14 recovered was small. Although muscle did not have the highest percentage of carbon-14 in the metabolite phase, muscle had more mass than any other tissue; thus, the total amount of metabolite was significant. The percentage of carbon-14 in the metabolites for muscle varied from 21.5 to 52%. The percentage of carbon-14 in the metabolites for brain-spinal cord was similar to muscle with a range of 16.1 to 46.8%. Stomachs and intestines had relatively low concentrations of metabolite--further indication that the isotope remaining in these tissues had not yet entered the blood.

For all tissues combined, a greater percentage of carbon-14 metabolites was obtained from ^{14}C toluene than from ^{14}C naphthalene (Table IV). All tissues in the freshwater fish had more metabolites of ^{14}C toluene than metabolites of ^{14}C naphthalene. For some tissues of seawater fish, this pattern varied but not in the more stable and important tissues such as muscle, brain-spinal cord, and liver.

TABLE 3. Total recovery of carbon-14 from eight tissues of Dolly Varden char, as a percentage of administered ±95% confidence interval (95% CI). ^{14}C toluene or ^{14}C naphthalene had been administered orally to char acclimated in seawater (SW) and freshwater (FW) 24 h before sampling. Significant difference at $P \geq 0.01$ between: a. freshwater naphthalene and saltwater naphthalene.

Tissue	Average weight (g)	Toluene				Naphthalene			
		FW (N=9)	95% CI	SW (N=9)	95% CI	FW (N=10)	95% CI	SW (N=10)	95% CI
Intestine	1.26	0.89	0.20	0.77	0.26	5.18	2.0	3.17	1.01
Stomach	1.01	7.69	1.67	6.71	1.74	8.61[a]	2.2	10.39	5.1
Heart	0.11	0.02	0.006	0.01	0.003	0.007[a]	0.001	0.15	0.05
Muscle	41.8	0.84	0.16	1.22	0.21	1.53[a]	0.34	12.98	1.7
Gill	1.27	0.65	0.18	1.02	0.24	0.87[a]	0.34	11.94	2.4
Gall bladder	0.08	0.17	0.09	0.25	0.13	0.84	0.01	0.12	0.02
Liver	0.69	0.11	0.02	0.07	0.007	0.21	0.08	0.36	0.06
Brain and spinal cord	0.24	0.015	0.004	0.028	0.006	0.015[a]	0.002	0.053	0.007
Total recovery (%)		10.4		10.1		16.5		39.21	

The tissues of freshwater fish retained more ^{14}C toluene
metabolites and slightly less ^{14}C naphthalene metabolites
than the tissues of seawater fish (Table IV). However, the
percentage of toluene and naphthalene metabolites varied
considerably from one tissue to the next. For example, the
gills and stomach of seawater fish has less toluene
metabolites and more naphthalene metabolites than the gills
and stomachs of freshwater fish. The brain-spinal cord of
freshwater fish, however, consistently had more of both
toluene and naphthalene metabolites than the brain-spinal cord
of seawater fish.

DISCUSSION

The total carbon-14 recovered from tissues and excretions
varied between 21.5% and 53% of the administered doses
(Table V). A number of factors contribute to the low recovery
values, but the most important source of error is probably the
estimate of carbon-14 remaining in the tissues after 24 h.
The estimate for tissues was definitely low because several
tissues were not sampled, i.e., reproductive organs, spleen,
kidney, bone, blood, eye, head, skin and associated mucus.
The mass of these organs is less than 20% of the total mass of
the fish, suggesting an estimate error of less than 20%. However, some of the tissues may have had relatively high
specific activities, and the error may have been larger.
Some of the excreted carbon-14 in the static cloacal
chamber may not have been recovered. Some toluene and
naphthalene, which are slightly volatile, may have evaporated,
even though the temperature was 10°C. Microbes probably
converted some of the ^{14}C naphthalene and ^{14}C toluene to $^{14}CO_2$.
The skin, exposed in the cloacal chamber, may have absorbed
some of the excreted carbon-14, but conditions were not
favorable for this because of the concentration gradient
across the skin after the oral exposure and the impervious
nature of the skin. Although some carbon-14 must have been
lost from the cloacal chamber, accumulation of excreted
carbon-14 in the chamber was almost linear, as determined at
8 h and 24 h; therefore, the losses were not extreme.
Although there was virtually no evaporative loss or
microbial degradation of excreted carbon-14 from the gill
chamber during the 15-min stop-flow sampling periods, the
extrapolated estimates of carbon-14 excreted from the gills
were probably low. During the first 8 h after exposure,
carbon-14 was recovered from the gill chamber for 15 min of
every 2 h. Although peak excretion for toluene was within

TABLE 4. Percentage of recovered carbon-14 activity present in the metabolite fractions of eight tissues from Dolly Varden char exposed in freshwater (FW) and seawater (SW). Char received ^{14}C toluene or ^{14}C naphthalene orally 24 h before sampling. Significant differences at $P > 0.01$ between: a. FW naphthalene and SW naphthalene and SW napthalene, and b. FW toluene and SW toluene.

Tissue	Average weight (g)	Toluene				Naphthalene			
		FW (N=9)	95% CI	SW (N=9)	95% CI	FW (N=10)	95% CI	SW (N=10)	95% CI
Intestine	1.26	38.9[b]	6.0	49.9	6.0	7.2[a]	1.9	18.5	2.6
Stomach	1.01	21.1[b]	5.7	7.6	1.1	9.2[a]	4.1	14.0	2.3
Heart	0.11	43.6	7.0	43.6	5.8	24.3	5.6	21.6	3.3
Muscle	41.8	62.0	7.5	60.8	7.3	22.8	3.4	21.5	2.9
Gill	1.27	23.9[b]	4.9	11.0	1.9	8.8[a]	2.6	17.7	2.8
Gall bladder	0.08	76.4[b]	7.7	52.6	7.5	76.8	6.2	74.1	6.2
Liver	0.69	42.3[b]	4.6	40.4	4.0	11.0[a]	1.9	26.3	3.6
Brain and spinal cord	0.24	46.8[b]	5.2	28.9	4.7	19.5[a]	3.1	16.1	1.7
Recovered carbon-14 appearing in metabolites (%)		44.4		36.8		22.5		26.2	

TABLE 5. *Comparison of excretion and tissue retention of carbon-14 by Dolly Varden char acclimated in seawater (SW) and freshwater (FW) at 24 h after oral administration of ^{14}C toluene and ^{14}C naphthalene.*

Total recovery of carbon-14 as percentage of administered

	Toluene		Naphthalene	
	FW	SW	FW	SW
Total excretion	22.7	22.6	11.4	13.7
Total tissue retention	10.4	16.5	10.1	39.2
Total accounted for	33.1	39.1	21.5	52.9

this 8-h period, the highest excretion rates were probably
not during out 15-min sampling period. When the water
flow was stopped, the rate the fish excreted carbon-14 may
have decreased (although the percentage of oxygen saturation
was monitored during the 15-min static period and was
determined not to have been reduced to stressful levels).

In our study, carbon-14 from ^{14}C toluene and ^{14}C
naphthalene was excreted primarily from the gills. Carbon-14
was recovered from the urine and the cloacal chamber during
the 24 h; but this amount, plus the carbon-14 in the gall
bladder awaiting excretion via the gut, was less than 20% of
the carbon-14 excreted via the gills. Fish exposed to crude
oil depurate hydrocarbons quite rapidly (Roubal et al., 1977;
Rice et al., 1977b); and, presumably, the aromatic hydrocarbons with low molecular weight, like toluene and naphthalene, are lost primarily via the gills as shown in our
study. High-molecular-weight aromatic hydrocarbons are
probably lost via the gills also; however, metabolism of high-
molecular-weight hydrocarbons by the liver and excretion of
other metabolites via the gall bladder-cloaca may be more
significant than excretion of the aromatic hydrocarbons via
the gills.

Although gill excretion of petroleum hydrocarbons and
metabolites was suggested by Varanasi et al. (1979), the bile
and urine of fish have been generally recognized as the major
avenues for the excretion of hydrocarbons and metabolites
(Lee et al., 1972; Varanasi et al., 1979). The evidence for
bile and urine as the major excretory pathways has not come
from studies measuring quantities of eliminated hydrocarbons
and their metabolites, but from the accumulation of a high
specific activity of labeled isotope in bile and urine,
primarily as metabolite (Lee et al., 1972; Roubal et al.,
1977; Melancon and Leach, 1978; Stathem et al., 1978; Collier
et al., 1978; Malins et al., 1979). Indeed our studies have
also revealed a high specific activity of labeled isotope in
liver, bile, and urine. However, when one considers the
volume of bile and urine (we collected all the urine for 24 h
and all the bile present at 24 h), the total amount of carbon-
14 excreted by the bile and urine was less than the total
amount of carbon-14 excreted by the gills, despite the high
specific activity in bile and urine. In fact, most of the
metabolites in the bile may not be excreted via the gut to the
cloaca. It is possible that the metabolites could be re-
absorbed into the blood from the intestines and redistributed
to the tissues or eliminated via the gills.

The skin and epidermal mucus of fish may also excrete
petroleum hydrocarbons. Naphthalene and its metabolites
accumulate in the skin and mucus of rainbow trout exposed

to naphthalene in water or fed naphthalene (Varanasi et al., 1978). Because mucus is continuously sloughed from the fish body, some excretion of petroleum hydrocarbons via mucus is implied. Varanasi et al. (1978) did not quantitate the significance of this pathway nor did we. In our study, however, excretion via the skin and mucus appeared to be of only minor significance compared to excretion via the gills.

In our studies, force feeding the hydrocarbon also favored its loss via the gills. When the fish is exposed to petroleum hydrocarbons in the water, the major uptake of hydrocarbons is probably via the gills (Lee et al., 1972). When the blood-water concentration gradient is reversed, which occurs when the hydrocarbon is in the food or is force fed, much of the parent hydrocarbon absorbed from the gut would be lost via the gills.

Twice as much carbon-14 from ^{14}C toluene as carbon-14 from ^{14}C naphthalene was excreted by all pathways. There are several reasons why toluene is excreted more rapidly than naphthalene. Because toluene is more polar than naphthalene, it is more soluble than naphthalene in the water environment of a cell. For this reason, toluene can probably travel relatively freely within a cell and between cells rather than being sequestered as tightly as naphthalene in lipid tissues throughout the body. Consequently, toluene is released into clean water by the gills more quickly than naphthalene.

In our study, Dolly Varden char metabolized the ^{14}C toluene and ^{14}C naphthalene. Some of the metabolites were excreted, and some were present in the tissues at 24 h. Metabolism was not surprising because the livers of several species of fish, including salmonids, contain aryl hydrocarbon hydroxylase (AHH), an enzyme system capable of metabolizing petroleum hydrocarbons (Pederson et al., 1974; Payne and Penrose, 1975; Gruger et al., 1977a, 1977b; Statham et al., 1978; Gerhart and Carlson, 1978). Recently, we found AHH activity in the livers of Dolly Varden char at our laboratory (Rice and Collodi, unpublished). With AHH activity in the liver, it is not surprising to find a high percentage of the carbon-14 from metabolites in the gall bladder and in the cloacal chamber.

Although urine excretion was small, the proportion of carbon-14 metabolites in the urine was quite high. There may be some AHH activity in the kidney or, as previously mentioned, the metabolites may be excreted by the kidneys. Low levels of AHH activity have been detected in the kidneys of fish (James et al., 1979). Although some carbon-14 from metabolites was found in the gill chamber, most recovered carbon-14 was still attached to the parent hydrocarbon; therefore, the metabolites leaving the gills were probably

extracted from the blood and were not generated at the gill
site.

More toluene was metabolized than naphthalene. The
percentage of carbon-14 from metabolites of toluene excreted
or in the tissues at 24 h was greater than the percentage of
carbon-14 from metabolites of naphthalene. Because toluene is
more polar than naphthalene, toluene may be metabolized more
easily than naphthalene, thus explaining some of these
differences. Indeed, the high levels of isotope activity
found in both the metabolite and nonmetabolite fractions of
the gill chamber, cloacal chamber, and urine of fish exposed
to toluene support this assumption.

Metabolites of aromatic hydrocarbons are probably toxic
to fish, but parent hydrocarbons are probably the major cause
of death in short-term exposures. Within a short-term
exposure, metabolites probably contribute to the toxic load
fish must endure, but apparently the metabolites are not as
toxic as the parent hydrocarbons, nor are metabolites
synthesized fast enough to accumulate in significant concentration within critical tissues. Sims and Grover (1974)
have reported long-term cytotoxic effects from aromatic
hydrocarbon metabolites formed in mammalian systems. Since
some metabolites persist in fish tissues longer than the
parent hydrocarbons (Varanasi et al., 1978; Melancon and Lech,
1978), the potential for long-term cytotoxic effects from the
metabolites exists in fish, especially in long-term exposures
where the tissues would be continuously exposed to
metabolites.

The most obvious difference between freshwater and seawater fish exposed to toluene and naphthalene appears to be in
the fish's ability to metabolize these compounds. In our
study, the muscle and neural tissues of freshwater fish
contained more carbon-14 metabolites of ^{14}C toluene and ^{14}C
naphthalene than muscle and neural tissues of seawater fish.
Also, more carbon-14 from toluene and naphthalene was recovered in the brain-spinal cord of seawater fish than in the
brain-spinal cord of freshwater fish. The increased toxicity
of toluene and naphthalene to fish in higher salinities
(Levitan and Taylor, 1979; Moles et al., 1979) may be caused
by the inability of the fish to metabolize the hydrocarbons;
therefore, the parent compounds are retained in these
important tissues at higher concentrations and for longer
periods of time.

The toxicity of aromatic hydrocarbons and water-soluble
fractions of oil to fish is rapidly apparent, suggesting the
nervous system is affected. Symptoms of toxicity develop
within 6-18 h, and most deaths occur during the first 24 h.
Reflex activity and equilibrium are lost prior to death

(Dixit and Anderson, 1977; Moles *et al.*, 1979). Apparently, the primary toxic action is in the neural tissue. Neuro-sensory organs undergo histopathological changes after sub-lethal exposure to naphthalene (DiMichele and Taylor, 1978). Dixit and Anderson (1977) have correlated the loss of equilibrium with concentrations of naphthalene in the brain of exposed Fundulus.

        Although in several studies aromatic hydrocarbons have been found in brains of fish (Korn *et al.*, 1976, 1977; Dixit and Anderson, 1977; Collier *et al.*, 1978; Varanasi *et al.*, 1979), hydrocarbon metabolites in the brain have been found in only a few studies (Roubal *et al.*, 1977; Collier, 1978; this study). Significant concentrations of hydrocarbon-metabo-lizing enzyme systems (AHH) have never been found in neural tissue and would not be expected, because brain tissue generally has a low level of metabolic activity. The hydro-carbon metabolites were probably formed elsewhere, presumably the liver, and were transported to the brain tissue via the blood. The blood-brain barrier may exclude some compounds, but apparently the metabolites of toluene and naphthalene were not excluded.

        Isolation of a fish in a metabolism chamber, such as used in this study, imposes some degree of unquantitated stress, even though our fish were "acclimated" overnight in the chambers before administering the toxicant and no fish died in the chambers. In a more recent study, we force fed isotope-labeled naphthalene and toluene to Dolly Varden char via capsules and then returned the fish to their normal holding tanks where they swam free for various periods prior to tissue sampling. Recovery of the isotopes from the tissues and the distribution of the isotopes between parent hydrocarbon and its metabolites (Thomas and Rice, in prep.) was very similar to the distribution of isotopes reported for this study. Because tissue deposition of hydrocarbons and metabolites was similar in these two studies, the excretory patterns are probably similar.

SUMMARY

        The gills were the most important pathway for excretion of carbon-14 from ^{14}C naphthalene. Most of the carbon-14 excreted by the gills was still attached to the parent compound. About 10% of the excreted carbon-14 appeared in the cloacal chamber, mostly as metabolites. Less than 1% of the total carbon-14 was excreted in the urine, predominantley as metabolites.

Tissues retained a significant amount of carbon-14 at 24 h. Although muscle contained large amounts of carbon-14 because of its mass, the gall bladder had the highest specific activity. The brain also retained significant quantities of carbon-14.

Although more ^{14}C toluene was excreted and metabolized than ^{14}C naphthalene, more ^{14}C naphthalene was retained in the tissues. A lower percentage of the carbon-14 was recovered in ^{14}C naphthalene metabolites than in ^{14}C toluene metabolites.

Seawater and freshwater Dolly Varden char excreted similar amounts of carbon-14; however, the percentage of metabolites in the excretions and tissues of seawater fish was lower than the percentage of metabolites in excretions and tissues of freshwater fish. For example, we recovered greater amounts of carbon-14 with a lower percentage of metabolites from the brain-spinal cord of seawater fish than from the brain-spinal cord of freshwater fish--possibly explaining why seawater Dolly Varden are more sensitive to aromatic hydrocarbons and the water-soluble fraction of oil than freshwater Dolly Varden.

LITERATURE CITED

Anderson, J. W. 1977. Effects of petroleum hydrocarbons on the growth of marine organisms. Rapp. P.-v. Réun. Cons. int. Explor. Mer, 155-165.

Armstrong, R. H. 1965. Some migratory habits of the anadromous Dolly Varden *Salvelinus malma* (Walbaum) in southeastern Alaska. 36 p. Seattle: Univ. Washington.

Caldwell, R. W., E. M. Caldarone, and M. H. Mallon. 1977. Effects of a seawater-soluble fraction of Cook Inlet crude oil and its major aromatic components on larval stages of the Dungeness crab, *Cancer magister* Dana. In: Fate and Effects of Petroleum Hydrocarbons in Marine Ecosystems and Organisms. pp. 210-220. Ed. by D. A. Wolfe, New York: Pergamon Press.

Collier, T. K. 1978. Disposition and metabolism of naphthalene in rainbow trout (*Salmo gairdneri*). M.S. Thesis. Seattle: Univ. Washington.

Collier, T. K., L. C. Thomas, and D. C. Malins. 1978. Influence of environmental temperature on disposition of dietary naphthalene in coho salmon *(Oncorhynchus kisutch)*: isolation and identification of individual metabolites. Comp. Biochem. Physiol. 61C: 23-28.

Dimichele, L., and M. H. Taylor. 1978. Histopathological and physiological responses of *Fundulus heteroclitus* to naphthalene exposure. J. Fish. Res. Board Can. 35: 1060-1066.

Dixit, D., and J. W. Anderson. 1977. Distribution of naphthalenes within exposed *Fundulus similus* and correlations with stress behavior. In: Proc. 1977 Oil Spill Conf. (Prevention, behavior, control, cleanup). pp. 633-636. Washington, D.C.: American Petroleum Institute.

Gerhart, E. H., and R. M. Carlson. 1978. Hepatic mixed-function oxidase activity in rainbow trout exposed to several polycyclic aromatic compounds. Environ. Res. 17: 284-295.

Gruger, E. H., Jr., M. M. Wekel, P. T. Numoto, and D. R. Craddock. 1977a. Induction of hepatic aryl hydrocarbon hydroxylase in salmon exposed to petroleum dissolved in seawater and to petroleum and polychlorinated biphenyls, separate and together, in food. Bull. Environ. Contam. Toxic. 17: 512-520.

Gruger, E. H., Jr., M. M. Wekell, and P. A. Robish. 1977b. Effects of chlorinated biphenyls and petroleum hydro-carbons on the activity of hepatic aryl hydrocarbon hydroxylase of coho salmon *(Oncorhynchus kisutch)* and chinook salmon *(O. tashawytscha)*. In: Fate and effects of petroleum hydrocarbons in marine organisms and ecosystems. pp. 323-331. Ed. by D. A. Wolfe. New York: Pergamon Press.

James, M. O., M. A. Q. Khan, and J. R. Bend. 1979. Hepatic microsomal mixed-function oxidase activities in several marine species common to coastal Florida. Comp. Biochem. Physiol. 62C: 155-164.

Korn, S., N. Hirsch, and J. W. Struhsaker. 1976. Uptake, distribution, and depuration of ^{14}C-benzene in northern anchovy, *Engraulis mordax*, and striped bass, *Morone saxatilis*. NOAA (U.S.), Fish. Bull. 74: 545-551.

Korn, S., N. Hirsch, and J. W. Struhsaker. 1977. The uptake, distribution, and depuration of ^{14}C benzene and ^{14}C toluene in Pacific herring, *Clupea harengus Pallasi*. NOAA (U.S.), Fish. BUll. 75: 633-636.

Korn, S., D. A. Moles, and S. D. Rice. 1979. Effects of temperature on the median tolerance limit of pink salmon and shrimp exposed to toluene, naphthalene, and Cook Inlet crude oil. Bull. Environ. Contam. Toxicol. 21: 521-525.

Lee, R. F., R. Sauerheber, and G. H. Dobbs. 1972. Uptake, metabolism and discharge of polycyclic aromatic hydro- carbons by marine fish. Mar. Biol. 17: 201-208.

Lee, R. F., C. Ryan, and M. L. Neuhauser. 1976. Fate of petroleum hydrocarbons taken up from food and water by the blue crab (*Callinectes sapidus*). Mar. Biol. 37: 363-370.

Levitan, W. M., and M. H. Taylor. 1979. Physiology of salinity-dependent naphthalene toxicity in *Fundulus heteroclitus*. J. Fish. Res. Board. Can. 36: 615-620.

Malins, D. C., T. K. Collier, L. C. Thomas, and W. T. Roubal. 1979. Metabolic fate of aromatic hydrocarbons in aquatic organisms: analysis of metabolites by thin-layer chro- matography and high-pressure liquid chromatography. Intern. J. Environ. Anal. Chem. 6: 55-66.

Melancon, M. J., Jr., and J. J. Lech. 1978. Distribution and elimination of naphthalene and 2-methylnaphthalene in rainbow trout during short- and long-term exposures. Arch. Environ. Contam. Toxicol. 7: 207-220.

Moles, A., S. D. Rice, and S. Korn. 1979. Sensitivity of Alaskan freshwater and anadromous fishes to Prudhoe Bay crude oil and benzene. Trans. Am. Fish. Soc. 108: 408-414.

Moore, S. F., and R. L. Dwyer. 1974. Effects of oil on marine organisms: a critical assessment of published data. Water Res. 8: 819-827.

Payne, J. F. 1976. Field evaluation of benzopyrene hydro- xylase induction as a monitor for marine petroleum pollution. Sci. 191: 945-946.

Payne, J. F., and W. R. Penrose. 1975. Induction of aryl
    hydrocarbon (benzo[a]pyrene) hydroxylase in fish by
    petroleum. Bull. Environ. Contam. Toxicol. 14: 112-116.

Pedersen, M. G., W. K. Hershberger, and M. R. Juchau. 1974.
    Metabolism of 3, 4-benzpyrene in rainbow trout (*Salmo
    gairdneri*). Bull. Environ. Contam. Toxicol. 12:
    481-486.

Rice, S. D., A. Moles, T. L. Taylor, and J. F. Karinen. 1979.
    Sensitivity of 39 Alaskan marine species to Cook Inlet
    crude oil and no. 2 fuel oil. In: 1979 Oil Spill
    Conference. pp. 549-554.

Rice, S. D., J. W. Short, and J. F. Karinen. 1977a. Com-
    parative oil toxicity and comparative animal sensitivity.
    In: Fate and effects of petroleum hydrocarbons in marine
    ecosystems and organisms. pp. 78-94. Ed. by D. A. Wolfe.
    New York: Pergamon Press.

Rice, S. D., R. E. Thomas, and J. W. Short. 1977b. Effect
    of petroleum hydrocarbons on breathing and coughing
    rates and hydrocarbon uptake-depuration in pink salmon
    fry. In: Physiological responses of marine biota to
    pollutants. pp. 259-277. Ed. by F. J. Vernberg, A.
    Calabrese, F. P. Thurberg, and W. B. Vernberg. New
    York: Academic Press.

Rossi, S. S., and J. M. Neff. 1978. Toxicity of polynuclear
    aromatic hydrocarbons to the polychaete *Neanthes
    arenaceodentata*. Mar. Poll. Bull. 9: 220-223.

Roubla, W. T., T. K. Collier, and D. C. Malins. 1977.
    Accumulation and metabolism of carbon-14 labeled benzene,
    naphthalene, and anthracene by young coho salmon
    (*Oncorhynchus kisutch*). Arch. Environ. Contam. Toxicol.
    5: 513-529.

Roubal, W. R., S. I. Stranahan, and D. C. Malins. 1978.
    The accumulation of low molecular weight aromatic hydro-
    carbons of crude oil by coho salmon *Oncorhynchus kisutch*)
    and starry flounder (*Platichthys stellatus*). Arch.
    Environ. Contam. Toxicol. 7: 237-244.

Sims, P., and P. L. Grover. 1974. Epoxides in polycyclic
    aromatic hydrocarbon metabolism and carcinogenesis. Adv.
    Cancer Res. 2: 165-274.

Statham, C. N., C. R. Elcombe, S. R. Szyjka, and J. J. Lech.
    1978. Effect of polycyclic aromatic hydrocarbons on
    hepatic microsomal enzymes and disposition of methyl
    naphthalene in rainbow trout *in vivo*. Xenobiotica 8:
    65-71.

Thomas, R. E., and S. D. Rice. 1979. The effect of exposure
    temperatures on oxygen consumption and opercular
    breathing rates of pink salmon fry exposed to toluene,
    naphthalene, and water-soluble fractions of Cook Inlet
    crude oil and no. 2 fuel oil. In: Marine pollution:
    functional responses. pp. 39-52. Ed. by W. B. Vernberg,
    A. Calabrese, F. P. Thurberg, and F. J. Vernberg. New
    York: Academic Press.

Varanasi, U., M. Uhler, and S. I. Stranahan. 1978. Uptake
    and release of naphthalene and its metabolites in
    skin and epidermal mucus of salmonids. Toxicol.
    Applied Pharmacol. 44: 277-289.

Varanasi, U., D. J. Gmur, and P. A. Treseler. 1979.
    Influence of time and mode of exposure of biotrans-
    formation of naphthalene by juvenile starry flounder
    (Platichthys stellatus) and rolk sole *(Lepidopsetta
    bilineata)*. Arch. Environ. Contam. Toxicol. 8: 673-692.

EFFECTS OF PETROLEUM HYDROCARBONS ON THE GROWTH AND ENERGETICS OF MARINE MICROALGAE

James E. Armstrong[1]

Florida Department of Environmental Regulation
Tallahassee, Florida

Steven W. G. Fehler

NOAA Environment Assessment Division
Washington, D.C.

John A. Calder

NOAA, Bering Sea-Gulf of Alaska Project
Juneau, Alaska

INTRODUCTION

Crude oil and refined petroleum products are complex hydrocarbon mixtures which interact with various cellular processes in different ways to produce toxicity to marine microalgae.

Previous studies in the field and laboratory have shown that the addition of petroleum and refined petroleum products severely impact the initially exposed populations of marine phytoplankton (Gordon and Prouse, 1973; Kauss et al., 1973; Pulich et al., 1974; Winters et al., 1976; Lee and Anderson, 1977).

[1]present address: TransAgra Corporation, Memphis, Tennessee.

In general, the relative molar toxicity of hydrocarbons varies inversely with water solubility such that the high molecular weight relatively water-insoluble hydrocarbons are actually more toxic on a molar basis (Calder and Lader, 1976). This general relationship holds very well in algal toxicity assays with pure compounds (Hutchinson *et al.*, 1979). However, variations in sensitivity of microalgal species to particular oils or refined products suggests that the initial toxicity of petroleum water-soluble fractions is probably a function of specific molecules, most likely the nitrogen, sulfur, and oxygen containing heterocyclics (Griffin and Calder, 1977). Parker *et al.* (1976) found that the asphaltics, oxygen and nitrogen containing compounds, which made up about 0.5% of the whole fuel oils analyzed, account for at least 20% of the water-soluble fraction.

The present studies were undertaken to investigate the sublethal effects of petroleum and component hydrocarbons on the growth and physiology of representative blue-green, green, and diatom species of marine microalgae. In order to gain insight into the mechanisms of petroleum toxicity to microalgae, we studied short and long term effects of oil water-soluble fractions and specific compounds on the growth, electron transport systems, and kinetics of selected enzymes. These experiments sugggest two different, although not mutually exclusive, biotoxic mechanisms of petroleum water-solubles to marine microalgae. Our studies indicate the toxic mechanisms of petroleum hydrocarbons to microalgae are an immediate inhibition of electron transport systems and/ or uncoupling of phosphorylation and a long term growth inhibition which may be caused by hydrocarbon uptake into the membranes.

MATERIALS AND METHODS

Organisms and Growth Conditions

Axenic cultures of microalgae used in this study were obtained from C. Van Baalen, The University of Texas Marine Science Institute. *Chlorella autotrophica* strain 580 and *Dunaliella tertiolecta* strain Dun are green microalgae originally isolated by R. R. L. Guillard. The estuarine diatom *Cylindrotheca* sp. strain N-1 was isolated by Morgan (1975) and the coccoid blue-greens *Agmenellum guadruplicatum* strain PR-6 and *Coccochloris elabens* strain Di were isolated by Van Baalen (1962).

Bacterized cultures of *Nannochloris* sp. strain Nan and *Cyclotella meneghiniana* strain CM-2 were obtained from R. L. Iverson, Department of Oceanography, Florida State University.

Algal strains were grown at $30\pm0.1°C$ in ASP-2 medium (Provasoli *et al.*, 1957; Van Baalen, 1962) containing 8 µg vitamin B_{12}/liter and 1 mg vitamin B_1/liter. The diatom N-1 was cultured in the same medium supplemented with 250 mg $Na_2SiO_35H_2O$/liter and 100 mg NH_4Cl/liter. Microalgae were cultured in 25 x 200 mm Pyrex culture tubes containing 30 ml of medium using the test tube method of Myers (1950). Cell suspensions were continuously bubbled with air enriched to approximately 1% CO_2 (v/v). Illumination from four F96T12/WW or F30T12/WW fluorescent lamps, two on each side of the water bath, provided 140 and 180 µEinsteins/m^2/sec of photosynthetically active radiation (400-700 nm) respectively, as estimated with a Li-Cor Model LI-185 photometer equipped with an LI-192S underwater quantum sensor (Lambda Instruments Corp.). Light/dark cycled cultures were isolated from ambient light in a black cloth-covered cabinet and automatically cycled (12 h light, 12 h dark) using an Intermatic timer (Scientific Products, Inc.).

Stock cultures were maintained under continuous illumination on the appropriate medium solidified with 1% agar (Difco 0140). For preparation of the experimental inocula, a loop of cells was transferred from agar slants, grown up, and back transferred a minimum of two times in liquid culture under both continuous and cycled illumination.

Growth was routinely monitored turbidimetrically with a Bausch and Lomb Spectronic 20 colorimeter set at 600 nm. A linear relationship between biomass and optical density was verified.

The specific growth rate constant, k, is expressed in terms of $\log_{10}$ units per day, as defined by the growth equation $\log_{10}N/N_o=kt$ (Kratz and Myers, 1955). In these units, a doubling time of 24 hours corresponds to k=0.301. The relative standard deviation, s/x, of k for any given culture treatment grown months apart was less than 0.10 (i.e., <10%).

Test Oils and Preparation of
Water-Soluble Extracts

No. 2 fuel oil, Southern Louisiana Crude and Kuwait Crude are American Petroleum Institute reference oils and were obtained from J. Anderson, Texas A&M University.

The water soluble components of each oil were extracted by addition of 80 ml of oil to 720 ml of culture medium in a 1000 ml equilibrating flask. Flasks were stirred for 24 hours at room temperature (23±2°C), at a rate which avoided formation of an oil-water emulsion. The mixture then was allowed to equilibrate without stirring for 12 hours. The aqueous phase was then drawn off into a Pyrex bottle by means of a stopcock at the base of the equilibrating flask.

1-Methylnaphthalene saturated media were prepared following the procedure of Eganhouse and Calder (1976). A ground-glass stoppered one-liter flask, containing a teflon-coated magnetic stirring bar and 500 ml of culture medium, was autoclaved and allowed to cool. Excess (100 µl) of filter-sterilized 1-methylnaphthalene (Aldrich Chemical Co.) was added to the flask and stirred for 24 hours at room temperature (23±2°C). After a further 12-hour unstirred equilibration period, the saturated solution was aseptically drawn through a stopcock at the base of the flask. Saturated 1-methylnaphthalene media were aseptically diluted to 500 ml volumes and distributed to autoclaved growth tubes. The concentration of 1-methylnaphthalene in saturated media was determined by the method of Calder and Lader (1976).

Biochemical and Chemical Assays

Oxygen exchange of stirred algal cell suspensions was measured with a Clark-type electrode (OX-15250, Gilson Medical Electronics) in a thermostatted (30°C) chamber. Changes in the electrode current were amplified with a Keithly 414S Picoammeter and recorded on a Houston Instruments Omniscribe recorder. Photosynthetically saturating light was provided by a Dukane projector with a Sylvannia CEW-150 watt projection bulb (the maximum output being 1900 µEinsteins/m^2/sec). The output of the electrode was calibrated by the method of Robinson and Cooper (1970).

ATP was extracted from algal cells according to Brezonik *et al.*, (1975). ATP concentrations in the algal extracts were estimated by luciferin-luciferase luminescence using desiccated firefly extract (Sigma FLE-250).

Alkaline phosphatase and phosphodiesterase activities were estimated in whole cell suspensions and in cell extracts. Hydrolysis of p-nitro-phenylphosphate and bis-p-nitro-phenylphosphate with 20 units of commercial alkaline phosphatase added, was measured kinetically at 400 nm in thermostatted (30°C) cuvettes at pH 9.3, to yield estimates of the alkaline phosphatase and phosphodiesterase activities, respectively. The enzymes were induced by phosphate starvation for 48 hours.

Protein in the cell extracts and enzyme preparations was estimated by the method of Lowry *et al.*, (1951) using bovine serum albumin (Sigma A-7511) as a standard.

RESULTS AND DISCUSSION

The effects of water-soluble hydrocarbons of No. 2 Fuel
Oil, Kuwait Crude, and Southern Louisiana Crude oils on the
growth rates of six marine microalgal strains are shown in
Table 1. The growth rates of the green microalga *Chlorella
autotrophica* strain Ind 580 and the diatom *Cylindrotheca* sp.
strain N-1 were not decreased by the petroleum water-solubles
at the concentrations tested. As previous studies have shown,
No. 2 Fuel Oil is one of the more toxic petroleum products to
microalgae (Pulich *et al.*, 1974). *Nannochloris* sp. strain Nan
and *Cyclotella meneghiniana* strain CM-2 were inhibited by
water-solubles of Southern Louisiana Crude as well as No. 2
Fuel Oil.

The cellular ATP content or pool may be considered a
sensitive indicator of pollutant stress (Brezonik *et al.*,
1975; Clegg and Koevenig, 1974) and nutrient deficiency (Holm-
Hanson, 1970). The data presented in Figure 1 represents

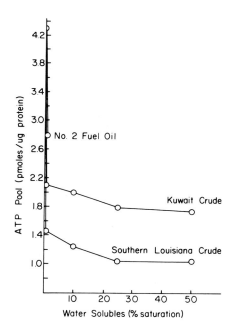

*FIGURE 1. The effect of water-solubles of No. 2 Fuel
Oil, Southern Louisiana Crude, and Kuwait Crude on the ATP
pools of Agmenellum quadruplicatum strain PR-6. Strain PR-6
failed to grow in cultures containing No. 2 Fuel Oil water-
solubles above 1% saturation.*

TABLE 1. The effect of petroleum water-solubles on the growth rates of marine microalgae.

% Saturation of Oil Water-Solubles		Growth Rate (log_{10} units/day)					
		Strain Designation[a]					
		PR-6	Di	Ind 580	Nan	N-1	CM-2
Control		2.34	1.86	1.16	0.61	1.02	0.82
No. 2 Fuel Oil:	1%	2.31	1.78	1.19	0.60	1.05	0.86
	10%	--b	1.67	1.21	0.55	1.01	0.77
	25%	--	0.85	1.24	--	1.02	--
	50%	--	0.85	1.27	--	0.99	--
Kuwait Crude	1%	2.40	1.76	1.15	0.57	0.98	0.88
	10%	2.36	1.88	1.21	0.61	1.04	0.84
	25%	2.36	1.77	1.21	0.62	1.06	0.85
	50%	2.35	1.85	1.18	0.58	1.06	0.85
Southern Louisiana Crude:	1%	2.37	1.86	1.25	0.58	1.04	0.85
	10%	2.33	1.86	1.21	0.50	1.03	0.74
	25%	2.35	1.86	1.24	0.28	1.04	0.63
	50%	2.42	2.08	1.09	--	1.01	--

Notes: a. PR-6, Agmenellum quadruplicatum; Di, Coccochloris elabens; Ind 580, Chlorella autotrophica; Nan, Nannochloris sp.; N-1, Cylindrotheca sp.; CM-2, Cyclotella meneghiniana;

b. Indicates no growth within 3 days.

the ATP pools of *Agmenellum quadruplicatum* strain PR-6 cells cultured in medium 1-50% saturated with the oil water-solubles. The growth of the blue-green coccoid *A. quadruplicatum* PR-6 was not affected by Kuwait or Southern Louisiana Crudes, however, the ATP pools were reduced to less than 50% and 25% of the control values in the presence of the respective crude oil extracts. The effect of No. 2 Fuel Oil on the growth rates of *A. quadruplicatum* PR-6, *Coccochloris elabens* Di, and *Nannochloris* sp. Nan as a function of the cell ATP pools are given in Figure 2. The blue-green strains PR-6 and Di suffered no retardation in growth rate until the ATP pools decreased to less than 2 pmoles/μg protein, less than one-half the normal cell ATP concentration. Likewise, the ATP pools of *Nannochloris* were decreased to less than one-half the control values before a decrease in the growth rate was observed. However, this generalization did not hold when applied to the other algal strains and oils. No oil specific or algal-strain specific correlation between ATP pool reduction and a decrease in growth rate was evident.

Photosynthetic oxygen evolution by various microalgae is quite sensitive to the toxic actions of particular oils (Pulich *et al.*, 1974; Armstrong and Calder, 1978). We followed the rapid oscillations in the cellular ATP pools upon

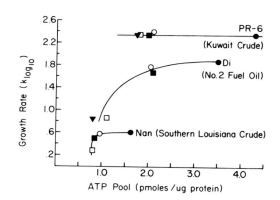

FIGURE 2.   *The effects of water-solubles of No. 2 Fuel Oil, Southern Louisiana Crude, and Kuwait Crude on the growth rates (k) of Agmenellum quadruplicatum strain PR-6, Coccochloris elabens strain Di, and Nannochloris sp. strain Nan as a function of the cellular ATP pools. Control (o), 1% water-solubles (●), 10% water-solubles (■), 25% water-solubles (□), and 50% water-solubles (▼).*

exposure of the treated cells to photosynthetically saturating
light to determine whether the immediate toxic effects of oil
water-solubles was uncoupling of phosphorylation or simply an
inhibition of electron transport (Fig. 3). The diatom
Cylindrotheca sp. strain N-1 was dosed with water-solubles of
No. 2 Fuel Oil 30 seconds prior to light exposure. After a
further 22 seconds in the dark, the ATP pool of the experimental cells was reduced to about 66% of the control cells.
Oxygen consumption of cell suspensions tested concurrently
was increased slightly suggesting an uncoupling of respiratory
phosphorylation. Upon exposure to light the ATP levels
oscillated with a periodicity of about 15 seconds in the control cultures which was decreased in the oil water-solubles
treated cells after one or two oscillation periods. We found
Chlorella autotrophica strain Ind 580 ATP levels to oscillate
with a periodicity of approximately 45 seconds. Lewenstein
and Schneider (1971) found the ATP pools of *Chlorella
pyrenoidea* to have a 40 second oscillation period. The
oscillations in algal ATP pools upon exposure to bright light
are presumed to be due to an imbalance in phosphorylation and
ATP utilization caused by the dark to light transition stress.
In general, the cell ATP levels found in the short-term experiments were about the same as the ATP levels of the algal
strains exposed to petroleum water-solubles for 24-36 hours
in the long-term growth studies. This suggests that both the

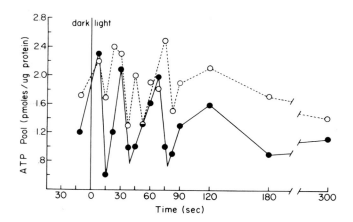

*FIGURE 3. Oscillations of the ATP pools of Cylindrotheca
sp. strain N-1 caused by dark-light transitions as effected
by exposure to No. 2 Fuel Oil water-solubles. Control (O),
treated cells (●).*

short-term and the long-term toxic mechanisms of the oils may be the same for microalgal strains which showed little or no decrease in the rate of growth yet had reduced ATP pools.

Alkaline phosphatase and phosphodiesterase hydrolyze extracellular phosphomonoesters and phosphodiesters, respectively. The released orthophosphate is then taken up by the cells. These enzyme systems are induced by inorganic phosphate deficiency. Water-solubles of Bunker C Crude, Kuwait Crude, and No. 2 Fuel Oil increased the initial reaction velocities of both enzymes from the diatom *Cyclotella meneghiniana* strain CM-2 while Southern Louisiana Crude decreased the reaction rate of alkaline phosphatase but had no measurable effect on phosphodiesterase activity (Fehler and Calder, 1976). Alkaline phosphatase activity assayed in whole cells of *Cyclotella meneghiniana* was 2.3 times greater than in extracted enzyme preparations but the reproducability of the assay was poor. The phosphodiesterase activities, however, were virtually identical *in vivo* and *in vitro*. Several compounds found in the water-soluble fraction of No. 2 Fuel Oils and shown to inhibit the growth of various microalgae (Winters *et al.*, 1976, Batterton *et al.*, 1978) were tested for their effects on these enzymes (Table 2). α-Isopropyl-phenol, 2,4-dimethylphenol, and N,N-dimethylaniline increased the initial velocity of the phosphodiesterase reaction (7.8-12.1%) when added at 5 mg/liter. None of the compounds were inhibitory, and none showed any effect on alkaline phosphatase activity. The significance of the inhibition or activation of specific enzymes or enzyme systems which results from exposure of the microalgae to petroleum water-solubles or individual compounds is unknown. However, we may assume that the change is undesirable.

Microalgae may exhibit growth lags when grown in artificial culture medium. The accepted explanation for these time lags in the initiation of cell division is that the culture medium is deficient in some respect or the medium components are acutely toxic to some of the inoculum cells. The microalgal strains used in these studies do not normally exhibit growth lags when transferred to fresh control ASP-2 medium, even when transferred at very low cell inoculum densities (Armstrong and Van Baalen, 1979). Algal cells incubated in medium containing 1-methylnaphthalene showed increasing lags in growth initiation with increasing concentrations of the hydrocarbon (Fig. 4) and a reduced ATP pool. Similar responses have been observed in studies of petroleum toxicity to various microalgal species (Prouse *et al.*, 1976; Soto *et al.*, 1975; Winters *et al.*, 1976; Pulich *et al.*, 1974). No trend in our data was easily discerned when the lags were plotted against the initial concentrations of

TABLE 2. The effect of petroleum hydrocarbon components on the alkaline phosphatase and phosphodiesterase activities of *Cyclotella meneghiniana* strain CM-2.

	Enzyme Activity (μmole/mg protein·min)	
	Alkaline Phosphatase	Phosphodiesterase
α-isopropylphenol	5.24 + 0.6%	1.79 + 0.5%
Control	5.30 + 0.4%	1.66 + 0.2%
2,4-dimethylphenol	4.94 + 0.3%	2.07 + 2.1%
Control	4.95 + 0.8%	1.82 + 1.7%
N,N-dimethylaniline	5.34 + 2.0%	1.92 + 0.2%
Control	5.05 + 0.6%	1.72 + 0.2%
p-toluidine	4.90 + 2.6%	1.53 + 0.9%
Control	4.77 + 0.8%	1.56 + 1.0%
m-cresol	4.94 + 0.6%	1.95 + 3.6%
Control	4.95 + 1.8%	1.71 + 5.2%
2,6-dimethylaniline	5.19 + 0.1%	1.77 + 0.0%
Control	5.19 + 0.8%	1.76 + 0.1%
2,4,6-trimethylaniline	5.22 + 0.2%	1.64 + 0.3%
Control	5.18 + 0.2%	1.66 + 0.9%

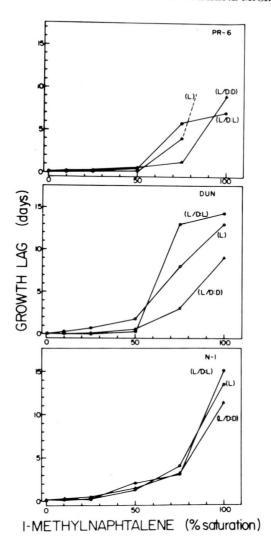

FIGURE 4. Increase in growth lags of Agmenellum quadruplicatum strain PR-6, Dunaliella tertiolecta strain Dun, and Cylindrotheca sp. strain N-1 as a function of the initial concentration of 1-methylnaphthalene in the culture medium. Cultures continuously illuminated (L); light/dark cycled cultures, exposed to 1-MN during the light period (L/D:L); light/dark cycled cultures, exposed to 1-MN during the dark period (L/D:D).

1-methylnaphthalene. It appears that the lag times increased
from a few hours to days between 25 and 50% saturation and the
three strains appear about equally sensitive. However, the
lag times plotted as a function of the theoretical dose per
cell show *A. quadruplicatum* PR-6 to be more sensitive than
Dunaliella tertiolecta Dun and the diatom *Cylindrotheca* sp
N-1 to be relatively resistant to 1-methylnaphthalene
(Fig. 5).

The immediate inhibition of photosynthetic oxygen
evolution plotted as a function of 1-methylnaphthalene dose
follows the same order of strain sensitivity and culture
light/dark regime (Fig. 6). The factors contributing to the
toxic effects of 1-methylnaphthalene in the short term
photosynthesis experiments and the long-term growth studies
may be identical. We found 1-methylnaphthalene to completely
volatilize from cell-free, air-bubbled culture tubes within 24
hours (Fig. 7). Soto *et al* (1975) convincingly demonstrated
that cell associated radiolabelled naphthalene was reduced by
cell division in *Chlamydomonas angulosa,* not volatilization
from the cells. The partition coefficient of 1-methyl-
naphthalene is estimated to be about 4000 using the equations
presented by Tulp and Hutzinger (1978). The cell associated
1-methylnaphthalene may then be assumed to intercalate in the
lipid-rich cell membranes and the effect on the initiation of
growth to be a result of the changes in the membranes similar
to the changes caused by temperature shifts of the algal
cultures (Oas, 1978).

SUMMARY

Water-soluble petroleum hydrocarbons in general can be
classified as non-specific metabolic inhibitors. Chronic
effects to marine microalgae include a reduction in the
cellular ATP pools with or without a concomitant reduction
in the rate of growth. Exposure of microalgae to oil water-
solubles inhibits photosynthetic electron transport as
measured by photosynthetic oxygen evolution and medium
alkalization and initial uncoupling of oxidative phosphorylation as indicated by an increased respiratory oxygen
consumption. Inhibition or an enhancement of the initial
reaction rate of specific enzyme or enzyme systems of
microalgae can be attributed to petroleum water-solubles and
specific hydrocarbons. The significance of these alterations
in specific enzyme activities are not clear at present. Oil
water-solubles inhibit photosynthetic reactions at very low
dose levels suggesting direct action on the electron transport

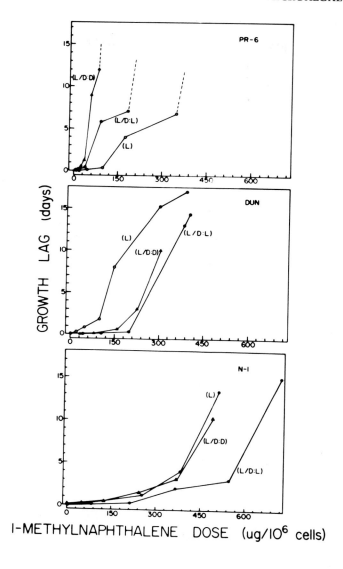

FIGURE 5. Increase in growth lags of Agmenellum quad-ruplicatum strain PR-6, Dunaliella tertiolecta strain Dun, and Cylindrotheca sp. strain N-1 as a function of the calculated dose of 1-methylnaphthalene per cell. Cultures continuously illuminated (L); light/dark cycled cultures, exposed to 1-MN during the light period (L/D:L); light/dark cycled cultures, exposed to 1-MN during the darl period (L/D:D).

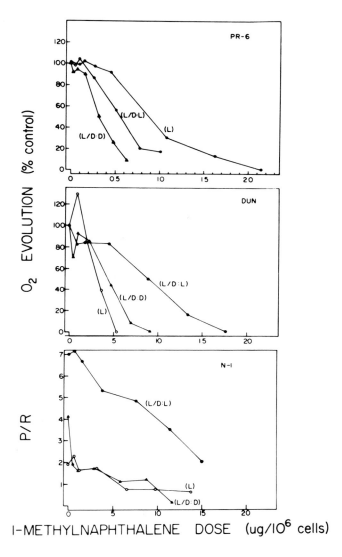

FIGURE 6. *Immediate effects of 1-methylnaphthalene on photosynthetic oxygen evolution of Agmenellum quadruplicatum strain PR-6, Dunaliella tertiolecta strain Dun and Cylindrotheca sp. strain N-1 as a function of dose per cell. Continuously illuminated cultures (L); light/dark cycled cultures, exposed to 1-MN during the light period (L/D:L); light/dark cycled cultures, exposed to 1-MN during the dark period (L/D:D).*

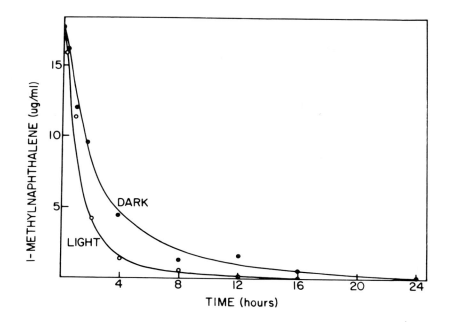

FIGURE 7. Evaporation of 1-methylnaphthalene from saturated ASP-2 culture medium as a function of time. The temperature was 30±0.1°C and the medium was bubbled with 1% CO_2-air at approximately 10 ml/min.

systems by specific compounds or classes of compounds, likely the N, S, O - substituted aromatics. Long-term effects such as reduced ATP pools and extended lags in growth are more likely caused by the uptake and incorporation of aromatic hydrocarbons into the cell membrane.

ACKNOWLEDGMENTS

    These studies were conducted at the Department of Oceanography, Florida State University and were supported by the National Science Foundation, International Decade of Ocean Exploration Grant OCE-76-82842. Our special thanks and gratitude are extended to Ms. Diane Miller for typing the manuscript.

LITERATURE CITED

Armstrong, J. E. and J. A. Calder. 1978. Inhibition of light-
    induced pH increase and oxygen evolution of marine
    microalgae by water-solubles of crude and refined oils.
    Appl. Environ. Microbiol. 35: 858-862.

Armstrong, J. E. and C. Van Baalen. 1979. Iron transport in
    microalgae: the isolation and biological activity of a
    hydroxamate siderophore from the blue-green alga
    Agmenellum quadruplicatum. J. Gen. Microbiol. 111:
    253-262.

Batterton, J., K. Winters, and C. Van Baalen. 1978.
    Anilines: selective toxicity to blue-green algae.
    Science 199: 1068-1070.

Brezonik, P. L., F. X. Browne, and J. L. Fox. 1975.
    Application of ATP to plankton biomass and bioassay
    studies. Water Res. 9 155-162.

Calder, J. A. and J. H. Lader. 1976. Effect of dissolved
    aromatic hydrocarbons on the growth of marine bacteria
    in batch culture. Appl. Environ. Microbiol. 32: 95-101.

Clegg, T. T. and J. L. Koevenig. 1974. The effects of four
    chlorinated hydrocarbon pesticides on ATP levels in three
    species of photosynthesizing algae. Bot. Gaz. 135:
    1368-1372.

Eganhouse, R. P. and J. A. Calder. 1976. The solubility of
    medium molecular weight aromatic hydrocarbons and the
    effects of hydrocarbon co-solutes and salinity. Geochem.
    Cosmochim. Acta. 40: 555-561.

Fehler, S. W. G. and J. A. Calder. 1976. Effects of the
    water-soluble components of petroleum on alkaline phos-
    phatase and phosphodiesterase from *Cycolotella men-
    eghiniana*. Am. Soc. Limn. & Ocn. Annual meeting,
    Savannah, GA.

Griffin, L. F. and J. A. Calder. 1977. Toxic effect of
    water-soluble fractions of crude, refined, and weathered
    oils on the growth of a marine bacterium. Appl. Environ.
    Microbiol. 33: 1092-1096.

Gordon, D. C. and N. J. Prouse.   1973.   The effects of three
    oils on marine phytoplankton photosynthesis.   Mar.
    Biol. 22:   329-333.

Holm-Hansen, O.   1970.   ATP levels in algal cells as influenced
    by environmental conditions.   Plant and Cell Physiol.
    11:   689-700.

Hutchinson, T. C., J. A. Hellebust, D. Mackay, D. Tam, and
    P. Kauss.   1979.   Relationship of hydrocarbon solubility
    to toxicity in algae and cellular membrane effects.   In:
    Proceedings of the 1979 Oil Spill Conference (Prevention,
    Behavior, Control, Cleanup).   pp. 541-547.   Los Angeles,
    CA: American Petroleum Institute.

Kauss, P., T. C. Hutchinson, C. Soto, J. Hellebust, and M.
    Griffiths.   1973.   The toxicity of crude oil and its
    components to freshwater algae.   Proceedings EPA/API/USGS
    Conference on Prevention and Control of Oil Spills, March
    13-15, Washington, DC.   p. 703-714.

Kratz, W. and J. Myers.   1955.   Nutrition and growth of
    several blue-green algae.   Amer. J. Bot.   42:   282-287.

Lee, R. F. and J. W. Anderson.   1977.   Fate and effect of
    naphthalenes: Controlled ecosystem pollution experiment.
    Bull. Mar. Sci. 27:   127-131.

Lewenstein, A. and K. Schneider.   1971.   The level of ATP in
    Chlorella.   In: Proc. 11. Intern. Congr. Photosynth.
    Res. Vol. 11. G. Forti, M. Avron, A. Melandri, (Eds.)
    pp. 1371-1378.   The Hague: Junk.

Lowry, O. H., N. J. Rosenbrough, A. E. Farr, and R. J. Randall.
    1951.   Protein measurement with the Folin phenol reagent.
    J. Biol. Chem. 193:   265-274.

Morgan, J. C.   1975.   M.S. Thesis.   The University of Texas
    at Austin, Austin, Texas.

Myers, J.   1950.   The culturing of algae for physiological
    research.   In: The Culture of Algae.   J. Brunel, G. W.
    Prescott, and L. H. Tiffany (Eds.).   Kettering Foundation:
    Ohio.   pp. 45-51.

Oas, T. G.   1978.   M.S. Thesis.   The Florida State University, Tallahassee, Florida.

Parker, P. L., K. Winters, C. Van Baalen, J. C. Batterson, and R. S. Scalan. 1976. Petroleum pollution: chemical characteristics and biological effects. In: Sources, effects and sinks of hydrocarbons in the aquatic environment. pp. 256-269. Proceedings of the Symposium, American University, Washington, DC: The American Institute of Biological Sciences.

Prouse, N. J., D. C. Gordon, Jr., and P. D. Keizer. 1976. Effects of low concentrations of oil accommodated in sea water on the growth of unialgal marine phytoplankton cultures. J. Fish. Res. Bd. Can. 33: 810-818.

Provasoli, L., J. J. A. McLaughlin, and M. R. Droop. 1957. The development of artificial media for marine algae. Arch. Mikrobiol. 25: 392-428.

Pulich, W. M., K. Winters, and C. Van Baalen. 1974. The effect of a No. 2 fuel oil and two crude oils on the growth and photosynthesis of microalgae. Mar. Biol. 28: 87-94.

Robinson, J. and J. M. Cooper. 1970. Method of determining oxygen concentrations in biological media, suitable for calibration of the oxygen electrode. Anal. Biochem. 33: 390-399.

Soto, C., J. A. Hellebust, and T. C. Hutchinson. 1975. Effect of naphthalene and aqueous crude oil extracts on the green flagellate *Chlamydomonas angulosa*. II. Photosynthesis and the uptake and release of naphthalene. Can. J. Bot. 53: 118-126.

Tulp, M. Th. M. and O. Hutzinger. 1978. Some thoughts on aqueous solubilities and partition coefficients of PCB, and the mathematical correlation between bioaccumulation and physico-chemical properties. Chemosphere 7: 849-860.

Van Baalen, C. 1962. Studies on marine blue-green algae. Bot. Mar. 4: 129-139.

Winters, K., R. O'Donnell, J. C. Batterton, and C. Van Baalen. 1976. Water-soluble components fo four fuel oils: chemical characterization and effects of growth of microalgae. Mar. Biol. 36: 269-276.

A STUDY OF THE RECOVERY OF A MARINE ISOPOD
(*SPHAEROMA QUADRIDENTATUM*) FROM
PETROLEUM-INDUCED SENSITIVITY

Wen Yuh Lee
J. A. C. Nicol

The University of Texas Marine Science Institute
Port Aransas Marine Laboratory
Port Aransas, Texas

INTRODUCTION

Numerous laboratory studies have been conducted on the
acute toxicity of petroleum oils, dispersants, and mixtures of
these substances to marine organisms. These studies provide
useful information on both the relative sensitivities of
local fauna and flora (Baker and Crapp, 1974; Nagell *et al.*,
1974; Lee and Nicol, 1977; Batterton *et al.*, 1978), and the
relative toxicities of various oils and dispersants (Anderson
et al., 1974; Gelder-Ottway, 1976; Byrne and Calder, 1977;
Donahue *et al.*, 1977). On the basis of this information, a
suitable dispersant can be chosen should an oil spill occur.
For additional information on the possible biological consequences at an oil spill site, chronic toxicity studies
carried out in both the laboratory and the field will be
necessary.

In general, toxicity studies on the entire life cycle of
an organism can yield significant data on growth rate,
fecundity, biochemical adaptation of the exposed populations,
and the viability of succeeding generations. By chronically
exposing successive generations of the annelid *Neanthes
arenaceodentata* to the water soluble fractions (WSF) of a No.
2 fuel oil, Rossi and Anderson (1978a) observed a significant
increase in the tolerance of adult worms and concluded that
genetic resistance to oil may have been involved. Lee (1978)
reported that juvenile isopods (*Sphaeroma quadridentatum*)

chronically exposed to sublethal concentrations of a WSF exhibited reduced growth rate, fecundity, and offspring
viability. Growth rate of the exposed isopods was adversely
affected at concentrations $\geq$ 3% WSF, while fecundity was depressed at lower levels ($\geq$ 1% WSF). Brood mortality (F_1 generation) in clean sea water was dependent largely on the
history of the parent populations. Offspring from the isopods
chronically exposed to WSF $\geq$ 1% suffered a mortality > 70%.
Since most toxicity studies terminate at the first generation,
and there are few data on the fate of the surviving offspring
and their responses to the oils, we have carried out further
studies on isopods. The objectives of this investigation are
to determine to what extent offspring of oil-contaminated
populations exhibit altered survival, fecundity and developmental rate, and to estimate the number of generations it
takes for these isopods to recover from such petroleum-
induced changes.

MATERIALS AND METHODS

     The present research is a continuation of a previous study
(Lee, 1978) and deals with oil resistance in offspring of the
isopods (*S. quadridentatum*) chronically exposed to the WSF of
a No. 2 fuel oil for 9 months. Basically, we performed
studies with two strains of offspring. One strain was from
the isopods (parent population) that had been chronically exposed to 0.1% WSF, and the other was from the 10% WSF-exposed
isopods. In both cases, only the parent populations had been
chronically treated with WSF. Then, beginning with the F_1
generation, the isopods were grown in WSF-free water and
juveniles of each generation were challenged with acute doses
of a No. 2 fuel oil WSF.
     To determine whether offspring of chronically exposed
isopods showed lower survival in WSF, young isopods (F_1) were
removed to clean sea water within 14h after their release from
the parent population. They were raised to five weeks old and
then, tested with 0 (control), 5, 15, 30, 45, 60 and 75% WSF
extracted from Baytown fuel oil (Exxon). For each concentration of WSF, twenty individuals (divided into two groups)
were exposed in 200 ml volumes per 10 individuals. Exposure
lasted for one month. During this period, the test medium
was renewed daily and dead animals were removed. Isopods were
fed with excessive food mixtures of powdered tropical fish
flakes and ground sea lettuce (*Ulva*). The containers were
randomly assigned on a bench, covered with glass plates and
gently aerated. F_2, F_3 and F_4 generations were treated

similarly. Responses of these offspring to WSF were determined by LC_{50} and further compared with that of offspring
derived from the controls. Values of LC_{50}, 95% confidence
intervals, and slope functions were estimated according to
Litchfield and Wilcoxon (1949).

Chronic effects of WSF on the reproduction of offspring
(F_1, F_2, and F_3) were evaluated in terms of the time interval
from larvae to larvae, and fecundity. At the beginning, the
newly hatched larvae (F_1) of WSF-exposed isopods (parent
populations) were collected and placed in a culture bowl containing 1.5 liter Millipore-filtered (0.45 µm) sea water,
which was renewed twice a week. The number of larvae in each
bowl varied and depended on the number of young produced on
that day. Individuals in each bowl were, therefore, of the
same age. They were grown to maturity in WSF-free sea water,
and allowed to reproduce. Once F_2 larvae appeared in a bowl,
the surviving isopods (F_1) were counted and sexed. Experiments lasted until all F_1 isopods in the bowl died. The total
number of larvae released during the life cycle was divided by
the number of surviving adult females and used as fecundity of
the F_1 generation. Three replicates were run for each concentration, as long as there were enough offspring. Reproductive
potential of the F_2 and F_3 generations was estimated in a
similar manner. Developmental rates were determined as the
time period between the appearance of isopod larvae of two
successive generations. Fecundity and developmental rates
obtained from each concentration were compared with that of
the controls and these data were tested by one-way analysis
of variance. Differences between generations were also in-
spected using least significant difference (LSD) as a
criterion (Snedecor and Cochran, 1968).

All experiments were conducted at 22 ± 2.3°C and a
salinity of 30 °/oo. Sea water was collected about 50 miles
off Port Aransas, Texas, filtered through a 0.45 µm Millipore
filter, and diluted to the desired salinity with demineralized
water. In order to suppress bacterial growth, antibiotics
were added to the sea water in amounts of 50 mg/l penicillin
G and 25 mg/l streptomysin sulphate. To reduce the possibility of contamination with WSF, all glassware used was
cleaned with organic solvents (chloroform and methanol),
soaked in water with detergent, and rinsed with distilled
water.

The WSF was prepared following the methods described by
Pulich et al., (1974). The total amount of organic content
in our 100% WSF stock solution of Baytown fuel oil was about
20 ppm. Dilutions of 10% and 0.1% of the stock solution were
equal to concentrations of 2 ppm and 20 ppb, respectively.
The test oil was kindly supplied by Dr. C. B. Koons of Exxon

Production Research Company, Houston, Texas.

Water soluble fractions of this fuel oil were analyzed by Winters et al. (1976). The percent changes in total organics, alkyl benzenes, alkyl phenols, anilines, naphthalenes and indoles, during a five days' exposure to air, were described by Lee et al. (1978). In general, phenols, indoles and anilines were lost at a much slower rate than naphthalenes and benzenes. Each test medium in the present study was renewed daily. We assumed that the WSF of the fuel oil in the test culture bowls behaved the same way as reported previously by Lee et al. (1978) and by Rossi and Anderson (1978b); therefore, it was assumed that the animals experienced concentrations from 100% to 25% over a 24h cycle.

RESULTS

Three successive control generations did not show significant changes in their responses to WSF (Table 1). Within the range of test concentrations (5% to 75% WSF), exposure of less than 48h did not cause 50% mortality of the test population. The average 96-h LC_{50} for those offspring was 56% WSF with 95% confidence intervals from 44.7 to 71.3% WSF. Larvae from 0.1%-exposed isopods had different responses (Table 2) from that of the controls. The 48-h exposure was long enough to induce 50% or higher mortality. The estimated 48-h LC_{50} was 68.5% WSF for the F_1 generation, and 63.5% for the F_2's. The 96-h LC_{50} values for the first three generations were less than the average value of the controls and also suggested a trend toward increasing resistance in later generations.

Offspring (F_1) of the 10%-exposed group were not tested for their responses to WSF, because the number of F_1 animals produced was small. Therefore, observations on survival in WSF were made only on F_2, F_3, and F_4 individuals. It is interesting to note that offspring of the 10%-exposed group were no more susceptible to WSF than those of the 0.1%-exposed isopods (Table 3). In fact, offspring of the 10%-exposed group seemed to be more resistant to oil than were offspring from the 0.1%-exposed group. For example, the LC_{50} for the F_2 generation of the 0.1%-treated isopods was about 60% WSF at 48h, but greater than 75% WSF for the same generation of the 10%-exposed isopods. One common feature in both groups is that values of LC_{50} increased with later generations.

Data on reproduction and developmental rate for the three successive generations are shown in Table 4. Fecundity of the controls ranged from 70 in F_1 to 43 in F_3, with an average of

TABLE 1. Effect of acute exposure to WSF (Baytown fuel oil, Exxon) on the survival of laboratory-cultured isopods (*S. quadridentatum*). Values of LC_{50}, 95% confidence intervals (CI), and slope functions (SF) were determined for three successive generations. All offspring were from unexposed populations.

Generation		24 h	48 h	96 h	120 h
Control (F_1)	LC_{50} (%WSF)	>75.0[a]	>75.0	49.0	42.5
	95% CI	—	—	43.0–55.5	35.5–51.0
	SF	—	—	0.75	0.75
Control (F_2)	LC_{50} (%WSF)	>75.0	>75.0	65.0	51.0
	95% CI	—	—	43.5–97.0[b]	44.0–59.5
	SF	—	—	0.82	0.78
Control (F_3)	LC_{50} (%WSF)	>75.0	>75.0	54.0	45.0
	95% CI	—	—	47.5–61.5	36.5–55.5
	SF	—	—	0.81	0.71

[a] LC_{50} > the highest concentration tested

[b] Data are heterogeneous

TABLE 2. Effect of chronic exposure to 0.1% WSF (Baytown fuel oil, Exxon) on four successive generations of the isopod, S. quadridentatum. The average values of LC50 for the offspring from the controls are shown at the bottom of this table.

Generation		24 h	48 h	96 h	120 h
0.1% BT(F_1)	LC_{50} (%WSF)	>75.0	68.5	38.0	34.5
	95% CI	–	54.0–86.5	32.5–44.0	26.0–45.5
	SF	–	0.58	0.78	0.64
0.1% BT(F_2)	LC_{50} (%WSF)	>75.0	63.5	43.0	40.0
	95% CI	–	54.5–74.0	38.0–49.0	35.0–46.0
	SF	–	0.65	0.74	0.80
0.1% BT(F_3)	LC_{50} (%WSF)	>75.0	>75.0	44.5	43.0
	95% CI	–	–	40.0–49.5	38.0–49.0
	SF	–	–	0.85	0.81
0.1% BT(F_4)	LC_{50} (%WSF)	>75.0	>75.0	65.0	60.0
	95% CI	–	–	60.0–70.5	56.0–64.5
	SF	–	–	0.82	0.85
Control	LC_{50} (%WSF)	>75.0	>75.0	56.0	46.2
	95% CI	–	–	44.7–71.3	38.7–55.3
	SF	–	–	0.79	0.75

TABLE 3. Effect of chronic exposure to 10% WSF (Baytown fuel oil, Exxon) on three successive generations of the isopod, *S. quadridentatum*. The average values of LC_{50} for the offspring from the controls are shown at the bottom of this table.

Generation		24 h	48 h	96 h	120 h
10% BT(F_2)	LC_{50} (%WSF)	>75.0	>75.0	47.0	40.5
	95% CI	–	–	39.5–56.0	35.0–47.0
	SF	–	–	0.74	0.78
10% BT(F_3)	LC_{50} (%WSF)	>75.0	>75.0	49.0	46.0
	95% CI	–	–	45.0–53.5	43.5–48.5
	SF	–	–	0.88	0.89
10% BT(F_4)	LC_{50} (%WSF)	>75.0	>75.0	>75.0	59.0
	95% CI	–	–	–	52.0–67.0
	SF	–	–	–	0.74
Control	LC_{50} (%WSF)	>75.0	>75.0	56.0	46.2
	95% CI	–	–	44.7–71.3	38.7–55.3
	SF	–	–	0.79	0.75

473

TABLE 4. Chronic effects of 0.1% and 10% WSF (Baytown fuel oil, Exxon) on the reproduction and developmental rates of isopods (*S. quadridentatum*). (M: mean; S.D.: standard deviation).

Generation	Total No. of adults	Total No. of young released	Total No. of females	Time interval from larvae to larvae (days)	Fecundity (young/female)
Control (F$_1$)	13	630	9	119	70
Control (F$_2$)	19	854	15	126	57
Control (F$_3$)	20	597	14	144	43
			M and S.D.	130 $\pm$ 12.9	57 $\pm$ 13.5
0.1%(F$_1$)A*	19	576	12	114	48
0.1%(F$_1$)B	20	488	13	121	38
0.1%(F$_1$)C	20	374	11	124	34
			M and S.D.	120 $\pm$ 5.1	40 $\pm$ 7.2
0.1%(F$_2$)A	29	669	19	133	35
0.1%(F$_2$)B	21	511	9	130	57
0.1%(F$_2$)C	21	430	11	133	39

TABLE 4 con't.

Generation	Total No. of adults	Total No. of young released	Total No. of females	Time interval from larvae to larvae (days)	Fecundity (young/female)
				M and S.D.	
0.1%(F$_3$)A	22	1037	12	132 $\pm$ 1.7	44 $\pm$ 11.7
0.1%(F$_3$)B	25	802	10	105	86
0.1%(F$_3$)C	26	666	11	98	80
				105	61
				M and S.D.	
10%(F$_1$)	18	185	13	103 $\pm$ 4.0	76 $\pm$ 13.1
10%(F$_2$)	12	553	8	63	14
				126	69
10%(F$_3$)A	8	108	2	143	54
10%(F$_3$)B	18	772	15	119	52
				M and S.D.	
				131 $\pm$ 17.0	53 $\pm$ 1.4

*Three replicates were run for each generation whenever possible, and they were assigned to A, B, and C.

475

57 larvae per surviving female, while the average number of
offspring produced by the 0.1%-exposed group increased from
40 in the F_1 generation to 76 in the F_3's. One-way analysis
of variance indicated that the three successive generations
of 0.1%-exposed *S. quadridentatum* did not differ from the controls in their mean fecundity. However, when comparisons
were made between pairs of means, the average fecundity of F_3
generation was significantly different from that of both F_1
and F_2 generations. A similar trend was also found for the
parameter of developmental rate; differences between means
were not detected at the 5% level for the two groups of the
controls and 0.1%-treated isopods. Based on the calculated
value of least significant difference (13.7), significant
difference between mean developmental rates were observed for
the following two pairs only: F_3 versus F_1 and F_3 versus F_2

The F_1 generation of 10%-exposed isopods had a developmental rate of 63 days and fecundity of 14. The developmental
rate was considerably faster, while fecundity was much lower
than the 0.1% and control groups. After the F_1 generation,
developmental rates and fecundity closely approximated values
observed for the controls.

DISCUSSION

Chronic exposure of the isopods *Sphaeroma quadridentatum*
to the WSF of a No. 2 fuel oil resulted in a less resistant
F_1 generation. There was also a consistent trend toward
increasing resistance with later generations. This pattern
was especially evident from consideration of 96-h LC_{50} values.
For example, the 95% confidence intervals of the F_4 generation
in the 0.1%-exposed group did not overlap with those of either
the F_1 or F_2 generation.

Rossi and Anderson (1978a) exposed three successive generations of polychaetes (*Neanthes arenaceodentata*) continuously to sublethal concentrations of No. 2 fuel oil WSF and
observed very different results from ours. Compared to control adults, the F_1, F_2 and F_3 adult worms in their studies
became more resistant, and this petroleum-induced resistance
did not change with succeeding generations. They postulated
that a genetic change must be involved in the petroleum resistance observed in the adult *N. arenaceodentata*. The different results between these two studies were probably due to:
1) species tested, 2) methods employed, and 3) the life stage
at which the animals were challenged with WSF. In our study,
5-week old juveniles were tested, and offspring of all generations were maintained in clean seawater. Despite these
differences, we did find one thing in common: F_1 juveniles

became more susceptible to petroleum hydrocarbons, and the susceptibility decreased with passing generations. Rossi and Anderson (1978a) did not report significant changes in larval worms.

Reproduction and developmental rates of offspring from WSF-exposed *S. quadridentatum* responded in a similar manner to that observed in the petroleum-induced resistance, except that the change in magnitude was more or less dependent on the concentration of WSF to which the parent populations were exposed. For example, at 0.1% WSF both fecundity and rate of development of the F_1 and F_2 generations were not significantly different from that of the controls. At 10% WSF F_1's took less time to mature, but released only 14 young per surviving female; this corresponded to 1/4 of the control. Under the conditions of higher concentrations (>10% WSF) and shorter exposure (<35 days), it appeared that only the fecundity of F_1 generation was significantly affected (Lee, unpublished data). This confirms a previous report (Lee, 1978) that fecundity is a more sensitive parameter as a measure of the effect of oil than rate of development. Rossi and Anderson (1978b) reached the same conclusion in a study on *N. arenaceodentata*.

This study yields two important results. First, the exposed isopods under the conditions of low concentrations of WSF and long-term exposure produce a more sensitive F_1 generation in terms of fecundity, developmental rate, and survival. Second, the petroleum-induced sensitivity does not persist long when the offspring of WSF-exposed isopods are held in clean sea water. It takes only 1 to 2 generations for them to recover from the oil exposure.

Although hydrocarbon levels in exposed and depurated isopods were not monitored, other studies have shown that uptake of naphthalenes or other petroleum-derived aromatic hydrocarbons by marine animals exposed to low concentrations is very rapid during the first 48h and reaches equilibrium at the end of 7 to 8 days (Lee *et al.*, 1972; Korn *et al.*, 1976; Harris *et al.*, 1977). Depuration rates are dependent on the manner in which the hydrocarbons are accumulated; the rate of depuration is much slower for those hydrocarbons taken up by way of food than those taken directly from solution (Corner *et al.*, 1976). However, the rate of loss of hydrocarbons from either contaminated tissues or animals during depuration is very similar to the rate of uptake; loss is rapid in the first three days and greatly reduced thereafter (Neff *et al.*, 1976; Harris *et al.*, 1977; Nunes and Benville, 1979). Therefore, it is not unusual that the petroleum-induced sensitivity observed in *S. quadridentatum* lasted for only one to two generations. Indeed, more rapid recovery of normal development, growth or other functions (e.g. behavioral patterns),

following the termination of exposure to WSF, has been
reported for the mud crab, *Rhithropanopeus harrissi*
(Laughlin *et al.*, 1978), the annelid worm, *N. arenaceodentata*
(Rossi and Anderson, 1978b), and the planktonic shrimp,
Lucifer faxoni (Lee *et al.*, 1978).

The experiment conducted in this study simulated one
kind of pollution. The combination of low concentrations of
WSF and long-term exposure may be encountered at a place near
a crippled oil well or in an oil port where hydrocarbons are
continuously added at low levels. Results from our study
suggest that isopod populations would take at least 5 months
to 1 year to recover if the WSF were fully removed from the
environment. However, in the field, petroleum oils take a
long time to evaporate, oxidize and be degraded by bacteria.
Parts of petroleum residues may even be embedded in sediments
(Dow, 1978) and periodically reenter the water long after the
oil spill (Vandermeulen and Gordon, 1976). The estimated
times of recovery determined in this laboratory study are,
therefore, much less than that observed for the fauna in an
oil-spill site. For example, 7 years after the spill of fuel
oil at West Falmouth, Massachusetts, recovery of the marsh is
still incomplete (Krebs and Burns, 1977). A study of a
benthic community in a polluted Antarctic environment indicated that the benthos took about 8 years to recover
following pollution abatement (Platt, 1978). Obviously,
when assessing the chronic effects of an oil spill on biota,
factors such as the chemical properties of the oil and the
physical characteristics of the environment itself must also
be taken into account.

SUMMARY

1. Experiments were designed to determine (a) whether
offspring of WSF-exposed isopods become less resistant to WSF
in terms of rates of development, reproduction, and survival,
and (b) the number of generations it would take to recover
from such exposure.

2. Chronic exposure of juvenile isopods to low levels of
WSF resulted in a less resistant F_1 generation. Petroleum-
induced susceptibility lasted for about 1 to 2 generations
when progeny of exposed isopods were grown in WSF-free sea
water.

3. Based on the life cycle of *S. quadridentatum* observed
in the laboratory, treated isopods required only 1/2 to 1 year
to recover in WSF-free sea water. This short time interval
was related to published data on the rates of accumulation and

depuration of petroleum compounds by marine invertebrates. The rapid loss of accumulated hydrocarbons in clean sea water was probably the main reason why petroleum-induced sensitivity did not persist over a long period of time.

4. Isopods in this study recovered more rapidly than the fauna in an oil-spill site. Since recovery of isopods took place in WSF-free sea water, the estimated time interval should be considered only as the minimum time period for a population to recover following an oil spill. However, when applying laboratory results to a field study, we suggest that other factors such as the chemical characteristics of oil and its persistence in the environment must be also taken into account.

ACKNOWLEDGMENTS

This study was supported by a grant from the National Science Foundation, IDOE, No. GX-37345. We thank Drs. A. Calabrese, C. Kitting, and M. Morgan for their helpful comments on the manuscript.

LITERATURE CITED

Anderson, J. W., J. M. Neff, B. A. Cox, H. E. Tatem, and G. M. Hightower. 1974. Characteristics of dispersions and water-soluble extracts of crude and refined oils and their toxicity to estuarine crustaceans and fish. Mar. Bio. 27: 75-88.

Baker, J. M. and G. B. Crapp. 1974. Toxicity tests for predicting the ecological effects of oil and emulsifier pollution on littoral communities. In: Ecological aspects of toxicity testing of oils and dispersants. pp. 23-40. Ed. by L. R. Beynon and E. B. Cowell. Essex: Applied Science Publishers Ltd.

Batterton, J. C., K. Winters, and C. Van Baalen. 1978. Sensitivity of three microalgae to crude oils and fuel oils. Mar. Environ. Res. 1: 31-41.

Byrne, C. J. and J. A. Calder. 1977. Effect of water-soluble fractions of crude, refined and waste oils on the embryonic and larval stages of the quahog clam *Mercenaria* sp. Mar. Biol. 40: 225-231.

Corner, E. D. S., R. P. Harris, C. C. Kilvington, and S. C. M.
    O'Hara. 1976. Petroleum compounds in the marine food
    web: Short-term experiments on the fate of naphthalene in
    Calanus. J. mar. biol. Ass. U.K. 56: 121-133.

Donahue, W. H., R. T. Wang, M. Welch, and J. A. C. Nicol.
    1977. Effects of water-soluble components of petroleum
    oils and aromatic hydrocarbons on barnacle larvae.
    Environ. Pollut. 13: 187-202.

Dow, R. L. 1978. Size-selective mortalities of clams in an
    oil spill site. Mar. Pollut. Bull. 9: 45-48.

Gelder-Ottway, S. V. 1976. The comparative toxicities of
    crude oils, refined oil products and oil emulsions. In:
    Marine ecology and oil pollution. pp. 287-302. Ed. by
    J. M. Baker. New York: Halsted Press.

Harris, R. P., V. Berdugo, S. C. M. O'Hara, and E. D. S.
    Corner. 1977. Accumulation of ^{14}C-1-naphthalene by an
    oceanic and an estuarine copepod during long-term ex-
    posure to low-level concentrations. Mar. Biol. 42:
    187-195.

Korn, S., N. Hirsch, and J. W. Struhsaker. 1976. Uptake,
    distribution, and depuration of ^{14}C-benzene in northern
    anchovy, *Engraulis mordax,* and striped bass, *Morone
    saxatilis*. Fish. Bull. 74: 545-551.

Krebs, C. T. and K. A. Burns. 1977. Long-term effects of an
    oil spill on populations of the salt-marsh crab *Uca
    pugnax*. Science 197: 484-487.

Laughlin, R. B., Jr., L. G. L. Young, and J. M. Neff. 1978.
    A long-term study of the effects of water-soluble
    fractions of No. 2 fuel oil on the survival, development
    rate, and growth of the mud crab *Rhithropanopeus harrisii*.
    Mar. Biol. 47: 87-95.

Lee, R. F., R. Sauerheber, and A. A. Benson. 1972. Petroleum
    hydrocarbons: Uptake and discharge by the marine mussel
    Mytilus edulis. Science 177: 344-346.

Lee, W. Y. 1978. Chronic sublethal effects of the water sol-
    uble fractions of No. 2 fuel oils on the marine isopod,
    Sphaeroma quadridentatum. Mar. Environ. Res. 1: 5-17.

Lee, W. Y. and J. A. C. Nicol. 1977. The effects of the water soluble fractions of No. 2 fuel oil on the survival and behaviour of coastal and oceanic zooplankton. Environ. Pollut. 12: 279-292.

Lee, W. Y., K. Winters, and J. A. C. Nicol. 1978. The biological effects of the water-soluble fractions of a No. 2 fuel oil on the planktonic shrimp, *Lucifer faxoni*. Environ. Pollut. 15: 167-183.

Litchfield, J. R., Jr. and F. Wilcoxon. 1949. A simplified method of evaluating dose-effect experiments. J. Pharmac. exp. Ther. 96: 99-113.

Nagell, B., M. Notini, and O. Grahn. 1974. Toxicity of four oil dispersants to some animals from the Baltic Sea. Mar. Biol. 28: 237-243.

Neff, J. M., B. A. Cox. D. Dixit, and J. W. Anderson. 1976. Accumulation and release of petroleum-derived aromatic hydrocarbons by four species of marine animals. Mar. Biol. 38: 279-289.

Nunes, P. and P. E. Benville, Jr. 1979. Uptake and depuration of petroleum hydrocarbons in the Manila clam, *Tapes semidecussata* Reeve. Bull. Environm. Contam. Toxicol. 21: 719-726.

Platt, H. M. 1978. Assessment of the macrobenthos in an Antarctic environment following recent pollution abatement. Mar. Pollut. Bull. 9: 149-153.

Pulich, W. M., Jr., K. Winters, and C. Van Baalen. 1974. The effects of a No. 2 fuel oil and two crude oils on the growth and photosynthesis of microalgae. Mar. Biol. 28: 87-94.

Rossi, S. S. and J. W. Anderson. 1978a. Petroleum hydrocarbon resistance in the marine worm *Neanthes arenaceodentata* (Polychaeta: Annelida), induced by chronic exposure to No. 2 fuel oil. Bull. Environm. Contam. Toxicol. 20: 513-521.

Rossi, S. S. and J. W. Anderson. 1978b. Effects of No. 2 fuel oil water-soluble-fractions on growth and reproduction in *Neanthes arenaceodentata* (Polychaeta: Annelida). Wat. Air, Soil Pollut. 9: 155-170.

Snedecor, G. W. and W. G. Cochran. 1968. Statistical methods. 593 pp. Ames: The Iowa State University Press.

Vandermeulen, J. H. and D. C. Gordon, Jr. 1976. Reentry of 5-year-old stranded Bunker C fuel oil from a low-energy beach into the water, sediments, and biota of Chedabucto Bay, Nova Scotia. J. Fish. Res. Board Can. 33: 2002-2010.

Winters, K., R. O'Donnell, J. C. Batterton, and C. Van Baalen. 1976. Water-soluble components of four fuel oils: Chemical characterization and effects on growth of microalgae. Mar. Biol. 36: 269-276.

AN ECOLOGICAL PERSPECTIVE OF THE EFFECTS OF MONOCYCLIC AROMATIC HYDROCARBONS ON FISHES

Jeannette A. Whipple
Maxwell B. Eldridge
Pete Benville, Jr.

National Marine Fisheries Service
Southwest Fisheries Center
Tiburon Laboratory
Tiburon, California

INTRODUCTION

Monocyclic aromatic hydrocarbons (MAH) constitute a major class of petrochemicals potentially affecting organisms in the aquatic environment, but they have not received much attention in pollution research. MAH are highly toxic and relatively water-soluble when compared with other classes of petroleum hydrocarbons. However, the assumption has been made for some time that they are so volatile they do not persist in the aquatic environment long enough to affect organisms. We think this assumption is unfounded, based upon inadequate field measurements, and arises from considering these compounds only in relation to oil spills when a single large input occurs. The potential chronic sources of these compounds and their effects on aquatic organisms, particularly in estuaries, have been largely ignored. In addition, most studies of monocyclic aromatics have been conducted in laboratories, testing the effects of high lethal and sublethal levels. The actual occurrence of these compounds in water and organisms in the field, and their potential effects at low chronic levels in relation to other environmental factors, are still relatively unknown.

This paper attempts to: 1) summarize present knowledge of the effects of monocyclic aromatics on fishes, 2) provide an ecological perspective of the effects of these compounds on

fishes, 3) hypothesize modes of action of monocyclic aromatics
on fishes and, finally, 4) suggest directions of research
needed to determine whether these compounds may represent a
threat to our fisheries resources.

Most laboratory studies of the effects of monocyclic
aromatics on fishes have been done at the National Marine
Fisheries Service laboratories at Tiburon, California, Auke
Bay, Alaska and Seattle, Washington and these studies are
emphasized in the following discussion. The authors' experience with monocyclic aromatic compounds centers around
our laboratory and field studies on fishes of the San
Francisco Bay area. We realize that other ecosystems will
differ on many points.

Measurements and observations made in our experiments
have led us to hypothesize some modes of action of monocyclic
aromatics on fishes. The discussion on their effects is
placed in the context of these hypotheses. When studies
substantiating our hypotheses are available, references are
provided. It should be realized, however, that this synthesis
is offered primarily to provide a research framework and
does not necessarily imply that the hypotheses have been
completely validated.

DEFINITION, PROBABLE SOURCES AND FATES OF MONOCYCLIC AROMATIC
HYDROCARBONS

We do not intend to provide a comprehensive discussion of
the chemistry of the monocyclic aromatic hydrocarbons (MAH),
but for the benefit of the reader unfamiliar with these
compounds, the following summary of their characteristics,
sources and probable fates is provided.

Structure and Chemical Characteristics

Aromatic hydrocarbons are designated aromatic primarily
because the earliest known representatives were distinguished
by marked aromatic odors. Monocyclic aromatics are substances
containing one benzene ring. The benzene ring is characterized by a cyclic arrangement of carbon and hydrogen atoms
with a resonant bonding structure: an unsaturated, symmetrical ring of six equivalent CH groups (Gerarde, 1960).
The simplest monocyclic compound is benzene itself, consisting
of one ring, with no substitutions (Fig. 1). Other common
MAH include toluene (one substituted methyl group), ethylbenzene (one substituted ethyl group) and the isomers of

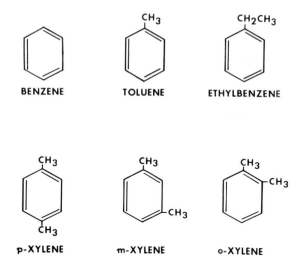

FIGURE 1. Six common monocyclic aromatic hydrocarbons. Unsymmetrical formulas (Kekulé) are used to represent the symmetrical molecules for simplicity.

xylene: *p*-xylene, *m*-xylene and *o*-xylene (three isomeric arrangements of two methyl groups). Some physical and chemical characteristics of these six monocyclics are summarized in Table 1. There are many other possible substituents on the benzene ring; these compounds will be referred to here, for purposes of simplification, as substituted benzenes.

MAH, for example benzene, are relatively soluble in water (Table 1) when compared with other petrochemicals. Since most research on the effects of MAH on fishes has been done with benzene and toluene, these particular compounds are most often discussed in this paper.

Potential Sources

Aromatic hydrocarbons have been used in industry for some time and benzene, toluene and xylene are three of the most important organic chemicals in industry. Recent data on the production of benzene, for example, showed an exponential increase in production in the last twenty years from less than 0.5 billion gallons in 1960 to about 1.5 billion gallons in 1979 (Davis and Magee, 1979).

The aromatic hydrocarbons obtained from petroleum and their derivatives come under the definition of petrochemicals-

TABLE 1. Physical - chemical properties of six prevalent monocyclic aromatic hydrocarbons. Data synthesized from Gerarde (1960) except where otherwise indicated.

Component	Molecular weight	Boiling point (°C) (760 mm Hg)	Melting point (°C)	Vapor pressure (mm Hg)	Index of refraction	Solubilities FW[a] 0 ppt (ppm)	SW[b] 25 ppt (ppm)
Benzene	78.11	80.1	+ 5.5	74.6 (20 C)	1.502 (20 C)	2026	1400
Toluene	92.13	110.6	−94.5	36.7 (30 C)	1.489 (24 C)	595	330
Ethyl-benzene	106.16	136.2	−94.9	10.0 (25.9 C)	1.493 (20 C)	175	180
para-Xylene	106.16	138.3	−55.9	10.0 (27.3 C)	1.500 (21 C)		180
meta-Xylene	106.16	139.1	−54.2	10.0 (28.3 C)	1.497 (20 C)		210
ortho-Xylene	106.16	144.4	−13.3	10.0 (32.1 C)	1.506 (20 C)	195	230

[a]Fresh water - McAuliffe, 1966.   25°C and 0 ppt.   Converted data to ml/liter.

[b]Sea water (estuarine) - Benville and Korn, 1977.   16°C and 25 ppt.

chemicals derived from petroleum or natural gas. The primary sources of MAH in the aquatic environment are from industrial petrochemicals and from crude oil. Some principal uses of MAH in industry are as: 1) starting materials and inter- mediates for synthesis of plastics, paints, pesticides, protective coatings, resins, dyes, drugs, flavors, perfumes, vitamins, explosives; 2) solvents for paints, dyes, resins, inks, lacquers, rubber, plastics and pesticides; and 3) con- stituents of aviation and automotive gasoline.

The MAH may also comprise from about 20 to 50% of the water-soluble fraction (WSF) of crude oil, depending upon the type of crude oil (Anderson et al., 1974b; Clark and MacLeod, 1977). Some crude oils are more aromatic and toxic in that they contain higher proportions of aromatics including monocyclics.

More specifically, we believe a major source of MAH petrochemicals in the San Francisco Bay estuary is from their use in pesticide mixtures as synergists, inhibitors, solvents, emulsifiers, wetting agents, etc. (Wiens, 1977). Petroleum products are also applied directly for control of insects, mites, weeds and fungus. A computer summary of the top 50 pesticides in terms of total pounds (extracted from the 1978 pesticide use reports for California)(State of California, 1978; Jung and Bowes, 1980) shows that aromatic petroleum hydrocarbons are heavily used in the ten counties forming the watershed draining into the San Francisco Bay-Delta. Six of the top ten chemicals listed include categories such as "petroleum hydrocarbons, petroleum oil, petroleum distillates, xylene, aromatic petroleum solvents, xylene-range aromatic solvents and petroleum distilled aromatics", totaling about 5 million pounds applied to 1.3 million acres.

There are other major sources of petrochemicals in the San Francisco Bay-Delta area, including municipal and industrial discharges, particularly in the Carquinez Straits area, which is along a major fish migratory pathway and also an area where fish kills consistently occur in summer (Kohlhorst, 1973). In fact, municipal effluents are probably among the major sources of petroleum input to the Bay, dis- charging a total of at least 72,400 pounds per day of "oil and grease" (Risebrough et al., 1978). Industrial discharges by refineries are apparently less-a total of 3,510 pounds per day (Risebrough et al., 1978). The total oil and grease measurement does not include, however, a large proportion of the more toxic water-soluble fraction such as the aromatic hydrocarbon components and, thus, inadequately estimates the relative toxicity of the discharges (Wolfe et al., 1979). Data from an API report (American Petroleum Institute, 1978) indicate that total MAH measured in refinery effluents can

range from only traces up to approximately 100 ppb.   Data
also indicate the concentration in intake water is sometimes
higher than in the effluents. The latter fact suggests that
there may be chronic water levels of MAH in the 100 ppb range.
The refineries listed in this report, however, are not
identified, and some San Francisco Bay refineries may have
lower or higher effluent concentrations.   In addition to
refineries, there are many other industrial dischargers of
petrochemicals. We need to examine the amounts and effects
of discharges more carefully, and further measurements of
toxic components in receiving waters are needed, not only in
San Francisco Bay, but also in other estuaries.

Other sources of MAH in San Francisco Bay include the
increasingly frequent spills associated with oil transport
and transfer activities. Petroleum refineries in the Bay
currently account for 3 to 4% of the total volume of crude
oil transported yearly by tankers throughout the world.   The
refineries in the northern area of the Bay process on the
order of a million barrels of crude oil daily (approximately
136 thousand tons) and the total volume processed yearly is
about 50 million tons (Risebrough *et al.*, 1978).

A minor source of MAH is from recreational boating
activity, with input of many toxic aromatics through outboard
motor gasoline effluent. At times this activity is
considerable (Hirsch *et al.*, ms in prep. (a)).

The relative contribution of various sources to the
concentration of MAH in the receiving waters of San Francisco
Bay has not yet been quantified, but a study has been
initiated with the cooperation of the State of California
Water Resources Control Board (Jung and Bowes, 1980; Whipple,
1979, 1980). Concentrations of individual MAH from other
areas are given below.

Fate of Monocyclic Aromatics in the
Aquatic Environment

Little is known about 1) the partitioning of MAH into
various compartments of an aquatic ecosystem, 2) the rates of
exchange among these compartments, or 3) the relative
importance of various pathways. The entire subject needs
further study for this class of compounds and will not be
discussed in detail here.

Table 2 summarizes some separate studies on various fates
of MAH in aquatic systems. There appears to be no study
attempting to bring these pathways into a single model for
this group of compounds. A comprehensive discussion of the
fate of petroleum hydrocarbons, in general, is given in a
review by Clark and MacLeod (1977) and in papers by Butler
et al., (1976), Gordon *et al.*, (1976) and Karrick (1977).

TABLE 2. Major pathways for the fate of monocyclic aromatics (MAH) in the aquatic environment. Rates probably vary with several environmental factors, including wind and wave action, temperature, and salinity.

Pathway	General conclusions	Selected references
Dissolution	Relatively soluble--152-1780 ppm in fresh water; 180-1400 ppm in seawater (Table 1).	Benville and Korn, 1977 Burwood and Speers, 1974 McAuliffe, 1966, 1977a Wasik and Brown, 1973
Evaporation	Rapid--from single spill a few hours to 10 days.	Gordon et al., 1976 Harrison et al., 1975 McAuliffe, 1977b
Atmospheric Input	Probably considerable.	Wiens, 1977
Concentration at Thermocline	Indication from vertical sampling profiles that volatile aromatics are at higher levels at the position of the thermocline.	Myers and Gunnerson, 1976
Emulsification Colloidal Dispersion	Little known.	Davis and Gibbs, 1975 McAuliffe, 1977a
Agglomeration Sorption Sinking Sedimentation	Monocyclics, including benzene, probably sorb to particulate matter; sorption of benzene to particulate matter in seawater was 4.8-28.4 μg/mg of particulates.	Zsolnay, 1972, 1977
Microbial Modification	In laboratory studies, monocyclics are biodegraded; including benzene, toluene, xylenes, tri- and tetra-methylbenzenes, alkylbenzenes, cyclo-alkyl benzenes. In the marine environment microbial degradation may be slower.	Gibson and Gibson et al., 1968 to 1977 (11 papers) Lee and Ryan, 1976 Marr and Stone, 1961
Photochemical Modification	There is preferential decomposition of aromatics.	Burwood and Speers, 1974 Hansen, 1977
Biological Ingestion and Excretion	Monocyclics are rapidly bioaccumu-lated from water to relatively high levels in adult fish and usually depurated rapidly. (Larval fish do not readily depurate, with high bio-accumulation resulting). Monocyclics are probably not bioaccumulated from food by fish.	See next few sections for references.

As previously discussed, MAH when compared to other
petroleum hydrocarbons are relatively soluble in water.
They also have high boiling points and vapor pressures (Table
1) and would be expected to volatilize from water relatively
rapidly. This pathway has been assumed to predominate, with
aromatics volatilizing too rapidly to affect aquatic biota.
Monocyclics also undergo photooxidation and microbial modification. However, there is some evidence that MAH, e.g.,
benzene, are sorbed to sediments and possibly to organic
aggregates. They may also be concentrated at thermoclines
and haloclines. Reversing loss through evaporation, MAH may
return to the water through atmospheric input. Finally, and
most significantly, MAH do occur in the water and are taken
up by aquatic organisms, including fish. The latter aspect
of the fate of monocyclic aromatics is emphasized in this
paper and discussed more fully below.

CONCENTRATIONS OF MONOCYCLIC AROMATICS IN WATER AND TISSUES
OF FISH SAMPLED IN THE FIELD

In lakes, rivers and estuaries, such as San Francisco
Bay, with continuous input of pollutants, there is some
evidence that MAH, including benzene, are present (e.g., Brown
et al., 1979; Whipple, 1979) in both water and fish tissues.
Table 3 summarizes some measurements of MAH in water samples
taken from the field. As can be seen from these data, few
measurements of individual monocyclic compounds have actually
been made. Many measurements of aromatic hydrocarbons found
in the literature, usually designated "total aromatics", do
not include benzene and toluene, and usually, by most
analytical methods, exclude ethylbenzene and xylenes.
One reason for the paucity of data on MAH may be the
assumption that they are not present in sufficient concentration to be harmful. Another reason is that the measurement
of these components requires special analytical techniques
which have only recently become feasible for low concentrations. Most previous field measurements of petroleum
hydrocarbons have been made of the alkane groups ($>C_7$) and a
few aromatics with higher boiling points (polycyclic aromatic
hydrocarbons, or PAH; Neff, 1979). The lower boiling point
MAH are lost in the concentration step used in the analysis
for polycyclic aromatics.
Table 3 includes only data where some estimate or measurement of the MAH is actually made. The range in open ocean
waters appears to be from approximately 0.01 to 5 ppb.
Closer to shore and in estuaries the range appears to be from

TABLE 3. Concentrations of monocyclic aromatics measured in the water column. Few data are available at the present time.

Location	Water Depth (meters)	No. of Samples	Hydrocarbon type	Hydrocarbon concentration (ppb)	References
Atlantic Ocean					
Skidaway River, GA	Surface	?	Benzene, toluene	3	Lee and Ryan, 1976
Tanker Route					
New York to	Surface		Total volatiles[a]	Med=0.1-0.3	Myers and
Gulf of Mexico	to 10	?		Range=0.08-2.4	Gunnerson, 1976
Pacific Ocean					
San Francisco Bay	2	8	Total aromatics	$\leq$5-59	DiSalvo and Guard, 1975
San Francisco Bay	Surface	5	Total 6 aromatics[b]	1-50	Benville (unpublished)
Tanker route					
San Francisco to	0-10	?	Total volatiles	Med=0.1-0.3	Myers and
Cook Inlet				Range=0.01-4.0	Gunnerson, 1976
GEOSECS					
Tanker routes	0-10	223	Total volatiles	Mean=0.22	Brown and
				Range=0.01-4.32	Huffman, 1976
Offshore oil seep	Surface	10	Benzene	ND-50	Koons and
Coal Oil Point, CA			Toluene	ND-80	Brandon, 1975
Baltic Sea and approaches					
Open water	1-200	6	Saturateds and monoaromatics	48-64	Zsolnay, 1972
Open water	1-200	40	Saturateds and monoaromatics	0-50	Zsolnay, 1977
Refineries[c]					
Refinery intake water	--	40			
			Benzene	Trace-40	American
			Toluene	Trace-15	Petroleum
			Ethylbenzene	Trace-20	Institute,
			Total	Trace-65+	1978
Refinery effluents	--	36			
			Benzene	Trace-30	American
			Toluene	Trace-60	Petroleum
			Ethylbenzene	Trace-40	Institute,
			Total	Trace-90+	1978
Fresh water lakes and rivers					
Fox River, Illinois	?	?	Benzene	100-200	Brown et al.,
(Industrial river)			Toluene	100	1979
Lake Chetek	?	?	Benzene	8-9	Brown et al.,
(Resort lake)			Toluene	90-100	1979
Lake of the Woods,	?	?	Benzene	ND	Brown et al.,
Canada (Pristine)			Toluene	Trace	1979

[a] Total volatiles measured are comprised of 75% (benzene + toluene + xylenes); 25% other.
[b] Total six monocyclics: benzene, toluene, ethylbenzene and p-, m-, and o-xylenes.
[c] Unidentified: some presumably on estuaries.
ND = Not detected.

1 to 100 ppb, and in fresh water up to 200 ppb. Measurement
units used for concentrations of MAH reported throughout
this paper are as follows:

ppm=μl/L, ppb=nl/L in water; ppm=nl/g in tissues.

Tables 4 and 5 summarize types and concentrations of low-
boiling point petrochemicals, including MAH, found in
tissues of striped bass collected from the Carquinez Straits
area of the San Francisco Bay-Delta (Whipple, 1979) during
their upward spawning migration. As far as we know, there
are no equivalent data for low-boiling point MAH for other
fishes. Although traces of naphthalenes were present, few
fish tissues examined contained significant levels of dicyclic
aromatics. Other polycyclic aromatics (PAH) may be present,
but tissue samples have not yet been analyzed for PAH. Our
measurements of MAH in tissues of striped bass show sur-
prisingly high concentrations (maximum in liver tissue=approx.
5.7 ppm; in ovary tissue=approx. 1.3 ppm; Table 5). The
predominant monocyclic aromatic found is benzene. These concentrations closely approximate those resulting in tissues of
fish exposed in the laboratory to 100 ppb of benzene (refer
to Table 10). The initial data from field-captured fish are
highly suggestive of a potential problem resulting from the
occurrence of these compounds in a chronically polluted estuary, particularly in view of the effects observed at similar
levels in laboratory studies.

RELATIONSHIP OF MONOCYCLIC AROMATIC HYDROCARBONS TO OTHER FAC-
TORS IN THE AQUATIC ENVIRONMENT

As shown above, analyses of tissue samples from striped
bass collected in the San Francisco Bay - Delta show the presence of MAH in relatively high concentrations (Whipple, 1979,
1980; Whipple *et al.*, 1979; Jung and Bowes, 1980). Although
studies are still underway, we find that there are high correlations between the presence of these compounds and poorer
condition of fish and their gametes. However, the cause and
effect relationships between MAH and certain deleterious effects in the field-captured fish are as yet unclear, and laboratory tests are being performed in an attempt to clarify
them.
      Traditionally, much of the work on effects of petroleum
hydrocarbons (including MAH) on aquatic organisms has been
restricted to controlled laboratory studies, testing single
compounds. This is primarily because of the difficulties in

TABLE 4. Low-boiling point petrochemicals scanned for in striped bass (*Morone saxatilis*) from San Francisco Bay; n=70 fish. Those compounds identified in striped bass liver tissue (total of 14 compounds) indicated with an asterisk. (Whipple, 1979, 1980; Whipple et al., 1979).

Aromatics	Alkyl cyclohexanes
*Benzene	*Methylcyclohexane
*Toluene	*1,4- and/or 1,1-Dimethylcyclohexanes
*Ethylbenzene	*1,2-Dimethylcyclohexane
*para-Xylene	
*meta-Xylene	
*ortho-Xylene	
Isopropylbenzene	
n-Propylbenzene	
*1, 3, 5-Trimethylbenzene	
1, 2, 4-Trimethylbenzene, tert-Butylbenzene	
1, 2, 3-Trimethylbenzene, sec-Butlybenzene	
Isobutylbenzene	
n-Butylbenzene, 1-Phenylbutene-2	
tert-Pentylbenzene	
1, 2, 4, 5-Tetramethylbenzene	
1, 2, 3, 5-Tetramethylbenzene	
1, 2, 3, 4-Tetramethylbenzene	
*Naphthalene (Trace only) Hexylbenzene	
Pentamethylbenzene	
2-Methylnaphthalene	
*1-Methylnaphthalene (Trace only)	
n-Heptylbenzene	
Hexamethylbenzene	
n-Octylbenzene, n-Nonylbenzene	
Fluorene n-Decylbenzene	

TABLE 5. Summary of low-boiling point petroleum hydrocarbons in liver and gonads[a] in striped bass (*Morone saxatilis*). Prespawning adults; n=70; gonads maturing (Whipple, 1979; Whipple et al., 1979)

	Mean samples w/hydrocarbons (ppm-WW)	No. fish w/hydrocarbons detectable	No. fish without hydrocarbons (ND)	Total no. fish analyzed	Range in mean Low — High (ppm-WW)
Liver tissue					
Females					
Total monocyclic aromatics[b]	0.430	46	10	56	0.020 - 3.285
Total alkyl cyclohexanes[c]	0.202	40	16	56	0.020 - 1.188
Total both	0.606	48	8	56	0.020 - 3.481
Males					
Total monocyclic aromatics	2.338	9	5	14	0.020 - 5.735
Total alkyl cyclohexanes	0.648	7	7	14	0.020 - 1.567
Total both	2.894	10	4	14	0.020 - 7.302
Gonadal tissue					
Females - Ovaries					
Total monocyclic aromatics	0.088	26	30	56	0.020 - 1.311
Total alkyl cyclohexanes	0.113	10	46	56	0.020 - 0.202
Total both	0.131	30	26	56	0.020 - 1.408
Males - Testes					
Total monocyclic aromatics	0.576	3	11	14	0.020 - 1.687
Total alkyl cyclohexanes	Trace	2	12	14	(Trace only)
Total both	0.589	3	11	14	0.020 - 1.707

ND = not detectable
[a] Data from 1978; fish from 1979 and 1980 being analyzed.
[b] Total monocyclic aromatics = Benzene + toluene + ethylbenzene + m-xylene + o-xylene; p-xylene undetectable.
[c] Total alkyl cyclohexanes = 1,4-dimethylcyclohexane + 1,1-dimethylcyclohexane + 1,2-dimethylcyclohexane.

testing complex mixtures of compounds such as occur in crude
oil and petroleum fractions. It has been pointed out by
Sprague (1971) and many others that laboratory studies are not
necessarily indicative of what actually happens in the natural
environment. The effect of any pollutant, or class of pollutants such as the MAH, on a fishery's population must ultimately be considered in relation to many other interacting
variables before we can estimate its relative contribution to
declines in the fishery. These variables may be naturally occurring (e.g., salinity) and/or man-introduced (e.g., pesticides). Fishes often migrate into or through estuaries with
chronic levels of several different pollutants, probably interacting additively and/or synergistically with MAH and with
natural factors.

An example of the probable interactions of various environmental factors affecting fish is shown in the conceptual
model of Figure 2. This figure shows a portion of a qualitative model for the striped bass population in the San Francisco Bay - Delta, and postulates the effective variables and

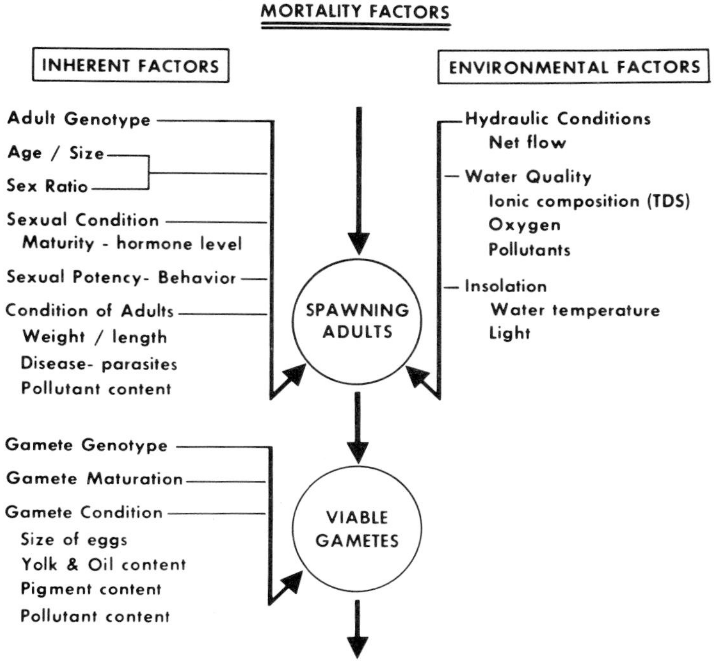

MORTALITY FACTORS

FIGURE 2. Conceptual model of factors affecting
mortality in striped bass (Morone saxatilis) during the
spawning adult and gamete life history stages (from Whipple,
1979, 1980; Whipple et al., 1979).

their interactions affecting survival (or alternatively growth
or reproduction) of a given life history stage (Whipple, 1979,
1980; Whipple *et. al.*, 1979). The stages shown in the figure
are the spawning adults and their gametes. We believe these
stages are among the most sensitive to the effects of pollutants. For this population, we are hypothesizing that pollutants, including MAH, are interacting with other environmental
variables to affect adults and their gametes. Our ultimate
goal is to make this model quantitative and predictive, determining the relative contribution of pollutants to reducing
survival, growth and reproduction in the population considered.
Very little has been done to quantify pollutant effects on a
population in the actual environment where pollutants may be a
probable factor in population persistence.

     Although we would like ultimately to be able to predict
effects of MAH on fishes in the aquatic environment, until
further field and laboratory work has been done on their interaction with other factors, such prediction will be difficult. Data are available, however, from several studies on
bioaccumulation and effects of MAH in fishes (primarily benzene and toluene). These studies enable us to make some preliminary predictions on the effects of MAH on fishes in the
natural environment.

     One of the difficulties in using data available from lab-
oratory experiments to predict effects in the field, is that
studies have been done under variable conditions. For example,
we believe that fishes possess considerable interspecific and
intraspecific inherent variability in their responses to pollutants, and environmental variability complicates this situation further. Table 6 lists some selected variables which we
hypothesize will influence bioaccumulation of MAH in both
field and laboratory studies and, subsequently, effects of
these compounds on fishes. The point of listing these sources
of variability is to provide a framework for future laboratory
studies, where variation can be controlled or eliminated.

     Further discussion of the influence of the variable listed
in Table 6 on uptake, retention and depuration of MAH follows.

UPTAKE AND ACCUMULATION OF MONOCYCLIC AROMATIC HYDROCARBONS

     The major route of uptake for MAH in fishes is probably
through the water column, although this is not the only potential source of exposure to MAH (Table 6). Few studies exist
comparing the routes of uptake and their relative importance
in the bioaccumulation of MAH in fishes. We have done studies
comparing the bioaccumulation of benzene in Pacific herring

TABLE 6. Some inherent and environmental variables affecting uptake, metabolism, retention and effects of MAH on fishes. Some factors (*) are still hypothetical, although currently being tested (Whipple, 1979, 1980; Whipple et al., 1979).

Major Inherent Factors	Major Environmental Factors
Interspecific variability (Table 9)	Source of exposure (Tables 7, 8) Water, food, particulates, or sediments, or combinations of these
Sex (Table 5)	Type of component or component mixture (Table 11)
Life history stage (Tables 7, 8, 9, 11)	Concentration, length of exposure (Tables 9, 11)
*Intraspecific genotypic variability Metabolism — functional Structural	
	Temperature (Table 11)
*Condition when exposed Degree of disease or parasitism Previous feeding regime — amount of fat Existing pollutant load	Salinity (Table 11)
	Alkalinity — pH (Table 11)
	*Presence of other pollutants Chlorinated hydrocarbons Heavy metals Others

Laboratory experiments can reduce this variability to a considerable degree by controlling (fixing) the variability and thus maximizing the variation due to effects of test (treatment) components.

In field experiments, more variables are uncontrollable and sample sizes must be larger, to reveal differences between exposed and unexposed populations.

larvae exposed from the water only, food only (contaminated rotifers) and water and food together (Eldridge and Echeverria, 1977). We have also compared the uptake and bioaccumulation of the components in the WSF of Cook Inlet crude oil in starry flounder exposed through water only, food only (contaminated clams) and both water and food (Whipple *et al.*, 1978a; also Yocom *et al.*, ms in prep.). Summaries of concentrations accumulated from different exposure sources are shown in Tables 7 and 8. Our general conclusion from studies done to date is that only trace amounts of MAH are taken up from contaminated food and that, in fish, bioaccumulation of these compounds probably does not occur from food. There are no studies of the potential bioaccumulation from other sources (e.g., particulates).

A number of studies (discussed below), however, show that MAH are readily and rapidly bioaccumulated from water. We believe that this is the major route of uptake in the field, and that the source of exposure is most likely to be a low concentration (1-100 ppb, occasionally higher) of MAH occurring chronically in polluted estuaries.

The rates of uptake of MAH, particularly benzene, are very rapid. Figure 3 shows uptake of benzene in the blood of striped bass exposed to approximately 1 ppm benzene (Benville *et al.*, 1980). Other studies show that uptake to maximum equilibrium levels in most tissues is reached within 2-24 hrs (Korn *et al.*, 1976a, 1977). Maximum levels in organs involved with metabolism of MAH are reached later (48-72 hrs). In longer exposures, such as in starry flounder exposed to the WSF of Cook Inlet crude oil in the laboratory, the order of magnitude of accumulation remained the same over the entire exposure period once equilibrium levels were reached (Figure 4). There was some indication of increased accumulation at about five weeks.

Tables 9 and 10 summarize the concentrations and bioaccumulation of benzene and toluene in fishes. Data are summarized to compare species, life history stages and tissues. Most data are from laboratory experiments (L); however, some field data (F) are also included for comparison.

Test concentrations of single components (benzene and toluene) varied from 1 ppb to approximately 10,000 ppb (0.001 - 10.000 ppm). The 10,000 ppb (10 ppm) level was near lethal for some species. Most test concentrations of both single components and total MAH, however, were approximately 100 ppb and were chosen to approximate field chronic levels. Data from field-captured striped bass (F) show that maximum accumulation of benzene is comparable to that in striped bass exposed to approximately 100 ppb in the laboratory. Except at the lowest and highest test concentrations (1 ppb and 10,000 ppb), the order of magnitude of accumulation in juvenile and

TABLE 7. Mean maximum concentrations of benzene and/or metabolites in herring larvae (*Clupea harengus pallasii*) exposed to ^{14}C-benzene in seawater and/or in food (rotifers; *Brachionus plicatilis*) for 72 hours (Eldridge and (Echeverria, 1977).

Initial Mean Benzene Concentration (ppm-μl/L)	Concentrations in Tissues - ppm; nl/g Wet Weight		
	Food + Water Exposure	Water Only	Food Only
0.144 - 2.10		0.49 - 8.16	
1.20	3.98		
1.20 (rotifers)			0.310

TABLE 8. Mean concentrations of different classes of petroleum hydrocarbons in some tissues of starry flounder (*Platichthys stellatus*) exposed to a total of 0.115 ppm MAH in the WSF of Cook Inlet crude oil and/or 8.0 ppm MAH in contaminated clams (*Tapes semidecussata*) for five weeks in the laboratory (Whipple et al., 1978a; also Yocom et al., ms in prep.). Control group (not shown) contained no detectable components.

Component Class Totals	Concentrations in Tissues (ppm; μg/g-WW)					
	Food + Water Exposure		Water Only		Food Only	
Liver	Mean	n	Mean	n	Mean	n
Alkyl cyclohexanes	26.45	22	30.38	20	1.22	21
*Monocyclic aromatics	19.53	22	21.44	20	0	21
Higher substituted benzenes	23.89	14	16.17	15	0	13
Dicyclic aromatics	10.15	14	8.37	15	0	13
Muscle						
Alkyl cyclohexanes	0.88	23	0.50	24	0	18
Monocyclic aromatics	1.04	23	0.60	24	0	18
Higher substituted benzenes	0.92	20	2.03	22	trace m-xylene	8
Dicyclic aromatics	0.30	20	0.31	22	0	8

*Monocyclic aromatics = Six most common in WSF of Cook Inlet Crude Oil; benzene, toluene, ethylbenzene, o-xylene, m-xylene, p-xylene.

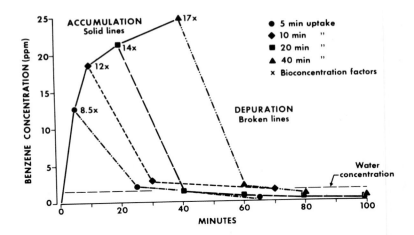

FIGURE 3. *Uptake, accumulation and depuration of benzene in blood of striped bass* (Morone saxatilis) *exposed to approximately 1.0 ppm benzene through the water column for one hour (from Benville* et al., *1980).*

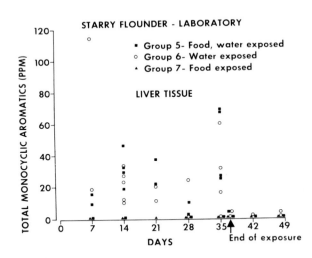

FIGURE 4. *Uptake, accumulation and depuration of six common monocyclic aromatic hydrocarbons in the WSF of Cook Inlet crude oil in the liver tissue of starry flounder* (Platichthys stellatus). *Exposed to 0.100 ppm total MAH in the WSF of Cook Inlet crude oil through the water column for five weeks (from Yocom* et al., *ms in prep.; Whipple* et al., *1978a).*

adult fishes was the same in most species (water concentration
from 10 to 5,000 ppb), being about 10 to 100 times the water
concentration (8.3 to 140 X). Maximum accumulation of
benzene, as a part of the WSF of crude oil, appeared to be of
the same order of magnitude as of benzene alone. Maximum
accumulation of total MAH in the WAF appeared to be approximately 100 times water concentrations in laboratory studies
(1 to 350 X). Toluene accumulated to higher levels than did
benzene in most laboratory studies, depending upon tissue.

When tissues and species were compared (Table 10),
northern anchovies accumulated the highest concentrations,
probably because these fish were more easily stressed by
laboratory conditions. Lowest accumulation occurred in
Pacific herring. The other species had approximately the
same maximum accumulations.

Accumulation of MAH varied with sex and proximity of males
and females to spawning (Whipple, 1979). In striped bass
during spawning season, benzene and other MAH were approximately equally distributed between liver and ovaries in
females. Accumulation in males was higher in liver and levels
in testes were usually undetectable, some with a trace of
toluene. During the nonspawning season, levels were higher
in female tissues, including liver, than in male. The levels
in liver and gonads were probably largely determined by the
amount of stored lipid, with increasing amounts of lipid
being transferred from the liver to the ovaries during
vitellogenesis. A corresponding change apparently does not
occur in males during the spawning season. The rest of the
year, female livers are higher in lipid than male livers.
When data are pooled for all species and stages of juveniles
and adults, females show higher accumulation (40 X) than
males (26 X).

Differences among life history stages for Pacific herring,
striped bass, starry flounder and coho salmon are also
summarized in Table 9. Accumulation in early life history
stages of striped bass was higher than in Pacific herring and
coho salmon, particularly in feeding larvae. Maximum
accumulation of ^{14}C-benzene and metabolites occurred in
feeding larvae of striped bass (1400 X at 384 hrs and still
increasing). Gonadal eggs also reached fairly high levels
when maturing (14-29 X). Generally, in striped bass, the
different stages bioaccumulate in the following, increasing
order: yolk-sac larvae, nonspawning adults and juveniles,
gonadal eggs, spawned eggs, prespawning adults, nonfeeding
post yolk-sac larvae and feeding post yolk-sac larvae. In
starry flounder, the order of increasing bioaccumulation is
as follows: immature ovaries, juveniles, nonspawning adults,

TABLE 9. Summary of mean concentrations and bioaccumulations of benzene and toluene in different species and life history stages. Most experiments were at an exposure level of approximately 0.100 ppm.

Variable Species and Life history stages	Water exposure level (ppm)	Benzene Range – Mean maximum concentrations[a] (ppm)	Benzene Maximum accumulation factor[b]	Benzene Hours to maximum accumulation	Toluene Range – Mean maximum concentrations[a] (ppm)	Toluene Maximum accumulation factor[b]	Toluene Hours to maximum accumulation	References
Starry flounder								
Gonadal eggs								
Immature	L 0.133 B[c] / L 0.088 T[c] / L 0.047 B[c]	ND	No accum.	--	0.200–0.830	9.4x	504	Whipple et al., 1978b
Mature	L 0.133 B[c] / L 0.088 T[c]	1.18–1.86	39x	72–96	2.16–5.21	93x	72–96	Whipple et al., 1978b
Juveniles	L 0.041 B[c] / L 0.043 T[c]	4.61–18.60	140x	504	14.24–34.71	395x	504	Whipple et al., 1978b
Nonspawning adults	L 0.047 B[c]	0.130–4.19	100x	840	0.420–8.90	205x	840	Yocom et al., ms in prep.; Whipple et al., 1978a
Prespawning adults	L 0.056 T[c]	1.44–1.57	33x	72–96	3.19–4.61	82x	72–96	Whipple et al., 1978b
Coho salmon								
Spawned eggs	*L 1.80 T				8.10	4.5x	unknown	Korn and Rice, in press
Alevins	*L 1.80 T				2.70	1.5x	unknown	Korn and Rice, in press
Emergent fry	*L 1.80 T				6.70	3.7x	unknown	Korn and Rice, in press
Pacific herring								
Gonadal eggs								
Immature	*L 0.100 B / *L 0.100 T	0.240	2.4x	24	0.440	4.4x	6	Korn et al., 1977
Mature	*L 0.100 B	1.40	14.0x	24				Struhsaker, 1977
Spawned eggs	*L 0.100 B	0.600	6.0x	12				Struhsaker, 1977
Feeding larvae								
Fed	*L 0.100 B	0.700	7.0x	not rchd 72				Struhsaker, 1977
Fed	*L 0.140 B	0.250	1.8x	not rchd 72				Eldridge and Echeverria, 1977
Not fed	*L 0.144 B	0.230	1.6x	6				Eldridge and Echeverria, 1977
Nonspawning adults	*L 0.100 B / *L 0.100 T	0.830	8.3x	6–48	3.90	39x	6–72	Korn et al., 1977

Striped bass

Variable	Water exposure level (ppm)	Benzene			Toluene			References
		Range - Mean maximum concentrations[a] (ppm)	Maximum accumulation factor[b]	Hours to maximum accumulation	Range - Mean maximum concentrations[a] (ppm)	Maximum accumulation factor	Hours to maximum accumulation	
Gonadal eggs (mature)	F (?)[d]	Trace only	No. accum.	chronic	Tr.-0.330	(3.3x)[d]	chronic	Whipple, 1979, 1980
Spawned eggs	*L 0.138	4.512	33x	33				Eldridge and Benville, ms in prep.
Yolk-sac larvae	*L 0.415	4.674	11x	120				Eldridge and Benville, ms in prep.
Feeding larvae Fed	*L 0.159	226.155	1400x	not rchd 384				Eldridge and Benville, ms in prep.
Not fed	*L 0.159	73.663	460x	360				Eldridge and Benville, ms in prep.
Juveniles	*L 0.088	1.302	15x	6				Korn et al., 1976a
Nonspawning adults	L 0.095	4.35	46x	336	0	No toluene, only benzene		Hirsch et al., ms in prep. (b)
	F(0.100)[d]	Tr.-4.58	(46x)[d]	chronic				Whipple et al., ms in prep.
Prespawning adults	L 0.750	2.0-20.0	27x	336				Hirsch et al., ms in prep. (b)
	F(0.100)[d]	Tr.-2.27	(23x)[d]	chronic	Tr.-0.330	?	chronic	Whipple et al., ms in prep.

*Indicates labeled compounds (accumulation of ^{14}C-benzene or ^{14}C-toluene + their metabolites, if any).

L = accumulation in laboratory-exposed fish.   B = benzene.

F = accumulation in field-captured fish.   T = toluene.

[a]Measurement of uptake in bile not included in mean maximum concentrations.

[b]Maximum accumulation factor = Mean maximum concentration of MAH in tissues divided by mean concentration in water column.

[c]WSF of Cook Inlet crude oil.

[d]If comparable to laboratory accumulation, approx. 0.100 ppm benzene in field.

503

TABLE 10. Summary of mean maximum concentrations and bioaccumulation of benzene in different species, tissues and sexes. Most experiments were at an exposure of approximately 100 ppb.

Tissues – Species[a]		Water Exposure Level (ppm)	Mean Maximum Concentration (ppm)	Maximum Accumulation Factors (Increasing order)
Gonads (Immature)				
Pacific Herring (A)	(*)	0.100	0.240	2.4 x
Striped Bass (A)	(−)	0.320	Trace	None
Stomach				
Striped Bass (J)	(*)	0.088	0.24	2.7 x
Testes (Mature)				
Striped Bass (A)	(−)	0.320	1.09	3.4 x
Heart				
Striped Bass (J)	(*)	0.088	0.26	2.9 x
Striped Bass (A)	(−)	0.320	1.82	5.7 x
Kidney				
Pacific Herring (A)	(*)	0.100	0.40	4.0 x
Striped Bass (A)	(−)	0.320	4.16	13 x
Pyloric Cecae				
Pacific Herring (A)	(*)	0.100	0.64	6.4 x
Colon				
Striped Bass (J)	(*)	0.088	1.32	15 x
Spleen				
Striped Bass (A)	(−)	0.320	6.40	20 x
Adrenal Gland				
Striped Bass (A)	(−)	0.320	6.72	21 x
Eye				
Striped Bass (A)	(−)	0.320	7.68	24 x
Blood				
Striped Bass (A)	(−)	0.320	8.64	27 x
Ovaries (Mature)				
Pacific Herring (A)	(*)	0.100	1.40	14 x
Striped Bass (A)	(−)	0.100	2.90	29 x
Muscle				
Striped Bass (J)	(*)	0.088	0.10	1.1 x
Pacific Herring (A)	(*)	0.100	0.63	6.3 x
No. Anchovy (A)	(*)	0.110	1.10	10 x
Striped Bass (A)	(−)	0.320	0.48	1.5 x

Table 10 con't.

Gills
Striped Bass (J)	(*)	0.088	0.49	5.6 x
Pacific Herring (A)	(*)	0.100	0.73	7.3 x
No. Anchovy (A)	(*)	0.110	4.60	42 x

Brain
Striped Bass (J)	(*)	0.088	0.63	7.2 x
Pacific Herring (A)	(*)	0.100	0.75	7.5 x
No. Anchovy (A)	(*)	0.110	4.60	42 x
Striped Bass (A)	(-)	0.320	7.36	23 x

Liver
Striped Bass (J)	(*)	0.088	0.86	9.8 x
Pacific Herring (A)	(*)	0.100	0.53	5.3 x
No. Anchovy (A)	(*)	0.110	6.01	55 x
Striped Bass (A)	(-)	0.320	6.72	21 x

Mesenteric Fat
Striped Bass (A)	(-)	0.320	35.0	109 x

Intestine
Striped Bass (J)	(*)	0.088	0.48	5.4 x
Pacific Herring (A)	(*)	0.100	0.83	8.3 x
No. Anchovy (A)	(*)	0.110	22.9	210 x

Gall Bladder (Bile)
Pacific Herring (A)	(*)	0.100	3.10	31 x
Striped Bass (J)	(*)	0.088	4.70	53 x
No. Anchovy (A)	(*)	0.110	480.	4400 x
Striped Bass (A)	(-)	0.320	3.80	12 x

[a]References: Korn et al., 1976a, 1977; Hirsch et al., ms. in prep.(b); Struhsaker, 1977.

(-) Benzene not labeled; (*) ^{14}C-Benzene and/or metabolites; (A) = Adult, (J) = Juvenile.

prespawning adults and mature ovaries (egg and larval stage
studies incomplete). Alevins of coho salmon accumulated more
than spawned eggs or emergent fry.

Tissues accumulating highest levels were those involved in
metabolism of compounds (^{14}C-benzene and metabolites measured)
and/or contained higher levels of lipids. The tissues are
arranged in order of minimum to maximum accumulations, the
least uptake occurring in testes and the highest in bile
(Table 10). In muscle, liver and bile, measurements were
made of both 1) ^{14}C-benzene and metabolites and 2) benzene
only. In the case of bile, a considerably higher proportion
of the measured concentration appears to be metabolites.

The uptake of benzene and toluene as a part of the WSF
of Cook Inlet crude oil by starry flounder appeared to be
slightly higher than uptake and accumulation of the individual
compounds, e.g., benzene in other species (Table 9). This
could be a species difference, however.

Figure 5A is a gas chromatogram showing the occurrence

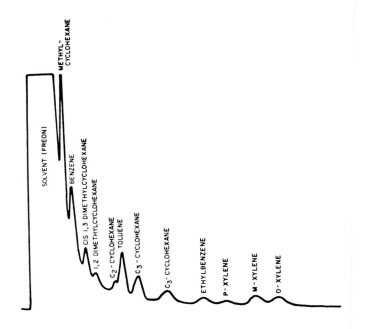

*FIGURE 5A. Chromatogram of low-boiling point hydro-
carbons, including monocyclic aromatic hydrocarbons, detected
in a maturing ovary of starry flounder (*Platichthys
stellatus*). Exposed to 0.115 ppm total MAH in the WSF of
Cook Inlet crude oil for seven days (from Whipple et al.,
1978b).*

of the low-boiling point compounds of methylcyclohexanes
and monocyclic aromatics in maturing ovarian tissue of
starry flounder exposed to the WSF of Cook Inlet crude oil
(Whipple *et al.*, 1978b). Highest concentrations were of
methylcyclohexane and benzene, but relative accumulations
differed because of the much lower concentrations of xylene
in the water. Figure 5B shows the relative uptake of
different classes of hydrocarbons in the WSF of Cook Inlet
crude oil in ovaries of starry flounder. Concentrations of
the six commonest monocyclics (M-1) were highest, followed
by methylcyclohexanes (CH), other substituted benzenes (M-2),
and finally by dicyclic aromatic hydrocarbons (D). Although
the MAH accumulate in fishes to relatively high concentrations
from the WSF, the relative toxicities of these classes of
compounds differ and effects cannot be determined on the basis
of concentration alone.

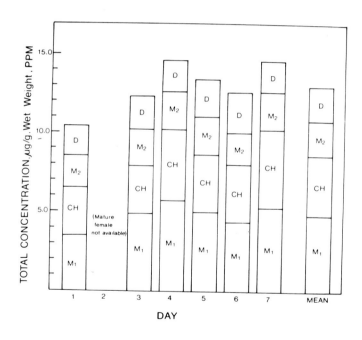

FIGURE 5B. Concentrations of monocyclic aromatics and
other compounds in maturing ovaries of starry flounder
(Platichthys stellatus). *M1 = six common monocyclics*
(benzene, toluene, ethylbenzene, p-, m-, and o-xylenes);
CH = total alkyl cyclohexanes; M2 = total higher substituted
benzenes; D = total dicyclic aromatics. Exposed to 0.115 ppm
total MAH in the WSF of Cook Inlet crude oil for seven days
(from Whipple et al., *1978b).*

METABOLISM

Although we know that uptake of MAH from water is very
rapid, and that bioaccumulation above water concentrations
occurs, we still do not know much about what happens to
these compounds within the fish. A considerable amount of
the MAH taken up is probably returned to the water through
the gills, unchanged. Some, however, is probably metabolized
(Korn *et al.*, 1976a, 1977).

Fish have oxidative enzyme systems for the metabolic
detoxification of xenobiotics, including the aromatic
petroluem hydrocarbons (Payne and Penrose, 1975). They are
associated with microsomes in the endoplasmic reticulum of
cells. These enzymes are NADPH-dependent, and called aryl
hydrocarbon hydroxylases (AHH). This area of research is
relatively new (1960's to present) and information on the
metabolic detoxification of petroleum hydrocarbons in fish
is still sparse. Research on mammals is summarized in the
papers of LaDu *et al.* (1971), White *et al.* (1973), Jerina and
Daly (1974) and Kappas and Alvares (1975). Recently, considerable work has been done on the metabolism of polycyclic
aromatics (PAH) in fishes (Lee, 1976, 1977; Stegeman and Sabo,
1976; Malins, 1977a, 1977b; Neff, 1978a, 1978b, 1979;
Varanasi and Malins, 1977). There is still little known, however, about the metabolism of MAH in fish. Gibson (1977)
summarized the differences in the metabolic processes used by
eucaryotic and procaryotic organisms to oxidize aromatic
hydrocarbons, including monocyclic aromatics such as benzene
and toluene. Bacteria apparently oxidize MAH to dihydrodiol
intermediates, followed by formation of catechols. The
catechols are then substrates for the enzymatic cleavage of
the aromatic ring. Fungi and higher organisms, on the other
hand, incorporate oxygen into the aromatic ring to form
arene oxides. The oxides undergo enzymatic addition of
water to yield *trans*-dihydrodiols. These, in turn, are
transformed into phenol, catechols and glutathione conjugates,
according to Jerina and Daly (1974).

Most of the studies indicate a basic similarity of
aromatic detoxification in mammals and fishes, including
evidence that the mixed function oxidases (MFO) are induced in
fish exposed to petroleum (Stegeman and Sabo, 1976; Payne and
Penrose, 1975).

Jerina and Daly (1974) summarized the hepatic metabolism
of benzene in mammals. Studies show that under physiological
conditions (*in vitro*), benzene oxide undergoes spontaneous
rearrangement to form phenol and react nonenzymatically with
the thiol group of glutathione. Benzene oxide also undergoes
enzymatic hydration to catechol. The general reactions are
as follows:

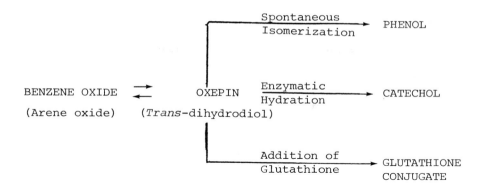

The rate of metabolism of benzene *in vitro* was relatively low and the arene oxides of benzene and alkylbenzenes very unstable. The hepatic metabolism of toluene and other aromatics is also described in Jerina and Daly (1974). The hepatotoxicity of these compounds in mammals is apparently caused by the arene oxides being covalently bound to hepatic protein, eventually resulting in necrosis. Addition of glutathione inhibits this process.

We do not know if the same process of metabolism of benzene occurs in fishes, but similarities in the MFO system suggest that it does. Roubal *et al.* (1977) compared the accumulation and metabolism of ^{14}C-labeled compounds (benzene, naphthalene and anthracene) in young coho salmon exposed through food and intraperitoneal injection. Although the metabolites of benzene were not identified, the results show that relative to naphthalene and anthracene, a lower proportion of the ^{14}C-activity of labeled benzene appeared in the form of metabolites at 24 hours in the tissues measured (brain, liver, gall bladder, flesh and carcass). Overall, the accumulation of ^{14}C-labeled benzene and metabolites was less than for ^{14}C-labeled naphthalene and anthracene. The paper also indicated that for naphthalene and anthracene, maximum ^{14}C-activity levels were lower and depuration slower when labeled compounds were administered via food as opposed to injection.

Studies of striped bass juveniles indicated that in liver and muscle tissue approximately half of the ^{14}C-labeled compounds were unchanged benzene and half in the form of metabolites when compared to gas chromatographic measurements of unchanged benzene (Korn *et al.*, 1976a, 1977; see also Table 10). In the gall bladder, however, most of the labeled material in bile was probably metabolites, as would be expected.

A recent study (Thomas and Rice, in press) examined the excretion of ^{14}C-labeled toluene and naphthalene from the gut of Dolly Varden char. Results showed that the major avenue for excretion of the parent hydrocarbons and their metabolites was via the gills, and minor portions were excreted via the gut and kidney. Most excretion via the gill was still in the form of the parent hydrocarbons; most excreted via the gut and kidney were in the metabolite phase.

The MAH bear structural similarity to many natural compounds such as steroid hormones, lipids, vitamins and neurotransmitters which also contain aromatic nuclei. This similarity may result in competition with other substrates for metabolism in the MFO system (LaDu *et al.*, 1971). MAH may also mimic effects of natural compounds, stimulating certain neuro-endocrine responses and behavior (Doggett *et al.*, 1977; LaDu *et al.*, 1971). More research in this area would be of con-siderable interest.

TRANSLOCATION

Based on the above studies, we propose that the trans-location of MAH through fish occurs as diagrammed in Figure 6. The major route of uptake is from water through the gills. Some uptake may be through the gut from swallowed water, particularly in salt water environments. A minor portion, if any, is taken through the intestinal wall. Most data indicate a minimal uptake of MAH through food (see Tables 7 and 8; Figure 4). The parent compounds readily solubilize in cell membranes (Roubal, 1974; Roubal and Collier, 1975) and are probably carried primarily via the erythrocytes (lipid cell membrane) through the general circulation in the blood (Benville *et al.*, 1980; Figure 3). Some of the monocyclics may also be carried by lipoproteins and leukocytes in the blood. Centrifugation of blood from benzene-exposed striped bass showed that the major portion of the benzene was associated with the erythrocytes, and not the serum fraction (Hirsch *et al.*, ms in prep. (b)). The blood circulates through the fish, and the aromatics apportion into target tissues and organs (Fig. 7) that have high lipid content (e.g., brain) (Korn *et al.*, 1976a, 1977). They are also transported to the liver where they are metabolized (some limited metabolic activity in other tissues is also possible). Benzene and metabolites are then both transported through the general circulation, where they reenter target tissues or are depurated. A considerable portion of the unchanged compounds probably diffuses back through the gills into the water. The

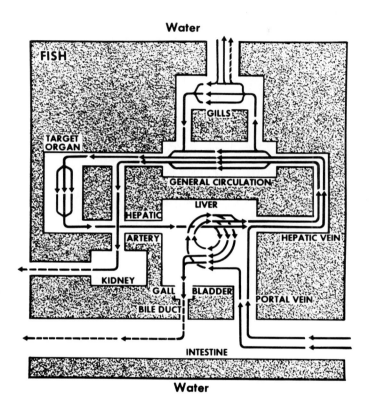

FIGURE 6. Translocation of monocyclic aromatic hydrocarbons and their metabolites in fish exposed through the
water column. Dashed lines indicate routes for excretion of
metabolites (adapted from Kappas and Alvares, 1975).

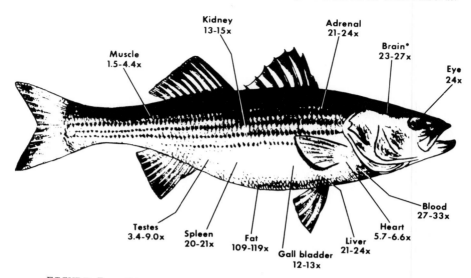

FIGURE 7. *Bioaccumulation of benzene in different tissues of the striped bass* (Morone saxatilis) *exposed to approximately 1.0 ppm benzene through the water column (from Benville et al., 1980).*

major route of excretion of metabolites appears to be via the bile into the intestine and out with the feces. Some metabolites are excreted through the gills. A minor portion of metabolites appears to pass through the kidney. At high exposure levels of MAH, the metabolic detoxification capacity of the liver may be exceeded and higher concentrations of unchanged compounds probably accumulate in the target organs.

DEPURATION

MAH are depurated relatively rapidly, parent compounds usually disappearing from most tissues within 48 hrs (e.g., Figs. 3 and 4), although the rate of disappearance is slower than the rate of uptake in most tissues (Fig. 3). The depuration of metabolites takes longer and persistence in tissues involved with metabolism is greater (Korn *et al.*, 1976a, 1977). In striped bass, Pacific herring and northern anchovy, residues of benzene/metabolites were detected in gills, liver and gall bladder seven days after termination of exposure. These organs, of course, are involved with metabolism and excretion of metabolites. In addition, residues were still detectable in fat tissue seven days after termination of exposure.

UPTAKE, ACCUMULATION AND DEPURATION IN LARVAL FISH

Adult and juvenile fishes appear to accumulate MAH very rapidly and to relatively high equilibrium levels within 24 hrs. The unchanged parent compounds are also rapidly depurated, although metabolites may persist for longer periods.

Larval fish, however, appear to have limited capacity to either metabolize or excrete ^{14}C-benzene. Table 9 and Figure 8 show some of the results of an experiment in which striped bass larvae were exposed to ^{14}C-labeled benzene from the water column in a semi-static system. Benzene was added daily to restore a test concentration of approximately 120 ppb benzene (Eldridge and Benville, ms in prep.). Figure 8 shows that the uptake of benzene in feeding larvae fed uncontaminated *Artemia* nauplii was higher than in starved larvae. The major route of uptake is still through water, but when larvae are actively feeding, they appear to continually accumulate benzene and/or metabolites more than when starved. Concentrations were still increasing 16 days after initiation of feeding (Day 24). Possible explanations for this are: 1) larvae have limited enzymatic capacity to metabolize benzene and thus accumulate benzene to high levels (at least 1400 X water concentration) or 2) larvae are unable to excrete accumulated metabolites or 3) both of these. The first explanation seems most plausible. Fetuses and young of mammals also have extremely limited capacity to metabolize aromatic hydrocarbons and other drugs (LaDu et al., 1971).

EFFECTS OF MONOCYCLIC AROMATICS

Acute Toxicity

The bioaccumulation of selected MAH in fishes was summarized previously (Tables 9 and 10). Generally, the effects of MAH will depend upon their concentration in fish and deposition in specific target organs and tissues. For example, the species, stages and tissues which accumulate highest levels, under most conditions, will be most sensitive to MAH; those bioaccumulating less generally will be least sensitive. The effects at a given concentration, however, will also vary according to other inherent and environmental variables as previously discussed and listed in Table 6.

The acute toxicity (96-hr LC_{50}'s) of MAH to fishes ranges from about 2.0 to 300 ppm (Table 11). When the toxicities of

Table 11. Acute bioassays of single monocyclics and for comparison, of water-soluble fraction (WSF) of oil containing monocyclics and other compounds.

Exposed Species/Stage	Salinity Temperature	System[a]	Compound	Concentration (ppm)	Effect (LC$_{50}$)[b] (TLm)[c]	References
Clupea harengus pallasi Pacific herring. Eggs through early cleavage (survival to hatching)	24 ppt SW 15.2°C	Semi-Static	Benzene	40-45	96h[b]	Struhsaker et al., 1974
Larvae 2 days after hatching	28 ppt SW 12.9°C	Semi-Static	Benzene	20-25	48h[b]	
Engraulis mordax Northern anchovy Eggs to hatching larvae (survival to hatching)	28 ppt SW 17.5°C	Semi-Static	Benzene	20-25	48h[b]	Struhsaker et al., 1974
Oncorhynchus kisutch Coho salmon Eggs (Hatching)	0 ppt FW 3.5-6°C	Semi-Static	Toluene	333 100	96h[b] 96h[b]	Korn & Rice, (In press)
Alevins Early				60	96h[b]	
Middle				20	96h[b]	
Late				9.36	96h[b]	

		System[a]	Compound	Value	Time	Reference
Eggs hatching	0 ppt FW 3.5-6°C	Semi-Static	Naphthalene	11.8	96h[b]	
Alevins						
Early				9	96h[b]	
Middle				8	96h[b]	
Late				4	96h[b]	
Emergent fry				2-3	96h[b]	Moles et al., 1979
Oncorhynchus tshawytscha						
Chinook salmon Juveniles	0 ppt FW 6°C	Semi-Static	WSF of Prudhoe Bay crude oil	3.59	96h[c]	
	9°C		Benzene	11.73	96h[c]	Moles et al., 1979
Oncorhynchus kisutch						
Coho salmon	0 ppt FW 8°C	Semi-Static	WSF of Prudhoe Bay crude oil	3.67	96h[c]	
	9°C		Benzene	14.09	96h[c]	Moles et al., 1979
Coho Salmon Juveniles	30 ppt SW 8°C	Static	Benzene	10-50	96h[b]	
			Toluene	10-50	96h[b]	
			Ethyl-benzene	10-50	96h[b]	
			Xylene	10-100	96h[b]	Morrow, 1974

[a] Systems: Static - Initial dose only, declining over test period; Semi-static - Daily addition to bring test concentration up to dose level over test period; Open flow - Constant exposure to open flow system of test concentrations. SW = seawater; FW = fresh water

Table 11 con't.

Exposed Species/Stage	Salinity Temperature	System[a]	Compound	Concentration (ppm)	Effect[b] (LC$_{50}$) (TLm)[c]	References
Oncorhynchus gorbuscha						
Pink salmon Juveniles	26-30 ppt SW 3.7-11[b]°C	Semi-Static	WSF of Cook Inlet crude oil	4.13 2.92	24h[c] 96h[c]	Rice et al., 1976
			WSF of #2 fuel oil	0.89 0.81	24h[c] 96h[c]	
Pink Salmon Smolts	0 ppt FW 4°C	Semi-Static	WSF of Prudhoe Bay crude oil	7.99	96h[c]	Moles et al., 1979
	28-30 ppt SW 4°C		WSF of Prudhoe Bay	3.73	96h[c]	
	0 ppt FW 4°C	Semi-Static	Benzene	17.09	96h[c]	
	28-30 ppt SW 4°C		Benzene	8.47	96h[c]	
Oncorhynchus nerka						
Sockeye salmon Smolts	0 ppt FW 6°C	Semi-Static	WSF of Prudhoe Bay crude oil	2.22	96h[c]	Moles et al., 1979
	28-30 ppt SW 6°C		WSF of Prudhoe Bay crude oil	1.05	96h[c]	
Sockeye Salmon Smolts	0 ppt FW 6°C	Semi-Static	Benzene	10.76	96h[c]	
	28-30 ppt SW 6°C		Benzene	5.55	96h[c]	

	Conditions	Exposure	Test substance	Value	Duration	Reference
Salvelinus malma						
Dolly Varden						
Juveniles	0 ppt FW 8°C	Semi-Static	WSF of Prudhoe Bay crude oil	2.75	96h[c]	Moles et al., 1979
			Benzene	11.96	96h[c]	
Smolts	0 ppt FW 8°C	Semi-Static	WSF of Prudhoe Bay crude oil	2.68	96h[c]	
	28-30 ppt SW 8°C	Semi-Static	WSF of Prudhoe Bay crude oil	1.38	96h[c]	
	0 ppt FW 8°C	Semi-Static	Benzene	11.90	96h[c]	
	28-30 ppt SW 8°C		Benzene	6.30	96h[c]	
(Dolly Varden) Smolts	26-30 ppt SW 3.7-11°C	Semi-Static	WSF of Cook Inlet crude oil	3.25	24h[c]	Rice et al., 1976
				2.94	96h[c]	
			WSF of #2 Fuel Oil	2.29	96h[c]	

Table 11 con't.

Exposed Species/Stage	Salinity Temperature	System[a]	Compound	Concentration (ppm)	Effect (LC$_{50}$)[b] (TLm)[c]	References
Morone saxatilis Striped bass Juveniles	25 ppt SW 16.0°C	Semi-Static	Benzene	6.9 5.8	24h[c] 96h	Benville and Korn, 1977
			Toluene	7.3 7.3	24h[c] 96h[c]	
			Ethyl-benzene	4.3 4.3	24h[c] 96h[c]	
			p-Xylene	2.0 2.0	24h[c] 96h[c]	
			m-Xylene	9.2 9.2	24h[c] 96h[c]	
			o-Xylene	11.0 11.0	24h[c] 96h[c]	
Striped Bass Juveniles	29 ppt SW 17.4°C	Open Flow	Benzene	10.9	96h	Meyerhoff, 1975

Species	Conditions	Test type	Compound	LC50	Time	Reference
Eleginus gracilis Saffron cod	26-30 ppt SW 3.7-11[b] C	Semi-Static	WSF of Cook Inlet crude oil	2.48 2.28	24h[c] 96h[c]	Rice et al., 1976
			WSF of #2 fuel oil	>4.56 2.93	24h[c] 96h[c]	
Aulorhynchus flavidus Tube-snout	26-30 ppt SW 3.7-11[b] C	Semi-Static	WSF of Cook Inlet crude oil	--- 1.34	24h[c] 96h[c]	Rice et al., 1976
			WSF of #2 fuel oil	--- ---	24h[c] 96h[c]	
Thymallus arcticus Arctic grayling	0 ppt FW 9°C	Semi-Static	WSF of Prudhoe Bay crude oil	4.40	96h[c]	Moles et al., 1979
	9°C		Benzene	14.71	96h[c]	
Cottus cognatus Slimy sculpin Juveniles	0 ppt FW 9°C	Semi-Static	WSF of Prudhoe Bay crude oil	6.44	96h[c]	Moles et al., 1979
	9°C		Benzene	15.41	96h[c]	
Gasterosteus aculeatus Three-spined stickleback Adults	0 ppt FW 5°C	Semi-Static	WSF of Prudhoe Bay crude oil	>10.45	96h[c]	Moles et al., 1979
	8°C		Benzene	24.83	96h[c]	

Table 11 con't.

Exposed Species/Stage	Salinity Temperature	System[a]	Compound	Concentration (ppm)	Effect (LC$_{50}$)[b] (TLm)[c]	References
Lepomis macrochirus Bluegill Adults	0 ppt FW 25°C	Static	Cyclohexane	42.33 40.00 34.72	24h[c] 48h[c] 96h[c]	Pickering and Henderson, 1966
			Benzene	22.49 22.49 22.49	24h[c] 48h[c] 96h[c]	
			Toluene	24.00 24.00 24.00	24h[c] 48h[c] 96h[c]	
			Ethylbenzene	35.08 32.00 32.00	24h[c] 48h[c] 96h[c]	
			Xylene	24.00 24.00 20.87	24h[c] 48h[c] 96h[c]	
			Phenol	25.85 23.88 23.88	24h[c] 48h[c] 96h[c]	
Carassius auratus Goldfish Adults	0 ppt FW 25°C	Static	Cyclohexane	42.33 42.33 42.33	24h[c] 48h[c] 96h[c]	Pickering and Henderson, 1966
			Benzene	34.42 34.42 34.42	24h[c] 48h[c] 96h[c]	
			Toluene	57.68 57.68 57.68	24h[c] 48h[c] 96h[c]	

Species	Conditions	Test type	Compound	Value	Time
Poecilia reticulata Guppy 6 months old	0 ppt FW 25°C	Static	Ethylbenzene	94.44	24h[c]
				94.44	48h[c]
				94.44	96h[c]
			Xylene	36.81	24h[c]
				36.81	48h[c]
				36.81	96h[c]
			Phenol	49.86	24h[c]
				49.13	48h[c]
				44.49	96h[c]
			Cyclohexane	57.68	24h[c]
				57.68	48h[c]
				57.68	96h[c]
			Benzene	36.00	24h[c]
				36.00	48h[c]
				36.00	96h[c]
			Toluene	62.81	24h[c]
				60.95	48h[c]
				59.30	96h[c]
			Ethylbenzene	97.10	24h[c]
				97.10	48h[c]
				97.10	96h[c]
			Xylene	34.73	24h[c]
				34.73	48h[c]
				34.73	96h[c]
			Phenol	49.86	24h[c]
				49.86	48h[c]
				39.19	96h[c]

Pickering and Henderson, 1966

521

Table 11 con't.

Exposed Species/Stage	Salinity Temperature	System[a]	Compound	Concentration (ppm)	Effect (LC$_{50}$)[b] (TLm)[c]	References
Pimephales promelas Fathead minnow Adults	0 ppt FW Soft water 25°C	Static	Cyclohexane	35.08 35.08 32.71	24h[c] 48h[c] 96h[c]	Pickering and Henderson, 1966
			Benzene	35.56 35.95 33.47	24h[c] 48h[c] 96h[c]	
			Toluene	46.31 46.31 34.27	24h[c] 48h[c] 96h[c]	
			Ethylbenzene	48.51 48.51 48.51	24h[c] 48h[c] 96h[c]	
			Xylene	28.77 27.71 26.70	24h[c] 48h[c] 96h[c]	
			Phenol	40.60 40.60 34.27	24h[c] 48h[c] 96h[c]	

Pimephales promelas
Fathead minnow
Adults

0 ppt FW
Hard water
25°C

Static

Cyclohexane	42.33	24h[c]		Pickering and Henderson, 1966
	42.33	48h[c]		
	42.33	96h[c]		
Benzene	34.42	24h[c]		
	32.00	48h[c]		
	32.00	96h[c]		
Toluene	56.00	24h[c]		
	56.00	48h[c]		
	42.33	96h[c]		
Ethylbenzene	42.33	24h[c]		
	42.33	48h[c]		
	42.33	96h[c]		
Xylene	28.77	24h[c]		
	28.77	48h[c]		
	28.77	96h[c]		
Phenol	38.62	24h[c]		
	38.62	48h[c]		
	32.00	96h[c]		

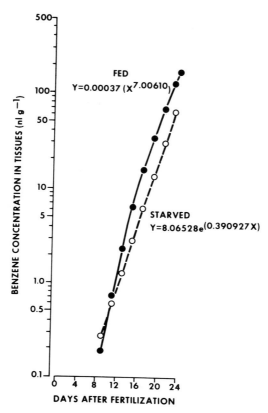

FIGURE 8. Uptake and accumulation of ^{14}C-labeled benzene in feeding and starved larvae of the striped bass (Morone saxatilis), *constantly exposed to a concentration of approximately 0.100 ppm ^{14}C-benzene through the water column (from Eldridge and Benville, ms in prep.).*

benzene to juveniles of different species are compared, salmonids appear slightly more sensitive than striped bass. Freshwater species such as the grayling, sculpin, stickleback, etc., appear less sensitive. The most sensitive to benzene was the sockeye salmon and the least sensitive, the guppy.

Among life history stages the most sensitive, at both acute and chronic levels, appears to be the feeding larvae, followed by gametes, prespawning adults, eggs, embryos and nonspawning adults. The least sensitive stage is the juvenile. This is in accord with the relative bioaccumulations at these stages (Table 9). Most acute bioassays are done with juvenile fish, although they appear to be the least

sensitive to MAH. More susceptible stages should be selected
for bioassays.

Recent work on striped bass (*Morone saxatilis*) and starry
flounder (*Platichthys stellatus*) indicates that there may be
intraspecific genotypic differences in the accumulation of
MAH. This may mean that there is also intraspecific
variability in susceptibility to the effects. In the striped
bass, for example, intraspecific differences in color
pattern and some meristic characteristics appear to correlate
positively with the MAH concentration in liver and gonads
(Whipple, 1979; Whipple *et al.*, ms in prep.). Factor analytic
results are being examined for correlations between genotypic
characters and concentrations of MAH (and other pollutants)
in striped bass. Right- and left-eyed starry flounder also
appear to vary in their relative uptake of MAH and other
components in the WSF of Cook Inlet crude oil (Yocom *et al.*,
ms in prep.). If laterality in starry flounder is genetically
based, the variants may also differ physiologically and metabolically in their susceptibility to MAH.

Fish may also vary in sensitivity to MAH if they are
already stressed by other environmental factors, or are low in
energy reserves due to spawning or other stress (see Table 6).
In striped bass, for example, there appears to be a strong
relationship between poor condition (estimated by condition
factors) and a higher concentration of MAH (Whipple, 1979;
Whipple *et al.*, ms in prep.). The degree of parasitism
(number of types, abundance, severity of host reactions) also
appears to correlate with the tissue concentration of MAH
(Whipple, 1979; Whipple *et al.*, ms in prep.). Cause and
effect relationships, however, are as yet unclear. For
example, does a heavily parasitized fish take up more MAH, or
does a fish stressed by MAH also become more heavily
parasitized, or do both events occur? Laboratory tests are
planned to clarify this issue. Moles (1980) found that coho
salmon fry (*Oncorhynchus kisutch*), infested with clam
glochidia of *Anodonta oregonensis* were significantly more
sensitive to toluene, naphthalene and the WSF of Prudhoe
Bay crude oil than uninfested fish.

Work is also being done to measure the pollutant load in
striped bass from San Francisco Bay, including not only MAH,
but heavy metals, PCB's, pesticides, etc. (Whipple, 1979,
1980; Whipple *et al.*, 1979; Whipple *et al.*, ms in prep.).
This work is still underway; however, results so far show
high loads of zinc, copper, mercury, PCB's and others (Jung
and Bowes, 1980). The interaction of the existing pollutant
load with MAH uptake in striped bass is being studied.

The relative toxicity of individual monocyclics (Table
11) varies with their solubility and bioaccumulation. The

position of the alkyl substitution on the aromatic ring also
appears to affect toxicity (Benville and Korn, 1977). In
general, *p*-xylene and benzene appear most toxic to fish, while
toluene, *m*- and *o*-xylene are least toxic. Ethylbenzene is
usually intermediate in toxicity. In comparison, cyclohexane
and phenol (phenol is a major metabolite of monocyclics) are
about as toxic as toluene (Pickering and Henderson, 1966;
Table 11).

Exposure periods in most acute bioassays (Table 11) are
24, 48 and 96 hrs. For most MAH the toxicity at 24 hrs
does not vary significantly from that at 96 hrs. We have
found that, although death may not occur at selected exposure
levels, delayed mortality often occurs after 96 hrs
(Struhsaker *et al.*, 1974). We feel that it is important to
examine surviving fish for some time after termination of
exposure. The length of exposure relative to MAH concentration is very important and should be considered in tests
of the effects of MAH on fishes, particularly when interested
in chronic low level effects, such as may occur in a polluted
estuary.

The source of exposure in most acute bioassays is through
the water column only. As discussed above, there appears to
be little accumulation through food and thus toxic effects
would not be expected through this route.

Studies done at Auke Bay, Alaska (Rice *et al.*, 1976;
Moles *et al.*, 1979) showed that temperature and salinity
affect the toxicity of MAH and the WSF of crude oil (Table
11). MAH are usually more toxic to fish in seawater than in
freshwater, and more toxic at lower temperatures. Freshwater
fish also appear to be slightly more sensitive to some MAH in
hard water (higher alkalinity) than in soft water (Pickering
and Henderson, 1966; Table 11).

The interaction of the MAH with one another, with other
hydrocarbons in the WSF of crude oil, and with other classes
of pollutants (such as chlorinated hydrocarbons and heavy
metals) is obviously complex. Little is known about whether
effects are additive, synergistic or antagonistic (Caldwell
et al., 1977). An acute bioassay of benzene and the WSF of
Prudhoe Bay crude oil on salmonids (Table 11) shows that the
toxicity of the WSF is greater than that of benzene alone.
Generally in acute bioassays, the polycyclic aromatics appear
to be more toxic and more persistent in tissues than the MAH
(Rice *et al.*, 1977; Neff, 1979). However, in the environment
their solubility in water is much less and the uptake probably
lower. Field data from San Francisco Bay also indicate that
dicyclics are not being bioaccumulated in striped bass (Tables
4 and 5). Alkyl cyclohexanes may be an important chronic
pollutant interacting with MAH, since they are bioaccumulated

to a higher degree and are more persistent (Tables 4, 5 and 8) and methylcyclohexane is about as toxic as toluene (Table 11). Competitive inhibition or stimulation between chlorinated hydrocarbons and petroleum hydrocarbons is probable, depending upon relative levels and their demand on the MFO system for detoxification (LaDu *et al.*, 1971). Heavy metals often act as inhibitors of the enzymes in the MFO system and could potentiate the effects of hydrocarbons, including monocyclics, by decreasing the rate of metabolic detoxification and increasing accumulated levels in target organs (Maines and Kappas, 1977). Further studies on the interactions of components within these major classes of pollutants are badly needed before we can determine their combined effects on fishes in the environment.

Hypothesized Modes of Action

We still know very little about the modes of action of MAH or other petrochemicals. We know that effects of MAH will differ according to several factors discussed previously, among them: 1) the life history, 2) the exposure level (lethal, high or low sublethal) and 3) the length of the exposure (short-term acute vs. long-term chronic). Similar effects probably occur whether fish are exposed to MAH, other aromatics (e.g., PAH), the total WSF of crude oils, or even other classes of hydrocarbons. Effects specifically attributable to single compounds are rare, or at least difficult to ascertain. From the literature it appears that effects of exposure to chlorinated hydrocarbons (pesticides) are similar to those observed in fishes exposed to petroleum hydrocarbons. A possible explanation is that the modes of action of most toxic hydrocarbons are similar, and that many effects are a result of a generalized enzymatic and hormone-mediated syndrome of responses to stress, similar no matter what the stressor. We hypothesize that four effect syndromes are generally seen: 1) effects due to inhibition or stimulation of enzymatic systems in metabolism, 2) hormone-mediated adaptive responses, 3) damage to organs due to hormone-mediated responses and 4) damage due to specific effects of an individual compound. Which responses occur probably depends primarily on the concentration and length of the exposure.

The hypothesized modes of action and effects are summarized in Table 12. Several of these modes of action are still hypothetical, and are meant to act only as temporary guidelines for study of the effects of MAH on fishes. Where observations or research tend to substantiate these hypotheses, references are noted.

Table 12. Some hypothesized modes of action and effects (functional and structural) of monocyclic aromatic hydrocarbons (MAH) at different life history stages and at different concentration levels of exposure. Hypothetical relationships are indicated by an asterisk.

LOW SUBLETHAL EXPOSURES OF MAH

Early Life History Stages
(Ref. Nos. 2, 11, 19-23, 49, 55, 77, 82, 92-94)

Organizational Level of Effect: (Effects often delayed)	Spawned eggs, embryos to newly hatched larvae	Feeding larvae
Cellular	*Stimulation of metabolism *Increased ATP, enzyme activity *Accelerated cellular respiration *Increased oxygen requirement	
Tissue and/or Organ System	Energy allocation altered Less yolk and/or oil Yolk-oil energy diverted to compensate for toxicant effect Accelerated heartbeat, development rate; increased respiration	Energy allocation altered Remaining yolk-oil rapidly utilized Accelerated development rate; increased respiration Accelerated organogenesis
Individual	Premature hatching; smaller, less viable larvae Less buoyant or active larvae	Smaller larvae, less active Desynchronization of first feeding
Population	Decreased growth, survival	Decreased growth, survival

528

HIGH SUBLETHAL EXPOSURES OF MAII

Early Life History Stages
(Ref. Nos. 19-23, 49, 55, 77, 82, 88, 92-94)

Organizational Level of Effect:	Spawned eggs, embryos to newly hatched larvae	Feeding larvae
Cellular	*Inhibition of metabolism *Decreased ATP levels, enzyme activity Reduced cellular respiration *Mutagenesis (DNA lesions) *"Sticky" chromosomes Abnormal cleavage	*Inhibition of metabolism *Decreased ATP levels, enzyme activity Reduced cellular respiration *Inhibited neurotransmission
Tissue and/or Organ System	*Inhibition neurotransmission *CNS depression Higher proportion yolk and/or oil Inhibited developmental rate Yolk-oil energy not fully utilized Decreased heartbeat (irregular) Decreased respiration rate Teratogenesis — Skeletal development abnormal, Muscle development abnormal, Deformities of eyes; Skeletal deformities of vertebral column, fins, cranial bones, jaw, Organogenesis, development inhibited	
Individual	Narcosis Delayed hatching; less viable and/or inactive and abnormal larvae More buoyant larvae Abnormal larvae	Narcosis; impaired swimming, feeding activity
Population	Decreased hatching, growth Decreased survival of yolk-sac larvae Decreased survival (Few recover)	Decreased growth, survival

Table 12 con't.

SUBLETHAL EXPOSURES OF MAH
(Initial Exposure Effects)[1]
Juveniles, Adults and Gametes

(Ref. Nos. 6, 7, 39, 50, 63-67, 75, 78-80, 83, 85, 86, 92-98)

Organizational Level of Effect: (Effects often delayed)

Have ability to detoxify (MFO system)
*Competitive inhibition of metabolism of toxicant, steroids, lipids, vitamins
(Metabolism of toxicant limited; insufficient enzyme levels)

Cellular

- *Inhibited metabolism and increased levels of adrenocorticosteroids
- *Inhibited metabolism and increased levels of sex hormones
- Increased accumulation of aromatic hydrocarbon

Tissue and/or Organ System

- *Increased gluconeogenesis in liver, protein catabolism, mobilization and utilization stored lipids
- *Neurotransmission altered
- *Affects electrolyte, water metabolism
- *Osmoregulation impaired; Edema
- Lymphocytopenia, thrombocytopenia, decreased mobilization of granulocytes; Increased erythrocytes
- *Haemopoietic organs affected
- Increased vitellogenesis
- Accelerated development of gonads, oogenesis, spermiogenesis
- Target tissues affected; cell membrane permeability altered

Individual

- Weakened; poor condition; Hypoactivity; Disequilibrium; Poor food localization
- *Less capacity to adapt to salinity changes
- *Decreased immunity to disease and parasites
- *Reduced inflammatory response, resistance to other toxicants
- Accelerated sexual maturation, reproductive behavior

Population

- Decreased growth, survival of adults, gametes
- Decreased growth, survival of adults, gametes
- Reduced reproductive success
- Desynchronization reproduction, spawning
- Reduced fecundity, fertilization, survival of eggs, larvae

[1]Also effects if homeostasis fails and negative feedback occurs.

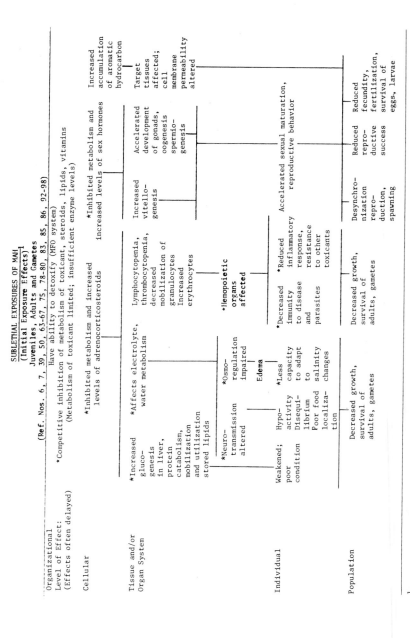

SUBLETHAL EXPOSURES OF MAH
(Adaptation to Exposure)
Juveniles, Adults and Gametes
(Ref. Nos. 6, 7, 39, 50, 63-67, 75, 78-81, 83, 85, 86, 92-98)

Organizational Level of Effect:

Neuroendocrine mediated response – Homeostasis – Return to pre-exposure condition

*Stimulation of pituitary hormones

	*Neurohypophyseal hormones stimulated		*Adenohypophyseal hormones stimulated			
Cellular	*Increased epinephrine (Norepinephrine)	*Increased oxytocin	*Increased thyrotropin	*Increased ATP, NADPH	*Increased RNA synthesis	*Induction enzyme synthesis in MFO system — *Increased metabolism of aromatic hydrocarbons
				*Increased metabolism of cholesterol	*Stimulation of metabolism of adrenocorticosteroid hormones	*Reduced levels of MAH in target tissues
Tissue and/or Organ System	Cardiovascular system; increased heartbeat, blood pressure	*Contraction of smooth muscles, gonads Accelerated release of gametes	Increased thyroxine Increased oxygen consumption tissues Increased respiration *Osmoregulatory stress reduced	*Reduced glucogenesis, protein catabolism, lipid mobilization	*Reduced levels of adrenocorticosteroid hormones in target organs	*Hemopoietic organs respond: Lymphocytosis Eosinophilia Neutrophilia
	*CNS neurotransmission affected			*Decreased sodium level; increased potassium		
Individual	*Avoidance responses Initial hyperactivity returns to normal	Premature spawning behavior, spawning	Accelerated respiration, oxygen consumption *Increased capacity for osmoregulation	*Osmoregulation, neurotransmission return to normal	*Intensification of inflammatory response Increased immunity to disease and parasites	
Population	Increased survival	Desynchronization of spawning Decreased survival of eggs, larvae	Increased survival			Increased survival

Table 12 con't.

LOW SUBLETHAL EXPOSURES OF MAH
(Negative feedback over longer chronic exposure)
Juveniles, Adults and Gametes
(Ref. Nos. 7, 39, 81, 92-97, 101)

Organizational Level of Effect:				
	*Increased competition for energy (ATP) required for metabolism and/or detoxification			
Cellular	*Decreased energy available for anabolism, respiration			*Decreased metabolism of MAH
Tissue and/or Organ System	Decreased vitellogenesis, Atresia of eggs	*Decreased lipid mobilization to liver	*Decreased muscle tissue, stored lipids, liver glycogen	*Increased levels of MAH in target tissues
	Reduced size, resorption of gonads		Reduced size of tissues, organs	Reversible and irreversible effects on target tissues
Individual	Inhibition of maturation, reproduction		Reduced growth, activity and capacity to respond to additional stress, e.g., disease, parasites	
			Poor condition	
Population	Reduced reproductive success, fecundity	Reduced growth (biomass)	*Reduced survival or life span	*Reduced survival or life span

532

SUBLETHAL EXPOSURES OF MAH

Toxic effects on target tissues possibly resulting from hormone-mediated, adaptive responses; chronic exposure (Ref. Nos. 6, 7, 39, 50, 63-67, 75, 78-81, 86, 92-98)

Organizational Level of Effect:

Cellular

Induction of MFO

*Increased epoxidation of aromatics with accumulation of toxic metabolites (Increased endoplasmic reticulum - ribosomes)

*Mutagenesis (DNA) and *Carcinogenesis (DNA)

*Induction of prophage

*Abnormal cell division

*Increased production leukocytes

*Abnormal leukocytes

*Abnormal erythrocytes

Tissue and/or Organ System

*Tumors: liver, kidney, epidermis, etc.

*Splenomegaly

*Leukemia   *Acute anemia

Individual

Weakened, poor condition

Weakened, poor condition

Population

Decreased survival

Decreased survival

Table 12 con't.

SUBLETHAL AND LETHAL EXPOSURES OF MAH

(Some toxic effects on target tissues resulting from increased levels of MAH and/or metabolites)

Most life history stages

(Ref. Nos. 4-7, 19-24, 39, 49, 50, 63-67, 74, 75, 77-81, 86-88, 92-98, 101)

Organizational Level of Effect: (Effects often delayed)					
Detoxification capacity of MFO exceeded and/or toxic metabolites not depurated					
Accumulation of aromatics and/or metabolites in target tissues and organs					
Tissue and/or Organ System	Liver: Lipid deposition *Change in type of lipid *Glycogen depletion Disruption of hepatic muralia Fibrotic infiltration Hypertrophy Hemorrhage; sinusoidal congestion with erythrocytes Necrosis	Ovaries: Increased vitellin in eggs, or atresia-resorption Egg membranes disrupted Egg vacuolization Reduced blood supply to eggs Erythrocyte destruction Hemorrhage Necrotic foci Testes: Resorption Hemorrhage Necrotic foci *Reduced sperm motility	Spleen: *Hypoplasia or *hyperplasia Fibrotic infiltration Granulomas Blood: Abnormal erythrocytes, leukocytes Destruction erythrocytes *Increased erythrocyte fragility *Cell membrane disruption *Increased lipoprotein	Heart: *Fatty infiltration Blood Vessels: Sloughing of endothelial lining *Vasodilation *Increased capillary permeability Gills: Increased mucus secretion Increased number mucus cells Lesions, vacuoles Epithelial sloughing	Brain & Nerve Tissue: *Lesions in hypothalamus, pituitary, optic nerve tract, olfactory bulbs Adrenals: Enlarged size Muscle: *Decrease in muscle tissue, less lipid; edema
Individual	Decreased metabolic efficiency Poor condition	Decreased reproduction	Decreased respiration, immunity, resistance to environmental extremes	Decreased cardio-vascular efficiency Decreased respiratory efficiency and osmoregulation	Impaired neuro-transmission and behavioral responses Neuroendocrine stress responses Reduced activity
Population	Decreased growth, survival	Decreased fecundity Reduced fertilization and reproductive success	Decreased growth, survival	Decreased growth, survival	Decreased growth, survival

534

LETHAL EXPOSURES OF MAH
All life history stages
(Ref. Nos. 5, 6, 19, 63, 65, 74, 75, 85, 88)

Organizational
Level of Effect: *Detoxification capacity of MFO system is exceeded

Cellular   *Decreased oxidation (metabolism) of aromatics   *Decreased conjugation of metabolites

Tissue and/or   Accumulation of aromatics and metabolites in lipid-rich tissues
Organ System

                    *Brain and nervous tissue affected,   Respiratory depression   Destruction of erythrocytes;
                    neurotransmission impaired;   reduced oxygen-carrying
                    CNS depression   capacity of blood

Individual   Narcosis, hypoactivity,   Anoxia   Acute anemia, Anoxia
                    disequilibirum

Population   No survival

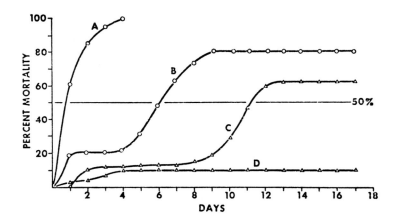

FIGURE 9. *Mortality curves of anchovies* (Stolephorus
purpureus) *exposed to different levels of osmoregulatory
stress in the laboratory. A = lethal (or acute) stress;
B and C = High sublethal (eventually lethal) stress; D =
Low sublethal (chronic) stress (Adapted from Struhsaker* et
al., *1974).*

To illustrate the responses of fish to stress, Figure 9
shows a family of mortality curves generated by exposing
anchovies to osmoregulatory stress (Struhsaker *et al.*, 1975).
The percent mortality over a 17-day period is shown at
different levels of stress (from lethal to low sublethal).
The configuration of two of these curves (B and C) is similar
to the general adaptation syndrome (GAS) described by Selye
(1950) (see also Rosenthal and Alderdice, 1976). A lethal
level of stress is represented by curve A, where continuous
mortality occurs until 100% of the fish are dead. At this
level, no homeostatic or adaptive response was adequate to
overcome the effects of the stress. In curves B and C,
there was an initial mortality, followed by an adaptive
response with zero mortality, followed by a negative feedback
where capacity to adapt was exceeded and mortality again
increased. Finally, mortality leveled off with some of the
fish surviving. The survivors probably represent a
genetically resistant portion of the population with greater
adaptive capacity. Curve D is equivalent to what may occur
with low sublethal exposures, where there may be a small
initial mortality (or none) followed by an adaptive response
of long duration. Ultimately, however, negative feedback may
occur. There is disagreement among researchers as to whether
there is a threshold level of "no effect" for many pollutants.

Some believe that there is no "safe level" of exposure if
organisms are exposed to pollutant stress for a sufficiently
long period of time. For example, the negative feedback may
shorten the life span, decrease growth or inhibit reproduction. This issue has not been resolved at the present time.

Chronic Effects

Many of the observed effects of MAH are probably due to a
generalized stress response. A possible specific toxic
effect may occur with the component benzene. In mammals,
benzene affects the blood-forming organs, resulting in
leukemia. This effect was observed to be specific to
benzene and was not induced by other aromatic compounds
(Gerarde, 1960). Preliminary studies of field-captured
striped bass indicate that there are effects on red and white
blood cells which are correlated with high levels of benzene
in the liver (Whipple, 1979). Further studies on hemopoietic
tissues are being done. The degree of a toxic effect will
depend largely on the levels of the monocyclics and/or
metabolites, such as phenol, in the target tissues. Our
research has also shown that prespawning females, their
ovaries, eggs and larvae experience irreversible damage to
tissues at very low levels of MAH in the water column
(50-100 ppb) and thus are potentially the most sensitive in
the actual environment (Struhsaker, 1977; Whipple *et al.*,
1978b).

Several other effects are shown in Table 12. Many of the
hypothesized modes of action and effects have been derived
from work on mammals (summarized by LaDu *et al.*, 1971). In
the table, hypothetical concepts are indicated by asterisks,
while those with studies substantiating the hypotheses are
referenced by numbers.

Many researchers have noted an apparent beneficial
effect of exposure of fishes to low levels of toxicants,
including the MAH (Eldridge *et al.*, 1977). This probably
results from the hormone-mediated adaptive responses of the
organisms to stress, at least temporarily stimulating
metabolic functions. This response is referred to by Smyth
(1967) as a "sufficient challenge". Under chronic conditions,
however, we feel that some negative feedback eventually
occurs (Table 12).

Hypothesized Effects at the Population Level

The effects of monocyclics summarized in Table 12
include potential effects at a population level. In essence,
these effects may result in the reduction in production of a

fishery as determined by fecundity, reproduction and larval
recruitment, growth and survival. For example, our work has
shown that such effects of MAH are possibly contributing to
the decline of the striped bass population and fishery in
the San Francisco Bay area (Whipple, 1979, 1980; Whipple *et
al.*, ms in prep.). Chronic levels of MAH occur in tissues of
striped bass which either alone, or in concert with other
pollutants identified (e.g., zinc), are sufficient to cause
negative effects. Although striped bass are relatively
hardy, many appear to be stressed beyond their capacity to
adapt. We hypothesize monocyclic aromatic petrochemicals
reduce the ability of fishes to adapt to natural stresses,
particularly during spawning migration and early life history
stages. Whether or not a population can genetically adapt to
this stress over longer periods depends upon the differential
selection of a hardier genotype. Although such intraspecific
diversity appears to exist in striped bass, present observations indicate that it is not sufficient to ensure
population persistence.

   Estimates of reduction in fecundity and larval survival
due to the uptake of monocyclic aromatics (both laboratory and
field data) indicate that a reduction of larval recruitment
as high as 40-50% may occur in some of the species studied,
for example, Pacific herring and striped bass (Struhsaker,
1977; Whipple *et al.*, ms in prep.) at the chronic levels now
occurring in San Francisco Bay.

SUMMARY

   Recent investigations of the sources of monocyclic
aromatic hydrocarbons (MAH) and their potential effects on
aquatic resources indicate that this hydrocarbon class is
prevalent in estuarine ecosystems such as the San Francisco
Bay-Delta, and may constitute a chronic pollution threat.
These toxic petrochemicals, interacting with other pollutants,
could cause quantitative reductions in the production of
fish populations by decreasing growth, reproduction, egg
viability, larval recruitment and survival. There are also
qualitative effects on fisheries, such as condition of fish
flesh and increased parasitism and disease. This paper
briefly discusses the sources and fates of MAH and summarizes
their effects on fishes.

   The chronic effects of monoaromatics are examined in the
context of a qualitative conceptual model, suggesting interactive effects of inherent environmental factors (abiotic
and biotic). Variability in uptake, bioaccumulation and

effects of monocyclics is considered in relation to inherent differences in fishes (e.g., interspecies variation, intra-specific variation in genotype, sex, age or life history stage and condition). The interaction of these inherent differences with certain environmental variables (e.g., temperature, salinity, pollutants and parasitism) is also discussed. The definition of variability is important in determining potential chronic effects on the biota, and in making selections of critical stages and effects to study and monitor.

Monocyclic aromatics discussed include: benzene, toluene, xylenes, ethylbenzene and substituted benzenes, in general. Examples of fish species are drawn from our research in the San Francisco Bay-Delta area. However, most of these species are widely distributed into other areas. Species discussed include: striped bass (*Morone saxatilis*), starry flounder (*Platichthys stellatus*), Pacific herring (*Clupea harengus pallasii*), northern anchovy (*Engraulis mordax*) and Chinook salmon (*Oncorhynchus tshawytscha*). The possible relationship of selected genotypic differences to variability in uptake, bioaccumulation and effects is discussed for striped bass and starry flounder. Differences among the following life history stages also occur: spawning adults, gametic eggs, spawned eggs, embryos, larvae, juveniles and nonspawning adults. Larvae, for example, appear to bioaccumulate very high levels of monocyclic aromatics.

Uptake, bioaccumulation, translocation and depuration of these components are discussed in relation to effects in terms of hypothesized modes of action and alterations of structure and function (morphology, physiology and behavior). Many responses appear to be nonspecific responses to pollutant stress, but some specific effects (e.g., blood parameters) may also occur.

ACKNOWLEDGEMENTS

We would like to acknowledge the assistance of the staff of the Physiological Ecology Program at the National Marine Fisheries Service Tiburon Laboratory, particularly Michael Bowers and Brian Jarvis, for assisting in several aspects of these studies. We also thank Rahel Fischer for assistance in editing the manuscript.

We would like to give special thanks to Stan Rice and Sid Korn of the Auke Bay National Marine Fisheries Service Laboratory for their cooperation and suggestions throughout all our research on monocyclic aromatic hydrocarbons.

Some of the work done at Tiburon Laboratory, NMFS was supported by funding allocated under the Marine Protection, Research and Sanctuaries Act, Title II, Section 202, P. L. 92-532. Partial support came from the Outer Continental Shelf Environmental Assessment Program, Environmental Research Laboratories, National Oceanic and Atmospheric Administration, with funding furnished by the Bureau of Land Management, U. S. Department of the Interior.

LITERATURE CITED

1. American Petroleum Institute. 1978. Analysis of refinery wastewaters for the EPA priority pollutants. Interim report. 139 pp. API Publication No. 4296. Washington, D.C.: American Petroleum Institute.

2. Anderson, J. W., J. M. Neff, B. A. Cox, H. E. Tatem, and G. M. Hightower. 1974a. The effects of oil on estuarine animals: Toxicity, uptake and depuration, respiration. In: Pollution and physiology of marine organisms, pp. 285-310. Ed. by F. J. Vernberg and W. B. Vernberg. New York: Academic Press.

3. Anderson, J. W., J. M. Neff, B. A. Cox, H. E. Tatem, and G. M. Hightower. 1974b. Characteristics of dispersions and water-soluble extracts of crude and refined oils and their toxicity to estuarine crustaceans and fish. Mar. Biol. 27: 75-88.

4. Benville, P. E., Jr., N. D. Hirsch, and J. A. Whipple. 1980. Benzene uptake and depuration by adult striped bass. Presented at 61st annual meeting, AAAS, Pacific Division. June 1980. Abstract in Newsletter of the San Francisco Bay and Estuarine Assoc., Aug. 1980. P. O. Box 160088, Sacramento, CA, 95816.

5. Benville, P. E., Jr. and S. Korn. 1977. The acute toxicity of six monocyclic aromatic crude oil components to striped bass (*Morone saxatilis*) and bay shrimp (*Crago franciscorum*). Calif. Fish Game 63(4): 204-209.

6. Brocksen, R. W. and H. T. Bailey. 1973. Respiratory response of juvenile Chinook salmon and striped bass exposed to benzene, a water-soluble component of crude oil. In: Proceedings Joint Conference on Prevention and Control of Oil Spills, pp. 783-791. Washington, D.C.: American Petroleum Institute.

7. Brown, E. R., E. Koch, T. F. Sinclair, R. Spitzer, O. Callaghan, and J. J. Hazdra. 1979. Water pollution and diseases in fish (An epizootiologic survey). J. Environ. Pathol. Toxicol. 2: 917-925.

8. Brown, R. A. and H. L. Huffman, Jr. 1976. Hydrocarbons in open ocean waters. Science (Wash., D.C.) 191: 847-849.

9. Burwood, R. and G. C. Speers. 1974. Photo-oxidation as a factor in the environmental dispersal of crude oil. Estuar. Coast. Mar. Sci. 2: 117-135.

10. Butler, J. N., B. F. Morris, and T. D. Sleeter. 1976. The fate of petroleum in the open ocean. In: Sources, effects and sinks of hydrocarbons in the aquatic environment, pp. 287-291. Proceedings of the symposium, American University, Washington, D.C., August 1976. American Institute of Biological Sciences.

11. Cairns, J., Jr., K. L. Dickson, and A. W. Maki, eds. 1978. Estimating the hazard of chemical substances to aquatic life. 278 pp. Philadelphia: American Society for Testing and Materials.

12. Caldwell, R. S., E. M. Caldarone, and M. H. Mallon. 1977. Effects of a seawater-soluble fraction of Cook Inlet crude oil and its major aromatic components on larval stages of the Dungeness crab, Cancer magister Dana. In: Fate and effects of petroleum hydrocarbons in marine ecosystems and organisms, pp. 221-228. Proceedings of a symposium, November 1976, Seattle. Ed. by D. A. Wolfe. New York: Pergamon Press.

13. California Department of Food and Agriculture. 1978. 1978 Pesticide use report. Annual. 187 pp. Sacramento: Calif. Dep. Food Agric.

14. Clark, R. C., Jr. and W. D. MacLeod, Jr. 1977. Inputs, transport mechanisms, and observed concentrations of petroleum in the marine environment. In: Effects of petroleum on arctic and subarctic marine environments and organisms, Vol. I, pp. 91-223. Ed. by D. C. Malins. New York: Academic Press.

15. Davis, D. L. and B. H. Magee. 1979. Cancer and industrial chemical production. Science (Wash., D.C.) 206: 1356-1358.

16. Davis, S. J. and C. F. Gibbs. 1975. The effect of weathering on a crude oil residue exposed at sea. Water Res. 9: 275-285.

17. DiSalvo, L. H. and H. E. Guard. 1975. Hydrocarbons associated with suspended particulate matter in San Francisco Bay waters. In: Proceedings of 1975 Conference on Prevention and Control of Oil Pollution, pp. 169-173. Washington: American Petroleum Institute.

18. Doggett, N. S., D. J. Bailey, and T. Qazi. 1977. Estrogen potentiating activity of two spiro compounds having approximately similar molecular dimensions to stilbestrol. J. Med. Chem. 20: 318-320.

19. Eldridge, M. B. and P. E. Benville, Jr. (Ms in prep.) Benzene uptake and accumulation and its effects in eggs and larvae of striped bass (*Morone saxatilis*).

20. Eldridge, M. B. and T. Echeverria. 1977. Fate of ^{14}C-benzene (an aromatic hydrocarbon of crude oil) in a simple food chain of rotifers and Pacific herring larvae. Am. Fish. Soc., CAL-NEVA Wildl. Trans. 1977: 90-96.

21. Eldridge, M. B., T. Echeverria, and S. Korn. 1978a. Fate of ^{14}C-benzene in eggs and larvae of Pacific herring (*Clupea harengus pallasi*). J. Fish. Res. Board Can. 35: 861-865.

22. Eldridge, M. B., T. Echeverria, and J. A. Whipple. 1977. Energetics of Pacific herring (*Clupea harengus pallasi*) embryos and larvae exposed to low concentrations of benzene, a monoaromatic component of crude oil. Trans. Am. Fish. Soc. 106: 452-461.

23. Eldridge, M. B., J. A. Whipple, D. Eng, M. J. Bowers, and B. M. Jarvis. 1978b. Laboratory studies on factors affecting mortality in California striped bass (*Morone saxatilis*) eggs. In: Proceedings of session on Advances in Striped Bass Life History and Population Dynamics. 108th Annual Meeting, Am. Fish. Soc., August 1978, Univ. Rhode Island.

24. Gerarde, H. W. 1960. Toxicology and biochemistry of aromatic hydrocarbons. 329 pp. New York: Elsevier Publishing Co.

25. Gibson, D. T. 1968. Microbial degradation of aromatic compounds. Science (Wash., D.C.) 161: 1093-1097.

26. Gibson, D. T. 1971. The microbial oxidation of aromatic hydrocarbons. Crit. Rev. Microbiol. 1: 199-223.

27. Gibson, D. T. 1972. Initial reactions in the degradation of aromatic hydrocarbons. In: Degradation of synthetic organic molecules in the biosphere: Natural pesticidal and various other man-made compounds, pp. 116-136. Proceedings of a conference, San Francisco, June 1971. Washington, D.C.: National Academy of Sciences.

28. Gibson, D. T. 1976. Microbial degradation of carcinogenic hydrocarbons and related compounds. In: Sources, effects and sinks of hydrocarbons in the aquatic environment, pp. 224-238. Proceedings of the Symposium, American University, Washington, D.C., August 1976. American Institute of Biological Sciences.

29. Gibson, D. T. 1977. Biodegradation of aromatic petroleum hydrocarbons. In: Fate and effects of petroleum hydrocarbons in marine ecosystems and organisms, pp. 36-46. Proceedings of a symposium, November 1976, Seattle. Ed. by D. A. Wolfe. New York. Pergamon Press.

30. Gibson, D. T. Microbial degradation of hydrocarbons. In: Physical and chemical sciences research report 1: The nature of seawater, pp. 667-696. Ed. by E. D. Goldberg. Dahlem Workshop Report.

31. Gibson, D. T., C. E. Cardini, F. C. Maseles, and R. E. Kallio. 1970a. Incorporation of oxygen-18 into benzene by *Pseudomonas putida*. Biochemistry 9: 1631-1635.

32. Gibson, D. T., M. Hensley, H. Yoshioka, and T. J. Mabry. 1970b. Formation of (+)-cis-2,3-dihydroxy-1-methyl-cyclohexa-4,6-diene from toluene by *Pseudomonas putida*. Biochemistry 9: 1626-1630.

33. Gibson, D. T., J. R. Koch, and R. E. Kallio. 1968. Oxidative degradation of aromatic hydrocarbons by microorganisms, I. Enzymatic formation of catechol from benzene. Biochemistry 7: 2653-2662.

34. Gibson, D. T., V. Mahadevan, and J. F. Davey. 1974. Bacterial metabolism of para- and meta-xylene: Oxidation of the aromatic ring. J. Bacteriol. 119: 930-936.

35. Gibson, D. T. and W. K. Yeh. 1973. Microbial degradation of aromatic hydrocarbons. In: The microbial degradation of oil pollutants, pp. 33-38. Ed. by D. G. Ahearn and S. P. Meyers. Baton Rouge, Louisiana: Center for Wetland Resources, Louisiana State University. Publ. No. LSU-SG-73-01.

36. Gordon, D. C., Jr., P. D. Keizer, W. R. Hardstaff, and D. G. Aldous. 1976. Fate of crude oil spilled on seawater contained in outdoor tanks. Environ. Sci. Technol. 10: 580-585.

37. Hansen, H. P. 1977. Photodegradation of hydrocarbon surface films. In: Petroleum hydrocarbons in the marine environment. Proceedings from ICES workshop, Aberdeen, September 1975. Ed. by A. D. McIntyre and K. J. Whittle. Rapp. P.-v. Réun. Cons. int. Explor. Mer 171: 101-106.

38. Harrison, W., M. A. Winnik, P. T. Y. Kwong, and D. Mackay. 1975. Crude oil spills. Disappearance of aromatic and aliphatic components from small sea-surface slicks. Environ. Sci. Technol. 9: 231-234.

39. Hirsch, N. D., R. Pimentel, and J. A. Whipple. (Ms in prep). Long-term fate and effects of chronic benzene exposure on striped bass.

40. Hirsch, N. D., J. A. Whipple, and P. E. Benville, Jr. (Ms in prep.). Introduction of monocyclic aromatics into seawater by an outboard engine.

41. Jerina, D. M. and J. W. Daly. 1974. Arene oxides: A new aspect of drug metabolism. Science (Wash. D.C.) 185: 573-582.

42. Jung, M. and G. W. Bowes. 1980. First progress report in cooperative striped bass study (COSBS). Sacramento: California State Water Resources Control Board.

43. Kappas, A. and A. P. Alvares. 1975. How the liver metabolizes foreign substances. Sci. Am. 232: 22-31.

44. Karrick, N. L. 1977. Alterations in petroleum resulting from physico-chemical and microbiological factors. In: Effects of petroleum on arctic and subarctic marine environments and organisms, Vol. I, pp. 225-299. Ed. by D. C. Malins. New York: Academic Press.

45. Kohlhorst, D. W. 1973. An analysis of the annual striped bass die-off in the Sacramento-San Joaquin estuary, 1971-1972. State of Calif., Resour. Agency, Dep. Fish Game, Anadromous Fish Branch, Admin. Rep. No. 73-7.

46. Koons, C. B. and D. E. Brandon. 1975. Hydrocarbons in water and sediment samples from Coal Oil Point area, offshore California. Paper No. OTC 2387, Seventh Annual Offshore Technology Conference, Houston, May 1975.

47. Korn, S., N. Hirsch, and J. W. Struhsaker. 1976a. Uptake, distribution and depuration of ^{14}C-benzene in northern anchovy, *Engraulis mordax,* and striped bass, *Morone saxatilis.* Fish. Bull., U.S. 74: 545-551.

48. Korn, S., N. Hirsch, and J. W. Struhsaker. 1977. The uptake, distribution, and depuration of ^{14}C-benzene and ^{14}C-toluene in Pacific herring, *Clupea harengus pallasi.* Fish. Bull., U.S. 75: 633-636.

49. Korn, S. and S. Rice. In press. Sensitivity to, and accumulation and depuration of, aromatic petroleum components by early life stages of coho salmon, *Oncorhynchus kisutch.* In: Early life history of fish, Vol. II. Ed. by K. Sherman and R. Lasker. Springer-Verlag.

50. Korn, S., J. W. Struhsaker, and P. Benville. 1976b. Effects of benzene on growth, fat content, and caloric content of striped bass, *Morone saxatilis.* Fish. Bull., U.S., 74: 694-698.

51. LaDu, B. N., H. G. Mandel, and E. L. Way. 1971. Fundamentals of drug metabolism and drug disposition. Baltimore: Williams and Wilkins.

52. Lee, R. F. 1976. Metabolism of petroleum hydrocarbons in marine sediments. In: Sources, effects and sinks of hydrocarbons in the aquatic environment, pp. 333-344. American Institute of Biological Sciences.

53. Lee, R. F. 1977. Accumulation and turnover of petroleum hydrocarbons in marine organisms. In: Fate and effects of petroleum hydrocarbons in marine ecosystems and organisms, pp. 60-70. Ed. by D. A. Wolfe. New York: Pergamon Press.

54. Lee, R. F. and C. Ryan. 1976. Biodegradation of petroleum hydrocarbons by marine microbes. In: Proceedings of the 3rd International Biodegradation Symposium, pp. 119-125. Ed. by J. M. Sharpley and A. M. Kaplan. London: Applied Science Publishers.

55. Lønning, S. 1977. The effects of crude Ekofisk oil and oil products on marine fish larvae. Astarte 10: 37-47.

56. Maines, M. D. and A. Kappas. 1977. Metals as regulators of heme metabolism. Science (Wash., D.C.) 198: 1215-1221.

57. Malins, D. C. 1977a. Biotransformation of petroleum hydrocarbons in marine organisms indigenous to the arctic and subarctic. In: Fate and effects of petroleum hydrocarbons in marine ecosystems and organisms, pp. 47-59. Ed. by D. A. Wolfe. New York: Pergamon Press.

58. Malins, D. C. 1977b. Metabolism of aromatic hydrocarbons in marine organisms. In: Aquatic pollutants and biologic effects with emphasis on neoplasia, pp. 482-496. Ed. by H. F. Kraybill, C. J. Dawe, J. C. Harshbarger, and R. G. Tardiff. Ann. N.Y. Acad. Sci. 298: 482-496.

59. Marr, E. K. and R. W. Stone. 1961. Bacterial oxidation of benzene, J. Bacteriol. 81: 425-430.

60. McAuliffe, C. 1966. Solubility in water of paraffin, cycloparaffin, olefin, acetylene, cycloolefin and aromatic hydrocarbons. J. Phys. Chem. 70: 1267-1275.

61. McAuliffe, C. D. 1977a. Dispersal and alteration of oil discharged on a water surface. In: Fate and effects of petroleum hydrocarbons in marine organisms and ecosystems, pp. 19-35. Ed. by D. A. Wolfe. New York: Pergamon Press.

62. McAuliffe, C. D. 1977b. Evaporation and solution of C_2 to C_{10} hydrocarbons from crude oils on the sea surface. In: Fate and effects of petroleum hydrocarbons in marine organisms and ecosystems, pp. 363-372. Ed. by D. A. Wolfe. New York: Pergamon Press.

63. Meyerhoff, R. D. 1975. Acute toxicity of benzene, a component of crude oil, to juvenile striped bass (*Morone saxatilis*). J. Fish. Res. Board Can. 32: 1864-1866.

64. Moles, A. 1980. Sensitivity of parasitized coho salmon to crude oil, toluene, and naphthalene. Trans. Am. Fish. Soc. 109: 293-297.

65. Moles, A., S. D. Rice, and S. Korn. 1979. Sensitivity of Alaskan freshwater and anadromous fishes to Prudhoe Bay crude oil and benzene. Trans. Am. Fish. Soc. 108: 408-414.

66. Morrow, J. E. 1974. Effects of crude oil and some of its components on young coho and sockeye salmon. 29 pp. EPA Report 666/3-73-018. U.S. EPA, Off. Res. Dev. Washington, D.C.

67. Morrow, J. E., R. L. Gritz, and M. P. Kirton. 1975. Effects of some components of crude oil on young coho salmon. Copeia 1975: 326-331.

68. Myers, E. P. and C. G. Gunnerson. 1976. Hydrocarbons in the ocean. U.S. Dep. Comm., NOAA, Mesa Special Report. 42 p.

69. Neff, J. M. 1978a. Accumulation and release of poly-cyclic aromatic hydrocarbons from water, food and sediment by marine animals. From: Symposium on carcinogenic polynuclear aromatic hydrocarbons in the marine environment. Sponsored by USEPA. August 1978, Pensacola, Florida.

70. Neff, J. M. 1978b. Polycyclic aromatic hydrocarbons in the aquatic environment and cancer risk to aquatic organisms and man. From: Symposium on carcinogenic polycuclear aromatic hydrocarbons in the marine environment. Sponsored by USEPA. August 1978, Pensacola, Florida.

71. Neff, J. M.   1979.   Polycyclic aromatic hydrocarbons in
    the aquatic environment.   Sources, fates, and biological
    effects.   262 p. London: Applied Sciences Publishers
    Ltd.

72. Payne, J. F. and W. R. Penrose.   1975.   Induction of aryl
    hydrocarbon (benzo(a)pyrene) hydroxylase in fish by
    petroleum.   Bull. Environ. Contam. Toxicol. 14:   112-226.

73. Pickering, Q. H. and C. Henderson.   1966.   Acute toxicity
    of some important petrochemicals to fish.   J. Water
    Pollut. Control Fed. 38:   1419-1429.

74. Rice, S. D., J. W. Short, and J. F. Karinen.   1976.
    Toxicity of Cook Inlet crude oil and No. 2 fuel oil to
    several Alaskan marine fishes and invertebrates.   In:
    Sources, effects and sinks of hydrocarbons in the aquatic
    environment, pp. 395-406.   American Institute of Bio-
    logical Sciences.

75. Rice, S. D., J. W. Short, and J. F. Karinen.   1977.
    Comparative oil toxicity and comparative animal sen-
    sitivity.   In: Fate and effects of petroleum hydrocarbons
    in marine ecosystems and organisms,   pp. 78-94.
    Proceedings of a symposium, November 1976, Seattle.   Ed.
    by D. A. Wolfe.   New York: Pergamon Press.

76. Risebrough, R. W., J. W. Chapman, R. K. Okazaki, and
    T. T. Schmidt.   1978.   Toxicants in San Francisco Bay
    and Estuary.   113 pp. Berkeley, CA: Association of Bay
    Area Governments.

77. Rosenthal, H. and D. F. Alderdice.   1976.   Sublethal
    effects of environmental stressors, natural and pol-
    lutional, on marine fish eggs and larvae.   J. Fish. Res.
    Board Can. 33:   2047-2065.

78. Roubal, W. T.   1974.   Spin-labeling of living tissues-
    A method for investigating pollutant-host interaction.
    In: Pollution and physiology of marine organisms,
    pp. 367-379.   Ed. by F. J. Vernberg and W. B. Vernberg.
    New York: Academic Press.

79. Roubal, W. T. and T. K. Collier.   1975.   Spin-labeling
    techniques for studying mode of action of petroleum
    hydrocarbons on marine organisms.   Fish. Bull., U.S.,
    73:   299-305.

80. Roubal, W. T., T. K. Collier, and D. C. Malins. 1977. Accumulation and metabolism of carbon-14 labeled benzene, naphthalene, and anthracene by young coho salmon (*Oncorhynchus kisutch*). Arch. Environ. Contam. Toxicol. 5: 515-529.

81. Selye, H. 1950. Stress and the general adaptation syndrome. Brit. Med. J. 1: 1383-1392.

82. Smith, R. L. and J. A. Cameron. 1979. Effect of water soluble fraction of Prudhoe Bay crude oil on embryonic development of Pacific herring. Trans. Am. Fish. Soc. 108: 70-75.

83. Smyth, H. F. 1967. Sufficient challenge. Fed. Cosmet. Toxicol. 5: 51-58.

84. Sprague, J. B. 1971. Measurement of pollutant toxicity to fish - III. Sublethal effects and "safe" concentrations. Water Res. 5: 245-266.

85. Stegeman, J. J. and D. J. Sabo. 1976. Aspects of the effects of petroleum hydrocarbons in intermediary metabolism and xenobiotic metabolism in marine fish. In: Sources, effects and sinks of hydrocarbons in the aquatic environment, pp. 423-436. American Institute of Biological Sciences.

86. Struhsaker, J. W. 1977. Effects of benzene (a toxic component of petroleum) on spawning Pacific herring, *Clupea harengus pallasi*. Fish. Bull., U.S.,: 75: 43-49.

87. Struhsaker, J. W., W. J. Baldwin, and G. I. Murphy. 1975. Environmental factors affecting stress and mortality of the Hawaiian anchovy in captivity. U. S. Dept. Comm., NOAA office of Sea Grant. Sea Grant Tech. Rep. UNIHI-SEAGRANT-TR-75-02.

88. Struhsaker, J. W., M. B. Eldridge, and T. Echeverria. 1974. Effects of benzene (a water-soluble component of crude oil) on eggs and larvae of Pacific herring and northern anchovy. In: Pollution and physiology of marine organisms, pp. 253-284. Ed. by F. J. Vernberg and W. B. Vernberg. New York: Academic Press.

89. Thomas, R. E. and S. Rice. In Press. Excretion of petroleum hydrocarbons and their metabolites by fish. In: Biological monitoring of marine pollutants. Ed. by F. J. Vernberg, A. Calabrese, F. P. Thurberg, and W. B. Vernberg. New York: Academic Press.

90. Varanasi, U. and D. C. Malins. 1977. Metabolism of petroleum hydrocarbons: Accumulation and biotransformation in marine organisms. In: Effects of petroleum on arctic and subarctic marine environments and organisms, Vol. II, pp. 175-270. Ed. by D. C. Malins. New York: Academic Press.

91. Wasik, S. P. and R. L. Brown. 1973. Determination of hydrocarbon solubility in seawater and the analysis of hydrocarbons in water extracts. In: Proceedings of 1973 Joint Conference on Prevention and Control of Oil Spills, pp. 223-227. Washington, D.C.: American Petroleum Institute.

92. Whipple, J. A. 1979. The impact of estuarine degradation and chronic pollution on populations of anadromous striped bass (*Morone saxatilis*) in San Francisco Bay-Delta, California. 6-month research report submitted to NOAA, Office of Marine Pollution Assessment, Oct. 1, 1979. To be published as NOAA Technical Memorandum (NOAA-TM-NMFS-SWFC-___)

93. Whipple, J. A. 1980. The impact of estuarine degradation and chronic pollution on populations of anadromous striped bass (*Morone saxatilis*) in San Francisco Bay-Delta, California. 6-month research report submitted to NOAA, Office of Marine Pollution Assessment, June 1, 1980. To be published as NOAA Technical Memorandum (NOAA-TM-NMFS-SWFC-___)

94. Whipple, J. A., M. B. Eldridge, P. E. Benville, Jr., M. J. Bowers, B. M. Jarvis, and N. Stapp. 1979. Effects of inherent parental factors, including pollutant uptake, on gamete condition and viability in striped bass (*Morone saxatilis*). Presented at a symposium on the Early Life History of Fish, Woods Hole, Mass., April 1979. In press. To be published by Springer-Verlag.

95. Whipple, J. A., B. M. Jarvis, M. J. Bowers, and P. E. Benville, Jr. (Ms in prep.). Effects of long-term chronic pollution on striped bass (*Morone saxatilis*) reared in the laboratory.

96. Whipple, J. A., T. G. Yocom, P. E. Benville, D. R. Smart, M. H. Cohen, and M. E. Ture. 1978a. Transport, retention, and effects of the water-soluble fraction of Cook Inlet crude oil in experimental food chains. In: Marine biological effects of OCS petroleum development, pp. 106-129. Ed. by D. A. Wolfe. NOAA Tech. Memo. ERL OCSEAP-1. Boulder, CO.

97. Whipple, J. A., T. G. Yocom, D. R. Smart, and M. H. Cohen. 1978b. Effects of chronic concentrations of petroleum hydrocarbons on gonadal maturation in starry flounder (Platichthys stellatus (Pallas)). In: Proceedings of Conference on Assessment of Ecological Impacts of Oil Spills, pp. 757-806. American Institute of Biological Sciences.

98. White, A., P. Handler, and E. L. Smith. 1973. Principles of biochemistry. 5th edition. 1296 pp. New York: McGraw-Hill Book Co.

99. Wiens, F. J. 1977. Reactive organic gas emissions from pesticide use in California. Report No. PD-77-02. Sacramento: California Air Resources Board.

100. Wolfe, D. A., J. W. Farrington, M. Katz, W. D. MacLeod, Jr., D. S. Miller, D. J. Reish, D. L. Swanson, J. A. Whipple, and T. G. Yocom. 1979. Oil and grease. In: A review of the EPA Red Book: Quality criteria for water, pp. 163-168. Ed. by R. V. Thurston, R. C. Russo, C. M. Fetterolf, Jr., T. A. Edsall, and Y. M. Barber, Jr. Water Quality Section, American Fisheries Society, Bethesda, MD.

101. Yocom, T. G., J. A. Whipple, D. R. Smart, M. H. Cohen, P. E. Benville, and M. E. Ture. (Ms in prep.). Uptake and retention of water-soluble components of crude oil from food (Tapes semidecussata) and water by starry flounder (Platichthys stellatus).

102. Zsolnay, A. 1972. Preliminary study of the dissolved hydrocarbons on particulate material in the Götland Deep of the Baltic. Kiel. Meeresforsch. 27: 129-134.

103. Zsolnay, A. 1977. Sorption of benzene on particulate material in sea water. In: Petroleum hydrocarbons in the marine environment, pp. 117-119. Proceedings from ICES workshop, Aberdeen, September 1975. Ed. by A. D. McIntyre and K. J. Whittle. Rapp. P.-v. Réun. Cons. Int. Explor. Mer 171: 117-119.

INDEX